HEPARIN

Structure, Function, and Clinical Implications

ADVANCES IN EXPERIMENTAL MEDICINE AND BIOLOGY

Recent Volumes in this Series

Volume 43
ARTERIAL MESENCHYME AND ARTERIOSCLEROSIS
Edited by William D. Wagner and Thomas B. Clarkson • 1974

Volume 44
CONTROL OF GENE EXPRESSION
Edited by Alexander Kohn and Adam Shatkay • 1974

Volume 45
THE IMMUNOGLOBULIN A SYSTEM
Edited by Jiri Mestecky and Alexander R. Lawton • 1974

Volume 46
PARENTERAL NUTRITION IN INFANCY AND CHILDHOOD
Edited by Hans Henning Bode and Joseph B. Warshaw • 1974

Volume 47
CONTROLLED RELEASE OF BIOLOGICALLY ACTIVE AGENTS
Edited by A. C. Tanquary and R. E. Lacey • 1974

Volume 48
PROTEIN-METAL INTERACTIONS
Edited by Mendel Friedman • 1974

Volume 49
NUTRITION AND MALNUTRITION: Identification and Measurement
Edited by Alexander F. Roche and Frank Falkner • 1974

Volume 50
ION-SELECTIVE MICROELECTRODES
Edited by Herbert J. Berman and Normand C. Hebert • 1974

Volume 51
THE CELL SURFACE: Immunological and Chemical Approaches
Edited by Barry D. Kahan and Ralph A. Reisfeld • 1974

Volume 52
HEPARIN: Structure, Function, and Clinical Implications
Edited by Ralph A. Bradshaw and Stanford Wessler • 1975

Volume 53
CELL IMPAIRMENT IN AGING AND DEVELOPMENT
Edited by Vincent J. Cristofalo and Emma Holečková • 1975

Volume 54
BIOLOGICAL RHYTHMS AND ENDOCRINE FUNCTION
Edited by Laurence W. Hedlund, John M. Franz, and Alexander D. Kenny • 1975

HEPARIN
Structure, Function, and Clinical Implications

Edited by
Ralph A. Bradshaw
Department of Biological Chemistry
Washington University School of Medicine
St. Louis, Missouri

and
Stanford Wessler
Department of Medicine
The Jewish Hospital
St. Louis, Missouri
and
Washington University School of Medicine
St. Louis, Missouri

PLENUM PRESS • NEW YORK AND LONDON

Library of Congress Cataloging in Publication Data

International Symposium on Heparin, St. Louis, 1974.
Heparin.

(Advances in experimental medicine and biology; v. 52)
Includes bibliographies and index.
1. Heparin — Congresses. 2. Heparin — Therapeutic use — Congresses. I. Bradshaw, Ralph A., 1941- ed. II. Wessler, Stanford, 1917- ed. III. Title. IV. Series.
QP702 H4157 1974 591.1'9'24 74-28408

DOI 10.1007/978-1-4684-0946-8

Proceedings of the International Symposium on Heparin held in St. Louis, Missouri, May 13-15, 1974

MyCopy version of the original edition 1975

A Division of Plenum Publishing Corporation
227 West 17th Street, New York, N.Y. 10011

United Kingdom edition published by Plenum Press, London
A Division of Plenum Publishing Company, Ltd.
4a Lower John Street, London W1R 3PD, England

Preface

The International Symposium on Heparin, held May 13-15, 1974, in St. Louis, Missouri, as a part of the dedication of the Shoenberg Pavilion of the Jewish Hospital of St. Louis, was conceived as a forum to bring together physicians and scientists with a basic interest in the structure, function and clinical usefulness of heparin. Few naturally occurring substances have commanded the breadth of interest among members of the biomedical research community as this compound has. Aspects of its covalent and three-dimensional structure, its biosynthesis, its interaction with and effect on physiologically important moieties and its use as a therapeutic agent in a variety of disease states have been actively studied for the past several decades. Thus, the present state of these studies seemed to be a timely subject for discussion, not only to gather together in one place representative samples of the myriad of data on heparin but also to underscore the ever increasing necessity for communication between basic research and clinical practice.

The content and organization of this monograph reflect the scope and importance of this substance. The first section deals with the molecular properties of heparin, including characterization of the basic structure, its biosynthetic origins, its determination and pharmacological standardization and the role of individual structural elements in its biological properties. The second section deals with the functional consequences of the interaction of heparin with various physiological entities including antithrombin III, platelets, and lipoprotein lipase. The last section deals with the several clinical uses of heparin, emphasizing not only the results of many trials but also the problems that remain to be elucidated.

The editors wish to thank the Board of Directors of the Jewish Hospital of St. Louis for their encouragement and financial backing without which this symposium would not have become a reality. We also wish to thank the National Heart and Lung Institute of the National Institutes of Health and the Council on Thrombosis of the American Heart Association for their financial support and the sponsorship of the Washington University School of Medicine. We also thank Mr. Norman Grossblatt for his editoral assistance in the preparation of this book. Finally, we believe it is appropriate to thank Ms. Judy Studebaker for her tireless efforts in shepherding this meeting from its conception to its termination in this monograph.

RALPH A. BRADSHAW
STANFORD WESSLER

St. Louis, Missouri
September, 1974

Contents

SECTION 1
STRUCTURAL ASPECTS

SECTION 1

STRUCTURAL ASPECTS

THE CHEMISTRY OF HEPARIN

Roger W. JEANLOZ

Laboratory for Carbohydrate Research, Harvard Medical School, Massachusetts General Hospital, Boston, Massachusetts (USA)

Although heparin was isolated in a pure, active form nearly six decades ago by McLean (41), the problems faced in solving its chemical structure have been complex enough that a recent circular (52) addressed to physicians states that "heparin is a heterogeneous substance whose exact structure is still unknown." Nevertheless, since I wrote (25), in 1958 that "the major complex chemical structure of heparin is still one of the least understood", much progress has been made, especially through the efforts of Cifonelli, Danishefsky, Dietrich, Lindahl, Linker, Perlin, Wolfrom, and their respective associates. Even if we do not have, as yet, a complete picture of the heparin molecule, and the relationship between structure and biological properties is still unclear, we have made enough progress to hope for a solution in the near future.

When heparin was first isolated (41), the concept of high-molecular-weight substances was as yet unknown, and it took two decades before heparin was recognized both as a high-molecular-weight substance and a polysaccharide. Since the techniques of purifying high-molecular-weight substances were developed mainly after the second World War, the attempts to obtain a compound with clearly defined chemical properties were hampered by unresolvable difficulties, as shown by the numerous preparations of crystalline, generally degraded salts and the lack of homogeneity of the resulting materials.

With the elucidation of its main carbohydrate components, heparin was found to be very similar to members of the class of compounds named "mucopolysaccharides" (now glycosaminoglycuronans), polysaccharides composed of amino sugar and uronic acid residues. The preliminary studies on the chemical structure of heparin suffered from the same misconceptions as the studies of other complex polysaccharides, namely that these compounds possess, like nucleic acids and proteins, homogeneous chemical structures. It is only in the last fifteen years that glycosaminoglycuronans, with the possible exception of hyaluronic acid, have been recognized to show variations in the nature of the uronic acid and hexosamine components, and of their substituents (acetyl and sulfate groups), as well as in the distribution of these components and substituents along the carbohydrate chains.

The hope of obtaining a chemically homogeneous molecule, with a well defined structure where it would be easy to establish a relationship between structure and biological properties, was definitely abandoned after the work of Lindahl (36) and associates showed that heparin is not a "pure" polysaccharide. These authors demonstrated that the carbohydrate chains are linked to a protein backbone of variable length, and that heparin exists, in the native state, mostly as a proteoglycan, as do other members of the class of glycosaminoglycuronans, such as chondroitin 4- and 6-sulfates and dermatan sulfate. Except for the length, lack of ramification, and repeating units of their carbohydrate chains, the proteoglycans resemble glycoproteins and may be considered as a special group of this large class of substances.

Any discussion of the chemical structure of heparin raises the question of the similarity of this structure with that of a family of compounds variously named "heparin monosulfate" (28), "heparitin sulfate" (37) or more recently "heparan sulfate." Although a common biosynthetic pathway for heparin and heparan sulfate has not been established as yet, the structures of these polymers present too many similarities and are too strongly differentiated from those of the other glycosaminoglycuronans, such as hyaluronic acid, chondroitin 4- and 6-sulfates and dermatan sulfate, to be considered separately. Thus, most of the observations reported here for heparin may be considered as valid also for heparan sulfate. In order not to confuse the reader with an excess of data, only the most relevant aspects of the chemical structure of heparin and heparan sulfate will be discussed on the following pages and, for further details, the reader may refer to one of the reviews that have been published in the past few years (2, 26, 29, 33).

Since heparin represents a family of related molecules where nature and distribution of components, length of carbohydrate and protein chains and total molecular weight vary, the source and modes of isolation and of purification are most important, especially when the characterization is based on the degree of anticoagulant activity. The importance of a defined source was not recognized in most of the earlier work, and this explains the difficulty in reproducing and interpreting past results. Because of the striking similarity of the structures of heparin and heparan sulfate to those of other glycosaminoglycuronans of animal origin, it is of interest to keep in mind variations between these structures (Table I).

Nature of the Carbohydrate Components

Like the structure of other glycosaminoglycuronans of animal sources, the carbohydrate chains of heparin are composed of an alternating sequence of 2-amino-2-deoxy sufars and hexuronic acids. Heparin, however, contains a much greater proportion of sulfate groups than any other sulfated glycosaminoglycuronans, reaching three groups per disaccharide unit (15), as compared to an average of one group per disaccharide unit for chondroitin 4- and 6-sulfates and dermatan sulfate.

D-Glucuronic acid was first recognized as the uronic acid component of heparin (and heparan sulfate). Since, at the time, the polysaccharide was considered to be a homogeneous polymer composed of identical repeating units (with eventually various lengths), the very low yield of D-glucuronic acid observed on isolation was attributed o the method of preparation, which led to degradation of most of the uronic acid component (54). The identification of D-glucuronic acid was confirmed by numerous procedures (26). It was only after L-iduronic acid was recognized as the major uronic acid component of dermatan sulfate (besides a small proportion of D-glucuronic acid) (50) that the nature of the uronic acid component of heparin was reinvestigated. The presence of L-iduronic acid residues in heparin was definitely established by the preparation of oligosaccharides containing both L-iduronic acid and D-glucuronic acid (31), by isolation (4) and characterization after reduction (53), by gas-liquid chromatography of the per(trimethylsilyl) derivative after acid hydrolysis (48) and methanolysis (23), and by nuclear magnetic resonance spectroscopy (45). The total relative proportion of D-glucuronic acid and L-iduronic acid residues per carbohydrate chain varies with the source of heparin (22, 51), and this relative proportion may also vary in relation to the distance from the extremity of the chain (51) (Table II). Similar observations have been made for heparan sulfate (51).

The identification of the 2-amino-2-deoxy hexose component has not presented the same difficulties as that of the uronic acid component, and the presence of only 2-amino-2-deoxy-D-glucose (D-glucosamine) has been established either by direct isolation (27), degradation with ninhydrin (25, 30), or oxidation to 2-amino-2-deoxy-D-gluconic acid (54).

TABLE I

Chemical Composition and Structure of Acidic Glycosaminoglycuronans

Properties	Heparin	Heparan Sulfate	Hyaluronic Acid	Chondroitin 4-Sulfate	Chondroitin 6-Sulfate	Dermatan Sulfate
Uronic acid component						
D-Glucuronic acid	+	+	+	+	+	+
L-Iduronic acid	+	+				+
Configuration α-L or β-D	+	+	+	+	+	+
Substituent:						
Sulfate at C-2 or L-iduronic acid	+	+				
Glycosyl residue at C-4	+	+	+	+	+	+
Amino sugar component						
2-Amino-2-deoxy-D-glucose	+	+	+			
2-Amino-2-deoxy-D-galactose				+	+	+
Configuration α-D	+	+				
β-D						
Substituents:						
N-Acetyl	+	+	+	+	+	+
N-Sulfate	+	+				
O-Sulfate	+	+				
0-3	+	+				
0-4				+		+
0-6	+	+			+	
Glycosyl residues at:						
0-3			+	+	+	+
0-4	+	+				

TABLE II

Distribution of Uronic Acids in Heparin and Heparan Sulfate from Various Sources[20]

Source	L-Iduronic acid (%)	D-Glucuronic acid (%)
Heparin		
Beef lung I	75	25
Beef lung II	84	16
Beef lung III	73	27
Hog mucosa I	73	27
Hog mucosa II	58	42
Beef mucosa	66	34
Whale tissue	60	40
Heparin sulfate		
Hog mucosa I	52	58
Hog mucosa II	33	67
Hog mucosa III	27	73
Beef lung I	41	59
Beef lung II	40	60
Beef lung III	37	63
Human umbilical cord	35	65

Sulfate and Acetyl Substituents of the Carbohydrate Components

The presence of sulfate groups as ester sulfates occur in various glycosaminoglycuronans and glycoproteins, but only in heparin and heparan sulfate) has a sulfate residue been found linked to the amino group of the 2-amino-2-deoxy sugar component. Thus, the N-sulfate (sulfoamido) group differentiates clearly the family of heparins and heparan sulfates from the chondroitin 4- and 6-sulfates, dermatan sulfate, keratan sulfate, and other sulfated glycoproteins. The N-sulfate group has been found only in heparin and heparan sulfate among all the known natural polysaccharides; in most other polysaccharides of animal origin, the amino group of the amino deoxy sugars is linked to an acetyl group, the exception being neuraminic acid, which is substituted either by an N-acetyl or an N-glycolyl group. It should be pointed out, however, that N-acetyl groups are found in various proportions in heparins having a low sulfate content (N-acetylglucosamine content 2-31%) (5, 7), and this variation exists even among heparins isolated from the same source (7).

Various methods -- such as alkaline hydrolysis (15), periodate oxidation, followed by reduction with borohydride, hydrolysis, and deamination with nitrous acid (55), and methylation (9) -- established the position of the O-sulfate groups at C-6 of the amino sugar residues and at C-2 of the uronic acid residues. The periodate oxidation of all the D-glucuronic acid residues, leaving a portion of the L-iduronic acid residues intact, clearly established the position of a sulfate ester at C-2 of the L-iduronic acid residues (35); this observation was confirmed by the isolation of oligosaccharides after nitrite degradation (18). This location of sulfate ester groups on the uronic acid component is also a characteristic that clearly differentiates the heparins from other sulfated glycosaminoglycuronans and glycoproteins.

It is evident that considerable variations of the chemical properties of the heparin molecule depend upon the number, location, and distribution of the strongly charged sulfate groups along the carbohydrate chain. Isolation of glycopeptides from heparin and heparan sulfates shows a greater proportion of N-acetyl groups near the protein-carbohydrate linkage in both heparin (6), and heparan sulfate (3), whereas N-sulfate (sulfoamido) groups are predominant near the nonreducing end, this increase being far more predominant in heparin than heparan sulfate. The remaining N-acetyl groups seem to be randomly distributed along the chain. A similar increase of the O-sulfate groups in the vicinity of the carbohydrate-protein linkage has been observed (51), and this accumulation of both N- and O-sulfate groups may reflect the influence of either the protein chain or the L-iduronic acid residues on the mechanism of sulfation.

Anomery of the Carbohydrate Components

The optical rotations of heparin (moderately positive) and heparan sulfate (slightly to moderately positive) are strikingly different from those of other glycosaminoglycuronans, owing to the presence of the α-anomer of the 2-amino-2-deoxy-D-glucopyranosyl (N-acetyl-D-glucosaminyl) residues, in contrast to the presence of the β-anomers of 2-acetamido-2-deoxy-D-glucopyranosyl residues in hyaluronic acid and of 2-acetamido-2-D-galactopyranosyl (N-acetyl-D-galactosaminyl residues) in the chondroitin 4- and 6-sulfates and dermatan sulfate. Variations of the values of optical rotation have been related to the various proportions of L-iduronic acid and D-glucuronic acid, both in heparin (8) and heparan sulfate (40) (Table III). Among all the known acidic polysaccharides of animal origin, only heparin and heparan sulfate contain the α-anomer of 2-acetamido-2-deoxy-D-glucose, and this property, like the presence

TABLE III

Correlation between D-Glucuronic Acid Content and Optical Rotation in Heparan Sulfate[a]

Source	D-glucuronic acid (%)	$[\alpha]_D^{25}$ (degrees) (c 1.0 in water)
Heparin by-products	90	+80
	80	+75
	30	+42
Beef lung	75	+68
Beef aorta	80	+60
Angloid liver	70	+70

[a]From Ref. 40.

of the N-sulfate group and of the O-sulfate group on the L-iduronic acid residue, strikingly differentiates this family of compounds from the other acidic glycosaminoglycuronans.

The uronic acid components, L-iduronic acid and D-glucuronic acid, however, have the same configurations at C-1 (α-L- and β-D, respectively) as those found in all other glycosaminoglycuronans containing these uronic acids. The determination of these various configurations has been based on the isolation of oligosaccharide fragments (44), deamination with nitrous acid (17, 44), and nuclear magnetic resonance (n.m.r.) studies (43, 44, 46).

Location of Glycosidic Linkages

Structure identification of complex polysaccharides is still based on the classical methods of degradation by chemical or enzymatic agents and structure identification of the resulting fragments, and on methylation of the polysaccharide, followed by identification of the methylated monosaccharides after hydrolysis. The peculiar properties of heparin, in which the sulfoamido group is very sensitive to acid hydrolysis, have suggested new methods of degradation based on the attack of the free amino group that is released by hydrolysis (42). Thus, the classical method of periodate oxidation, followed by sodium borohydride reduction coupled with acid hydrolysis and nitrous acid degradation (16), has proved particularly useful in giving fragments still containing sulfate groups at C-2 of the L-iduronic acid and at C-6 of the hexosamine component. Since all the D-glucuronic acid residues were oxidized by the periodate agent, it was concluded that these residues are linked glycosidically at C-4 and were not substituted by sulfate groups. The formation of a disaccharide containing a 2,5-anhydro-D-mannose 6-sulfate residue after nitrous acid degradation is clear evidence that the hydroxyl group at C-6 of the hexosamine component is not involved in the glycosidic linkage, thus confirming previous results indicating the linkage to be at C-4 (53). Identification of erythritol and L-threitol as fragments of the hydrolysis of periodate oxidized-sodium borohydride reduced heparin was another confirmation of a linkage at C-4 of the uronic acid residues (47).

Isolation of enzymes from adapted Flavobacterium heparinum by Dietrich and associates (10-14, 44, 49) and Linker and associates (20, 21, 38, 39) have also contributed to our understanding of the chemical structure of heparin and heparan sulfate. Earlier suggestions that the enzymes isolated by the first group of investigators were glycosidases (10-12) have not been sustained (14, 43) and the enzymes are, in fact, eliminases (10-14, 49). Thus, an unsaturated disaccharide trisulfate was obtained in about 70-80% yield; it was degraded first to D-glucosamine N,O-disulfate, and then to D-glucosamine 6-sulfate or D-glucosamine N-sulfate (14). Study of the unsaturated disaccharide trisulfate by n.m.r. spectrometry (44) established the repeating unit as (1-4)-O-(α-L-idopyranosyluronic acid 2-sulfate)-(1-4)-(2-deoxy-2-sulfoamino-α-D-glycopyranosyl 6-sulfate) (Fig. 1). The (1-4) linkage to the uronic acid component was suggested by the formation of the 4,5-double bond through water elimination. Recently, up to five oligosaccharides with similar structures have been isolated from the enzymic degradation of heparan sulfate (21).

No other work on the methylation of heparin and heparan sulfate has been reported since the isolation of 3-O-methyl and 2,3-di-O-methyl-D-glucose and 2-acetamido-2-deoxy-3-O-methyl- and 3,6-di-O-methyl-D-glucose from the hydrolysate of methylated, carboxyl-reduced heparin, which demonstrated the linkages at C-4 of both the D-glucuronic acid and D-glucosamine residues (9). The location of the glycosidic linkage at C-4 of the L-iduronic acid residues by the methylation procedure has not, as yet, been demonstrated, although the required, standard methylated derivatives are available (1).

Protein-Carbohydrate Linkage Region

One of the most important developments in the recent studies of the chemical structure of heparin (32, 36) and heparan sulfate (24) was the recognition that the carbohydrate chains are generally

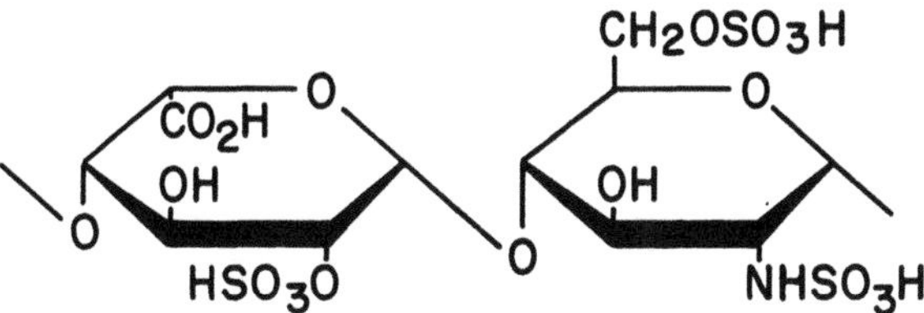

Fig. 1. Repeating unit of heparin.

linked to a protein backbone, and that the chemical structure of this "linkage region" is identical to that found for the "linkage region" of the chondroitin sulfates and dermatan sulfate. The amino acid component involved is serine, one of the amino acids that are found in the linkage of the carbohydrate chains of the mucin-type glycoprotein. In contrast to this type of compound where a 2-acetamido-2-deoxy-α-D-galactopyranosyl residue is involved in the linkage to serine (or threonine), the carbohydrate-protein linkage of the chondroitin sulfates, dermatan sulfate, heparin, and heparan sulfate involves a β-D-xylopyranosyl residue. This residue is linked to two other β-D-galactopyranosyl residues in a typical structure (Fig. 2) not found in any other known complex carbohydrate. Thus, despite the great differences in the structure of their main chains, which include residues of D-glucuronic acid, L-iduronic acid, 2-acetamido-2-deoxy-D-glucose, 2-sulfoamido-2-deoxy-D-glucose, and 2-acetamido-2-deoxy-D-galactose having various degrees and positions of O-sulfation, all the acidic polysaccharides of animal tissue, with the exception of hyaluronic acid and keratan sulfate, have in common an identical linkage region. This common feature suggests that an identical mechanism controls the initiation of the biosynthesis of these various polysaccharides.

Up to now, only D-glucuronic acid, linked to the trisaccharide of the linkage region, has been found as initiator of the carbohydrate chain of heparin or heparan sulfate, but the next uronic acid component can be either D-glucuronic acid or L-iduronic acid (Fig. 2).

No correlation has been established, as yet, between the amino acid sequence in the vicinity of the linkage to the serine residue

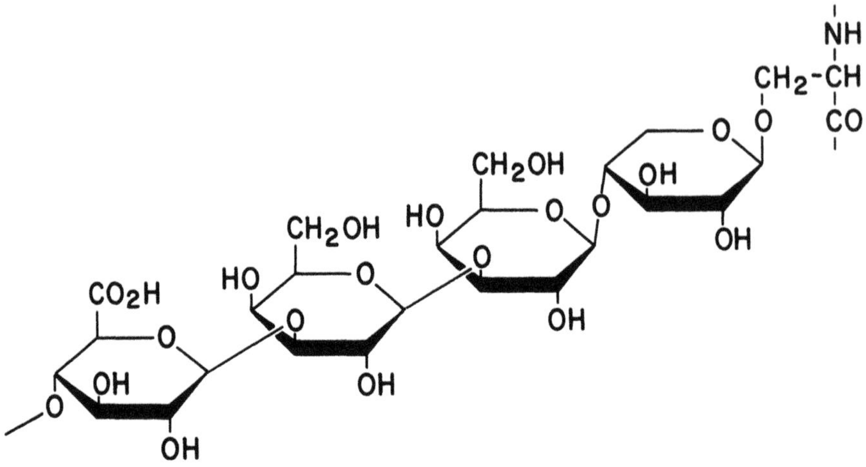

Fig. 2. Carbohydrate-protein linkage region.

and the variations in the structure and composition of the carbohydrate chains, and it will be of great interest to established whether these variations (which influence greatly the physiochemical properties of the proteoglycan) are coded in the protein backbone, in the organization of the enzymes building the chains, or depend only on the availability of the nucleotide precursors.

Protein Backbone

Our knowledge of the length, amino acid terminals, and sequence of amino acids of the protein backbone of the native heparin and and heparan sulfate proteoglycans is still very limited. There is strong evidence that the protein chain of the heparan sulfate proteoglycan is of rather high molecular weight and has numerous chains attached to it, whereas only one carbohydrate chain would be attached to a short peptide in heparin (33). The presence of "high molecular heparin" has been reported, however, and thus the presence, in heparin, of numerous carbohydrate chains attached to a high-molecular-weight protein backbone cannot be excluded (19). Since heparins containing carbohydrate chains devoid of amino acid components have also been isolated (34), it seems logical to admit the presence of a family of molecules varying in structure from numerous carbohydrate chains attached to a single protein backbone to a single carbohydrate chain without amino acids, with all the intermediate variations possible between these two extreme examples. The respective proportion of each structure could depend on numerous physiological conditions (source, location, age, hormonal influence, etc.).

Conclusions

From the brief review of the present literature, the reader will be easily convinced of the difficulties in studying the chemical structure of such a family of compounds, the heparins and heparan sulfates, where only a few of the components and other structure elements are (or seem until now to be) constant, such as the β-D-glucopyranosyluronic and α-L-idopyranosyluronic acid residues linked at C-4, the 2-amino-2-deoxy-α-D-glucopyranosyl residues linked also at C-4, and the trisaccharide linkage region. The remaining parameters, however, show great variations, such as distribution of the two uronic acid components, the degree and location of O- and N-sulfation (but restricted to C-2 of the L-iduronic acid and to C-4, C-6, and the amino group of the 2-amino-2-deoxy-D-glucose component), the length of the carbohydrate and protein chains, and the number of carbohydrate chains attached to the same protein backbone. Since any of these changes would be reflected in a change of the properties of the whole molecule (40), a study of the mechanism controlling all these parameters represents quite a formidable task.

REFERENCES

1. BAGGETT, N., STOFFYN, P.J. and JEANLOZ, R.W., J. Org. Chem., 28 (1963) 1041.
2. BRIMACOMBE, J.S. and WEBBER, J.M., Mucopolysaccharides, Elsevier Press, Amsterdam, 1964, p. 92.
3. CIFONELLI, J.A., Carbohyd. Res., 8 (1968) 233.
4. CIFONELLI, J.A. and DORFMAN, A., Biochem. Biophys. Res. Comm., 7 (1962) 41.
5. CIFONELLI, J.A. and KING, J., Carbohyd. Res., 12 (1970) 391.
6. CIFONELLI, J.A. and KING, J., Carbohyd. Res., 21 (1972) 173.
7. CIFONELLI, J.A. and KING, J., Biochim. Biophys. Acta, 320 (1973) 331.
8. CIFONELLI, J.A. and MATHEWS, M.B., Connect. Tissue Res., 1 (1972) 121.
9. DANISHEFSKY, I., STEINER, H., BELLA, A. and FRIEDLANDER, A., J. Biol. Chem., 244 (1969) 1741.
10. DIETRICH, C.P., Biochem. J., 108 (1968) 647.
11. DIETRICH, C.P., Biochemistry, 8 (1969) 2089.
12. DIETRICH, C.P., Biochem. J., 111 (1969) 91.
13. DIETRICH, C.P., NADER, H.B., BRITTO, L.R.G. and SILVA, M.E., Biochim. Biophys. Acta, 237 (1971) 430.
14. DIETRICH, C.P., SILVA, M.E. and MICHELACCI, Y.M., J. Biol. Chem., 248 (1973) 6408.

15. DURANT, G.J., HENDRICKSON, M. and MONTGOMERY, R., Arch. Biochem. Biophys., 99 (1962) 418.
16. FOSTER, A.B., HARRISON, R., INCH, T.D., STACEY, M. and WEBBER, J.M., J. Chem. Soc., (1963) 2279.
17. HELTING, I. and LINDAHL, U., J. Biol. Chem., 246 (1971) 5442.
18. HÖÖK, H., LINDAHL, U. and IVERIUS, P.-H., Biochem. J., 137 (1974) 33.
19. HORNER, A.A., J. Biol. Chem., 246 (1971) 231.
20. HOVINGH, P. and LINKER, A., J. Biol. Chem., 245 (1970) 170.
21. HOVINGH, P. and LINKER, A., Carbohyd. Res., 38 (1974) in press.
22. INOUE, S., Biochim. Biophys. Acta, 329 (1973) 264.
23. INOUE, S. and MIYAWAKI, M., Biochim. Biophys. Acta, 300 (1973) 73.
24. JACOBS, S. and MUIR, H., Biochem. J., 87 (1963) 38; KNECHT, J., CIFONELLI, J.A. and DORFMAN, A., J. Biol. Chem., 242 (1967) 4652.
25. JEANLOZ, R.W., Fed. Proc., 17 (1958) 1082.
26. JEANLOZ, R.W., in Carbohydrates, (Ed. Pigman, W. and Horton, D.) Academic Press, New York, 1970, vol. IIB, p. 615.
27. JORPES, J.E. and BERGSTRÖM, S., Z. Physiol. Chem., 244 (1936) 253.
28. JORPES, J.E. and GARDELL, S., J. Biol. Chem., 176 (1948) 267.
29. JORPES, J.E., Ann. N.Y. Acad. Sci., 115 (1964); Handb. Exp. Parmakol., 27 (1971) 143.
30. KORN, E.D., J. Amer. Chem. Soc., 80 (1958) 1520.
31. LINDAHL, U., Biochim. Biophys. Acta, 130 (1966) 308.
32. LINDAHL, U., Biochim. Biophys. Acta, 130 (1966) 388; Arkiv Kemi, 26 (1967) 101.
33. LINDAHL, U., in Chemistry and Molecular Biology of the Intercellular Matrix, (Ed. Balazs, E.A.) Academic Press, London, 1970, vol. 2, p. 943.
34. LINDAHL, U., Biochem. J., 116 (1970) 27; FRACASSINI, A., WHITE, C.J. and HUNTER, J.C., FEBS Lett., 32 (1973) 116.
35. LINDAHL, U. and AXELSSON, O., J. Biol. Chem., 246 (1971) 74.
36. LINDAHL, U., CIFONELLI, J.A., LINDAHL, B. and RODÉN, L., J. Biol. Chem., 240 (1965) 2817.
37. LINKER, A., HOFFMAN, P., SAMPSON, P. and MEYER, K., Biochim. Biophys. Acta, 29 (1958) 443.
38. LINKER, A. and HOVINGH, P., Biochim. Biophys. Acta, 165 (1968) 89.
39. LINKER, A. and HOVINGH, P., Biochemistry, 11 (1972) 563.
40. LINKER, A. and HOVINGH, P., Carbohyd. Res., 29 (1973) 41.
41. McLEAN, J., Amer. J. Physiol., 41 (1916) 250.
42. MEYER, K.H. and WEHRLI, H., Helv. Chim. Acta, 20 (1937) 353.
43. PERLIN, A.S., CASU, B., SANDERSON, G.R. and JOHNSON, L.F., Can. J. Chem., 48 (1970) 2260.
44. PERLIN, A.S., MACKIE, D.M. and DIETRICH, C.P., Carbohyd. Res., 18 (1971) 185.
45. PERLIN, A.S., MAZUREK, M., JAQUES, L.B. and KAVANAGH, L.W., Carbohyd. Res., 7 (1968) 369.

46. PERLIN, A.S., NG YING KIN, N.M.K. and BHATTACHARJEE, J.S., Can. J. Chem., 50 (1972) 2437.
47. PERLIN, A.S. and SANDERSON, G.R., Carbohyd. Res., 12 (1970) 183.
48. RADHAKRISHNAMURTHY, B., DALFERES, E.R. and BERESON, G.S., Anal. Biochem., 24 (1908) 397.
49. SILVA, M.E. and DIETRICH, C.P., Biochem. Biophys. Res. Comm., 561 (1974) 965.
50. STOFFYN, P.J. and JEANLOZ, R.W., J. Biol. Chem., 235 (1960) 2707.
51. TAYLOR, R.L., SHIVELY, J.E., CONRAD, H.E. and CIFONELLI, J.A., Biochemistry, 12 (1973) 3633.
52. VALCOURT, A.J., Pharmacy Newsletter, the Massachusetts General Hospital, Boston, April 1974.
53. WOLFROM, M.L., HONDA, S. and WANG, P.Y., Chem. Comm., (1968) 505; Carbohyd. Res., 10 (1969) 259.
54. WOLFROM, M.L. and RICE, R.A.H., J. Amer. Chem. Soc., 68 (1946) 537.
55. WOLFROM, M.L. and WANG, P.Y., Chem. Comm., (1967) 241; WOLFROM, M.L., WANG, P.Y. and HONDA, S., Carbohyd. Res., 11 (1969) 179.

DISCUSSION OF JEANLOZ PAPER

JAQUES

Dr. Jeanloz, would you comment on macro-heterogeneity as opposed to micro-heterogeneity, and whether you consider it to be equally important in what a particular investigation is reporting.

JEANLOZ

By micro-heterogeneity I meant a variation along the chain, for example in the distribution of the components or the substituents not so much a variation of the molecular size. I think it is difficult to make any clear difference between micro- and macro-heterogeneity. Macro-heterogeneity is very relative here in the field of carbohydrates. We have been thinking more or less in terms of micro-heterogeneity when the difference is small, lets say the difference between distribution of N-acetyl and N-sulfate groups, and in terms of macro-heterogeneity between different structures, for example heparin and hyaluronic acid. But I think there is no real definition of micro-and macro-heterogeneity, I am sorry to say.

JAQUES

Perhaps we used the wrong term then. Possibly gross-heterogeneity is the problem.

JEANLOZ

I will not use any adjective, and just say heterogeneity.

LINDAHL

I'd like to make a brief comment with regard to Dr. Jeanloz's description of the molecular weight of the various heparin preparations. You showed a slide with gel chromatograms of native heparin and of commercially available heparin. The "native" heparin in this particular case was isolated in our laboratory from a mouse mastocytoma, a rodent tissue; the polysaccharide was very heterogeneous and essentially of higher molecular weight than the commercial heparin. It should be noted that the differences in molecular weight are probably not due to drastic isolation conditions during preparation of the commercial heparin; rather, the high molecular weight material is a precursor which is degraded to smaller molecules fairly soon after synthesis of the polysaccharide. The macro-molecular precursor is apparently not an extended polysaccharide chain, but appears to resemble the highly branched and complex heparin first isolated by Dr. Horner from rat skin. We believe, as previously suggested by Dr. Horner, that the low molecular weight heparins isolated from various mammalian tissues are degradation products of such a precursor. In fact, we have recently been able to demonstrate the degradation process by pulse-labeling of neoplastin mouse mast cells with inorganic ^{35}S-sulfate _in vivo_.

JEANLOZ

I have one question for Dr. Lindahl. You mean by branched structure, attached to a protein core?

LINDAHL

This macromolecular branched heparin contains peptides, but it is not a proteoglycan in the general sense. The molecule includes several polysaccharide chains joined by a core structure in a so far unknown way. The core is not protein but appears to consist of polysaccharide. As this was originally suggested by Dr. Horner I think he is more competent than I am to answer this question.

HORNER

I think I should speak now as I was the person who perhaps added further confusion to the heparin field by describing macromolecular heparin. It is rewarding to find that Dr. Lindahl counts this as worthy of further pursuit. My own feeling, after working with rat skin heparin for a while, was that if it was just a peculiarity of rodents we would not spend too much time on it. However, after finding traces of this macromolecular heparin in humans, but having the problem of postmortem degradation, we proceeded to look at monkey tissues and to find lots of macromolecular and low molecular weight heparin in monkey tissues. This has led to my working hypothesis, which has been published, that the macromolecular form is an inactive precursor of physiologically active low molecular weight heparin. I would like to speak more about that tomorrow, but point out at this stage that macromolecular heparin does not seem to be a peculiarity of the rodents and may have relevance to human physiology.

CRYSTALLINE STRUCTURE OF HEPARIN

Edward D.T. ATKINS and Ian A. NIEDUSZYNSKI

H. H. Wills Physics Laboratory, University of Bristol, Bristol (UK)

Current trends in molecular biology argue convincingly in favor of a precise relation between biologic function and shape of macromolecules. This is especially relevant in the case of heparin; the evidence points to a specific inhibition by this highly charged polysaccharide of protein molecules involved in blood coagulation. Thus, information concerning the molecular structure and shape of the heparin molecule should be considered as an essential step forward in understanding the biologic mechanisms involved in coagulation.

X-ray diffraction is often used for exploring precise molecular structure because the wavelength of the radiation is roughly the same as the interatomic distances in molecules. The technique is most useful when the substance under investigation is in an ordered crystalline state, sometimes termed "solidstate," although often a considerable amount of water or mother liquor (often as high as about 100%) is present. In attempts to elucidate the molecular structure of an essentially linear polymer like heparin, with a relatively simple covalent repeating sequence, as opposed, say, to the globular proteins, usually the best that can be achieved is to induce the molecular chains to orient into a approximate parallel array and then to encourage disciplined juxtaposition of chains and precise interchain association. Such three-dimensional crystalline domains, often termed "micelles," will invariably be limited in extent and will exist with a variety of azimuthal orientations about the molecular chain direction - called the fiber axis, because the situation is closely analogous to the molecular morphology exhibited by natural fibers.

It is usual, when x-ray diffraction analysis is used on fiber patterns, for the chemical covalent repeating sequence to be established, and the x-ray diffraction data are used to elucidate the three-dimensional molecular structure of the polymer chain. When the present investigation concerning the molecular structure of heparin was started, this was not the case, for three main reasons: First, there were doubts as to whether the true covalent repeat was a tetrasaccharide or a disaccharide, or indeed, whether a unique repeating sequence could describe the character of this relatively short molecule. Second, there was much confusion concerning the anomeric configuration of the sugar residues, which, of course, have an important bearing on the glycosidic linkage geometry and, therefore, a substantial influence on the whole three-dimensional shape of the polymeric backbone. Third, much discussion centered around the chemical identity of the uronic acid units. Originally, the uronic acid composition was thought to be exclusively D-glucuronic acid alternating with equimolar quantities of D-glucosamine (40) and joined together through α 1-4, glycosidic linkages (13). Later, Cifonelli and Dorfman (12) showed the presence of the C(5) epimer, L-iduronic acid, and more recent evidence favors L-iduronic acid as the major uronic acid in heparin (26, 34, 39). The reported sulfate content of heparin averages five per tetrasaccharide, and the sulfates are thought to be distributed as illustrated in Fig. 1. The anomeric configuration of the uronic acid units has yet to be established, although nuclear magnetic resonance results by Perlin _et al._ (33) suggest the α-L configuration for the iduronic acid residues.

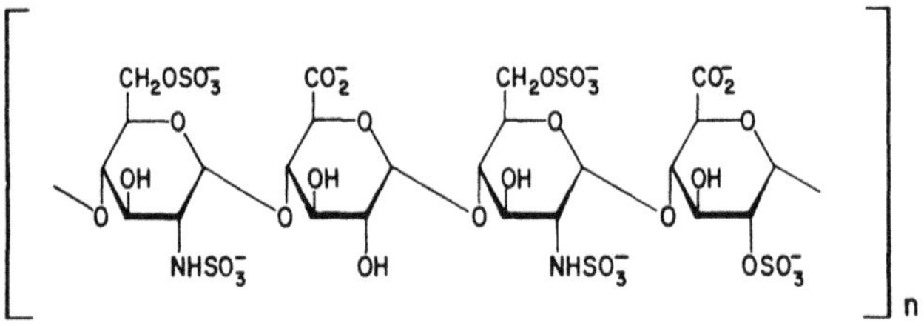

Fig. 1. Generalized covalent repeating sequence for heparin, at least until recent years, showing the α 1-4 glycosidic linkages between N-acetyl-D-glucosamine and D-glucuronic acid. Such a model would have a backbone geometry similar to that of amylose (poly-α-D-glucose). Current evidence indicates that D-glucuronic acid residues in heparin constitute only about 20-30% of the total uronic acid content and are in the β-D-configuration. The major uronic acid component is thought to be its C(5) epimer, α-L-iduronic acid. The distribution of sulfate groups, on the average five per tetrasaccharide, is still in some doubt and is probably a function of the uronic acid composition (after Lindahl, 25).

The α-D configuration previously assigned to the glucuronic acid residues has recently been discounted by Höök (18), and it is thought that most, and probably all, of the glucuronic acid residues have a β-D configuration (16, 28).

X-ray diffraction patterns from oriented material offer an independent method of examining the issues raised above and, in addition, by placing constraints on the sterochemically permissible conformations allowed by the x-ray criteria, help to establish more definitively the molecular structure of heparin.

Materials and Methods

Heparin. - The sodium salt of heparin extracted from hog intestinal mucosa was supplied by Professors Ulf Lindahl and Torvard Laurent, University of Uppsala. Sample H140 CpIII was a purified form of heparin stage XlV obtained from the Wilson Laboratories, Chicago, and is formally similar to preparation I-CPI of Lindahl *et al*. (29). It corresponds to a viscometric molecular weight of 12,000 and a sulfur: hexosamine ratio of 2.3:1. In addition, commerical Evans Pularin heparin and samples of type B heparin as used by Perlin *et al*. (33) were kindly supplied by Dr. B. Casu, Milan.

Macromolecular Heparin. - Samples of macromolecular heparin were kindly supplied by Dr. Alan Horner, Toronto. The molecular weight of the material was _ 10^6, and it was prepared from rat skin by the published method (20), except that Bio-Gel A-50m agarose gel granules, 100-200 mesh, were used for gel filtration. This macromolecular species of heparin is thought to consist of long chains, essentially similar to those of heparin, but covalently bound to a core. The sulfate content and anticoagulant activity appear to be similar to pig mucosal heparin (20, 21).

Polysaccharide Films and Fibers. - Specimens suitable for x-ray diffraction were prepared by allowing aqueous solutions (1% w/v) to dry down to make films approximately 40 μm thick, which were then cut into strips, stacked, and stretched with hanging weights at controlled relative humidity in a manner already described in detail (10).

X-ray Diffraction. - X-ray diffraction photographs were obtained by using CuK α radiation from an Elliott Rotating Anode X-ray generator. Various pin-hole collimators were used, 150-500 μm in diameter with specimen-to-film distances of approximately 4 cm. The cameras were filled with hydrogen during the exposure, and the specimens were kept at a particular relative humidity with various saturated salt solutions. The arrangement to obtain fiber-type x-ray patterns is shown schematically in Fig. 2.

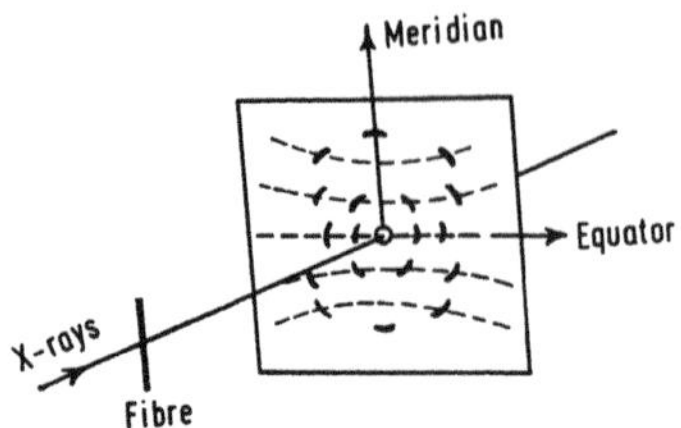

Fig. 2. Schematic arrangement for obtaining an x-ray fiber-type diffraction pattern. The oriented specimen is presented with its fiber axis perpendicular to the incident x-ray beam, and the diffraction pattern is collected on a photographic film. The specimen is sometimes tilted toward the incident x-ray beam to enhance particular meridional reflections. The layer line spacings correlate with the crystallographic repeat of the structure.

Molecular Model Building and Calculation. - Molecular models of all the structures considered were built by using both CPK space-filling models and accurate wire models with a scale of 4 cm = 0.1 nm. The models were checked by measurement and conformed to known stereochemical criteria. With the averaged coordinates of Ramachandran *et al.* (38) and Arnott and Scott (2), trial models were generated on a University ICL-4 computer, which fitted the x-ray data and were stereochemically permissible, involving no short contacts between atoms.

Results

The x-ray diffraction photograph obtained from the sodium salt of heparin (H140 Cp III) at a relative humidity of 80% is shown in Fig. 3. The reflections index on a triclinic unit cell containing only one chain segment (31). The measured spacing of the first layer line is 1.65 nm, and the simplest interpretation correlates this value directly with the crysyallographic repeating sequence along the molecular chain axis. (The value of 1.65 nm has an associated estimated error of ± 0.1 nm due to the excessive line broadening of reflections on this layer line (32). However, more recent x-ray diffraction photographs allow a more precise measurement of the spacing, and will take 1.65 nm as a firm value to avoid confusion). This value of 1.65 nm cannot be reconciled with fewer than four saccharide residues, and these must constitute the minimal cell contents. A noticeable feature of the x-ray fiber diagram (Fig. 3) is the breadth of the 001 reflection, which, after making allowance for dispersion, corresponds to a crystallite length of approximately 16 nm along the chain axis. X-ray powder diffraction patterns of the same crystalline phase were also obtained from the commercial Evans

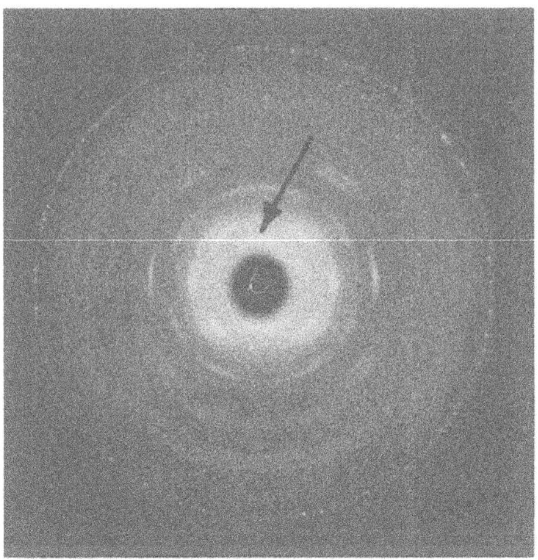

Fig. 3. X-ray fiber-diffraction photograph obtained from the sodium salt of pig mucosal heparin at a relative humidity of 80%. The fiber axis is vertical. The unit cell is triclinic, with a fiber repeat corresponding to 1.65 nm. Note the relative sharpness of the reflections in the radial direction on the equator (imaginary horizontal line), compared with 001 reflection (arrowed). The width of this reflection corresponds to a crystallite length along the chain axis of $\simeq$ 16 nm. The dotted ring is for calibration purposes.

Pularin material, which is widely used in medicine and comes from hog intestinal mucosa, and type B heparin, which is almost free of N-acetyl groups and has been characterized by the nuclear magnetic resonance data of Perlin _et al._ (33).

The x-ray diffraction pattern shown in Fig. 4 was obtained from the macromolecular heparin sample kept at a relative humidity of 78%. The reflections can be indexed on an orthorhombic unit cell with a fiber repeat of 1.73 nm (7). When the same sample is raised to a relative humidity of 84% and kept there for 24 hr, a molecular transition results, giving rise to the new diffraction pattern illustrated in Fig. 5. During this interconversion process, both patterns are observed, with the original slowly giving rise to the new one. This final pattern (Fig. 5) is a more highly oriented version of the x-ray pattern from normal heparin (Fig. 3), as originally obtained by Nieduszynski and Atkins (31). The breadth of the 001 reflection is less than that observed for normal heparin and indicates a noticeable increase in the number of coherently scattering tetrasaccharide repeats and therefore the molecular weight of an individual chain.

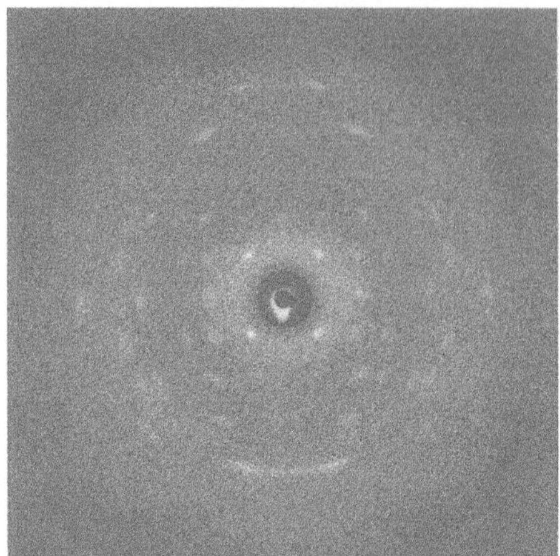

Fig. 4. X-ray fiber-diffraction photograph obtained from the sodium salt of macromolecular heparin at relative humidity of 78%. The fiber repeat in this case is 1.73 nm, and the reflections can be indexed on an orthorhombic unit cell.

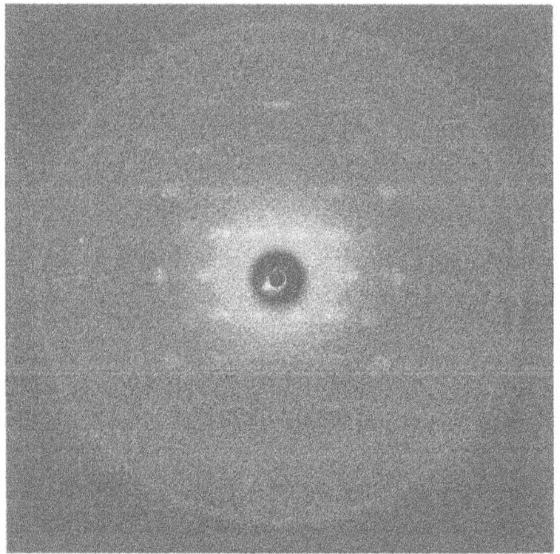

Fig. 5. X-ray-diffraction photograph obtained from the same sample as in Fig. 4, after remaining at a relative humidity of 84% for 24 hr. This x-ray pattern is a more highly oriented version of that obtained from pig mucosal heparin (Fig. 3).

Discussion

The crystallographic periodicity associated with the repeating sequence along the molecular chains has been observed to range from 1.65 to 1.73 nm, which is consistent with a tetrasaccharide repeating unit. This does not rule out a disaccharide covalent repeat for the backbone; we might have, for instance, a two-fold helix, in which each "fold" is equivalent to a disaccharide, but the different environments of successive disaccharide give rise to a crystallographic tetrasaccharide repeat. More complicated arrangements involving intertwining helices are unlikely, in that the density of the triclinic form favors only one molecular chain per unit cell, and the simplest interpretation is that heparin has a tetrasaccharide repeat. We may go a stage further since the length of the calculated crystallite size of approximately 16 nm gives a coherent scattering block of about nine tetrasaccharide repeats. The chemical composition of a tetrasaccharide with five sulfate groups and associated Na^+ counterions indicates a total of 1,200 atomic units, and the ratio of molecular weight (12,000) to this number is 10:1. The agreement supports the conclusion that a crystallographic tetrasaccharide repeat is involved and, furthermore, that it accounts for a major proportion of the molecule - at least for molecules that crystallize.

The observed axial periodicity of 1.65-1.73 nm gives a useful constraint on possible molecular models. The older and generally accepted model for heparin is based on a backbone of α-D-glucose units with exclusively α 1-4 glycosidic linkages, as shown schematically in Fig. 1 and redrawn to show the shape of the sugar rings and distribution of glycosidic bonds in Fig. 6. On the basis of standardized, average coordinates by Ramachandran _et al_. (38) and

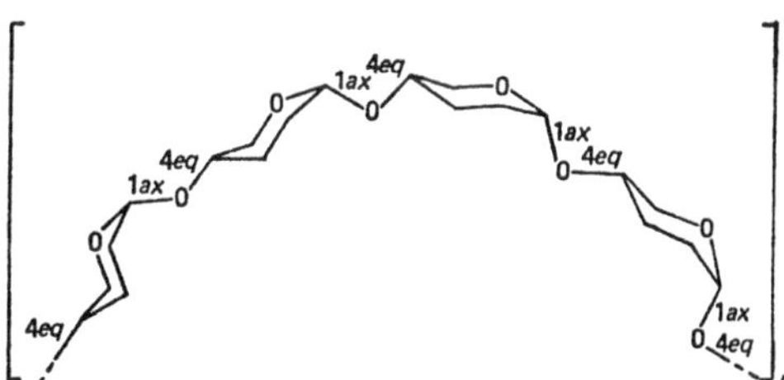

Fig. 6. Schematic diagram of model with a backbone based on N-acetyl-D-glucosamine and D-glucuronic acid and glycosidically linked α 1-4. ax, an axially disposed bond; eq, an equatorially disposed bond; D-GlcNac, represents N-acetyl-D-glucosamine, probably di- or mono- sulfated; UA, uronic acid.

Arnott and Scott (2), the maximal length of an α-D-glucose residue is 0.45 nm, which would permit theoretical tetrasaccharide periodicities of 1.80 nm. The value could be extended further, if bond-angle distortions were involved; but, at present, we will base our discussion on the standard chair shapes for the sugar residues. By considering the rotation about the glycosidic bonds in amylose (poly-α-D-glucose) designated by the angles φ and ψ in Fig. 7, Blackwell _et al_. (11) and Sundararajan and Rao (36) computed a conformational map, as shown in Fig. 8. In a comparable case with heparin, the rise per residue (h) would need to be $\geqslant 1.65/4 = 0.41$ nm, which is outside the sterically allowed region, unless, as mentioned earlier, the precise geometry of the molecule is relaxed. Extended four-fold helical conformations for amylose-inorganic salt complexes have been reported (23, 35) with axial periodicities of 1.61 nm (i.e., 0.4 nm of rise per monomer) for KBr amylose. It could be that the high ionic forces stablize such an extended chain conformation. In addition, Senti and Witnauer (35) reported layer line spacing of up to 2.26 nm, which is impossibly high for a sequence of four monomer residues, so serious consideration must be given to more complex interpretations, perhaps involving multistrand helices. On more general considerations, the structure shown in Fig. 6 is unlikely in that no good chemical evidence exists to support the presence of α-D-glucuronic acid residues. The uronic acid possibilities that appear to be available are β-D-glucuronic acid and α-L-iduronic acid, as illustrated in Fig. 9 (5, 16, 28, 33).

There is no reason to suppose that the α-D-glucosamine residues

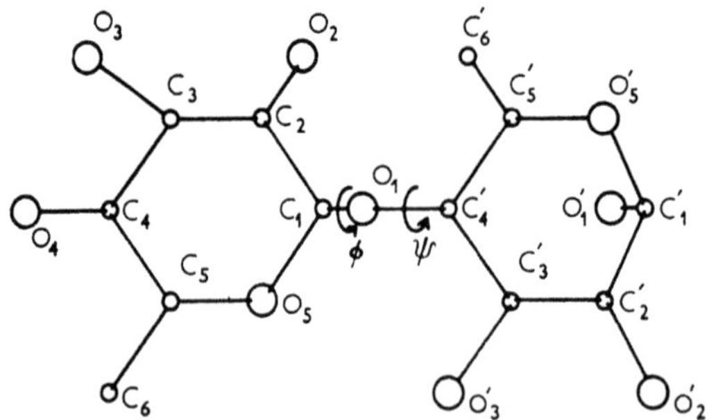

Fig. 7. Schematic representation of an α-D-glucose structure showing the two angles of rotation, φ and ψ. The $(\varphi, \psi) = (0, 0)$ starting position corresponds to C(1), O(1), O(4) and C(4'), O(1') all lying in the same plane. H(1) will be cis to H(4'). By rotations through various values of φ and ψ, a whole range of helical conformations can be generated, but only some will be stereochemically "feasible."

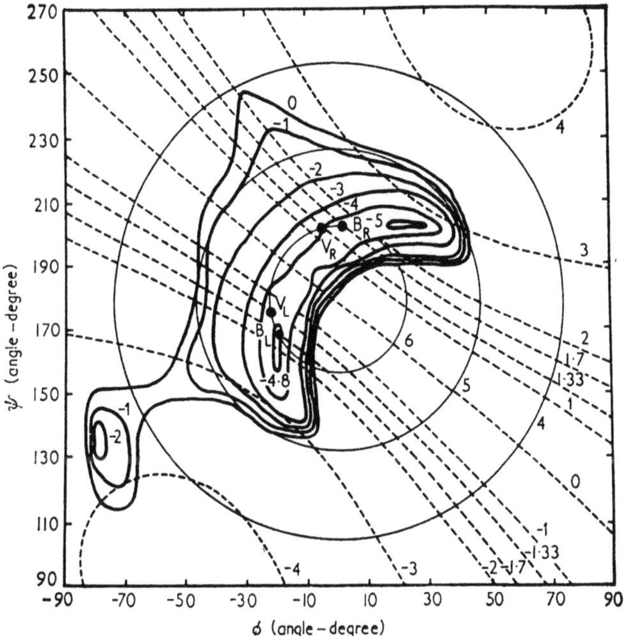

Fig. 8. Energy contour map for the unit shown in Fig. 7, with a bridge angle of 117°. Heavy line, potential energy in Kcal/residue. Thin lines, contours of constant n, the number of residues per turn of helix. Dashed lines, contours of constant h, the rise per residue in Å units parallel to the axis of the helix. Positive values of h indicate a right-hand helix, negative values a left-hand helix (after Blackwell et al., (11). Note that the potential energy is positive for values of h greater than 4Å (0.4 nm).

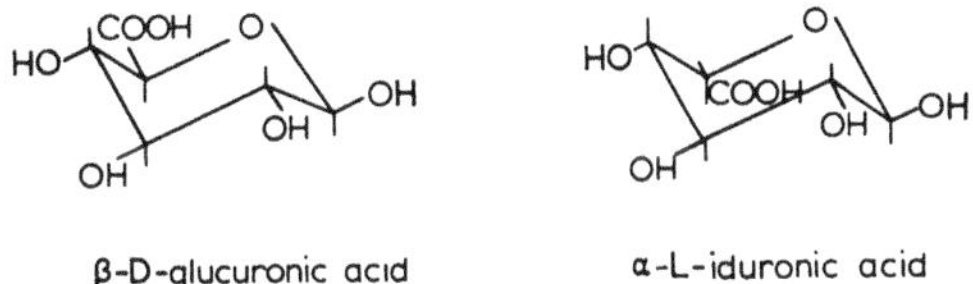

Fig. 9. Diagrammatic representation of both possible uronic acid residues in the normal C1 chair form. α-L-iduronic acid is the C(5) epimer of β-D-glucuronic acid.

are in other than the normal C1 chair conformation, linking through their 1-positions axially and their 4-positions equatorially. However, the precise uronic acid composition is in some doubt, and the β-D-glucuronic acid residues are more likely to exist in their energetically favorable C1 chair conformation and be diequatorially linked. Further consideration needs to be given to the chair shape of the α-L-iduronic acid residue. Atkins and Laurent (5) and Atkins and Isaac (3), in studies on dermatan sulfate, were able to show that their results were consistent with a C1 chair conformation for α-L-iduronic acid residues may adopt a solid-state conformation different from that found in dermatan sulfate as a consequence of sulfation in the 2-position (it could be in the 3-position without affecting the argument at this stage). In chondroitin 4-sulfate (22), dermatan sulfate (4, 22), and the plant polysaccharide carrageenins (1), the sulfate groups adjacent to the linkage positions are all axially placed, and inspection of molecular models suggests that axial sulfate groups adjacent to the linkage positions are sterically less cumbersome. Thus, if α-L-iduronic acid adopted the alternate 1C chair conformation, it would dispose its large side groups favorably - the sulfate group axial and the carboxyl equatorial. Lindahl *et al.* (27 and elsewhere in this volume) and Höök *et al.* have proposed that β-D-glucuronic acid is C(5) epimerized to α-L-iduronic acid at the polymer level and that the process requires concomitant sulfation of the polymer. The process may be considered as analogous to (but more complex than) the plant polysaccharide alginates. Here, β-D-mannuronic acid is C(5) epimerized to α-L-guluronic acid at the polymer level (15, 24). In this case, x-ray diffraction firmly shows that the sugar ring changes from the normal C1 chairs to the alternate 1C on epimerization (6, 8, 9). Returning to heparin, the elegant work of Lindahl and collaborators suggests that β-D-glucuronic acid and α-L-iduronic acid are in equilibrium at the polymer level, but that later sulfation of the α-L-iduronic acid residues (there is no evidence of sulfation of β-D-glucuronic acid) removes them from the equilibrium and drives the epimerization process (18, 27 and elsewhere in this volume). It is tempting to speculate that this process may be coupled with a change in the chair conformation of the sulfated α-L-iduronic acid, as illustrated in Fig. 10. It could then be argued that the reverse epimerase could not operate on this radically altered molecular shape. Indeed, Fransson (14), in periodate oxidation experiments, has put forward a similar suggestion for the blocks of sulfated α-L-iduronic acid in dermatan sulfate.

It is reasonable to assume, taking all the current evidence into account, that the heparin molecule is first polymerized as N-acetyl-α-D-glucosamine alternates with β-D-glucuronic acid. The molecular structure would be expected to be similar to that shown in Fig. 11. Each sugar ring is in the C1 chair form, and the glycosidic linkages are alternately 1(ax)-4(eq) between N-acetylglucosamine and uronic acid and 1(eq)-4(eq) between uronic acid and

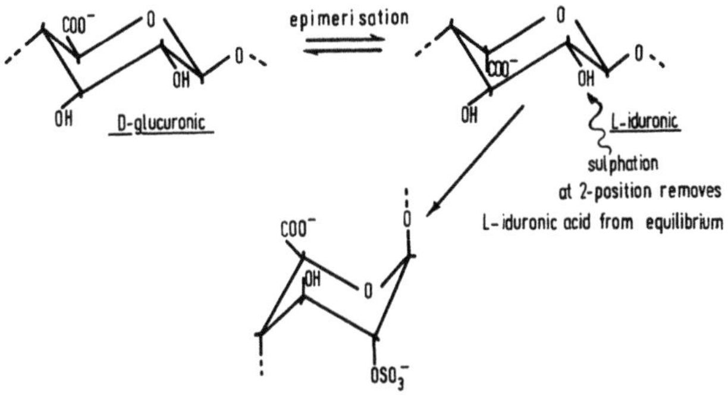

Fig. 10. A possible scheme for driving the epimerization process of β-D-glucuronic acid to α-L-iduronic acid. It is envisaged that there is an equilibrium between the two epimers. Sulfation of the L-iduronic acid in the 2-position might result in a change in the chair conformation to the alternate 1C form. This new shape effectively removes L-iduronic acid residues, and further epimerization must occur to maintain the equilibrium.

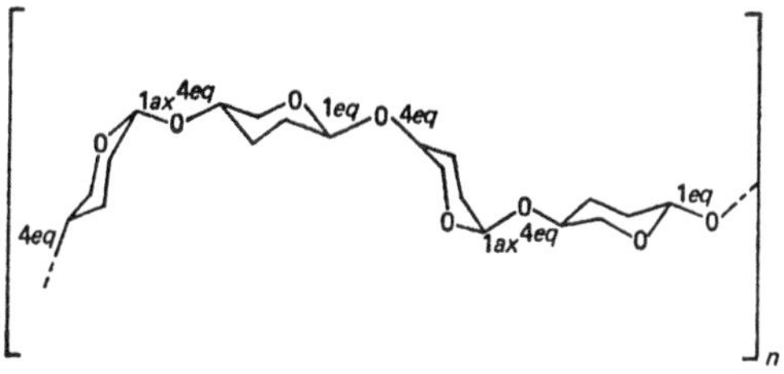

Fig. 11. Schematic representation of a tetrasaccharide repeat in which both uronic acid residues adope the normal chair conformation.

N-acetylglucosamine (i.e., alternating α and β linkages). The maximal theoretical extension, on the basis of published standard coordinates and an assumption of no bond angle (or length) distortion, is 1.98 nm, and space-filling trial structures can be built with measured repeats ≥ 1.9 nm. Such a model is consistent with the structure proposed for heparan sulfate with an observed repeat of 1.86 nm (5), as illustrated by the computer-generated two-fold helix shown in Fig. 12. This backbone conformation could be used as the basis for heparin by replacing the β-D-glucuronic residues, either just one per

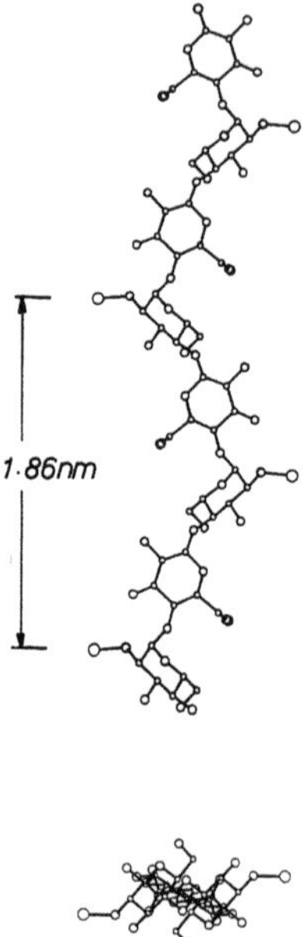

Fig. 12. Computer-drawn projections of a two-fold helix consistent with the 1.86 nm measured repeat for heparan sulfate (5). This backbone, consisting of β-D-glucuronic acid residues alternating with N-acetyl-α-D-glucosamine, probably represents the structure of heparin before epimerization and sulfation at the polymer level. The top diagram represents a view perpendicular to the chain axis, the bottom diagram, a view looking down the chain axis. All the sugar residues are in the normal C1 chair conformations.

tetrasaccharide, or both, with α-L-iduronic acid and appending sulfate groups at the appropriate positions. Figure 13 shows a computer-generated model designed to fit the observed 1.73 nm repeat. A range of variations between the covalent repeat shown in Fig. 12 and that shown in Fig. 13 are possible without at this stage invoking changes to the normal C1 chair conformation of the sugar residues.

However, as mentioned previously, incorporation of sulfated α-L-iduronic acid could well result in a change in conformation to the alternate 1C chair (see Fig. 10) for these particular residues. Figure 14 illustrates schematically the effect of such a change on the polymer backbone. Here we have a more complex glycosidic linkage geometry with one of the uronic acids becoming 1-4 diaxially linked.

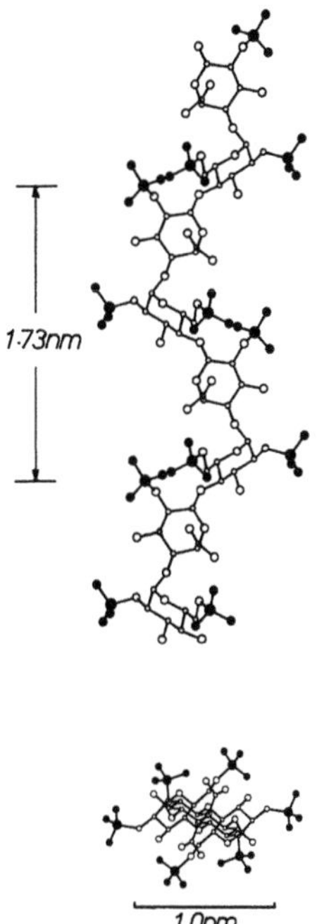

Fig. 13. Trial computer-drawn projections of a two-fold helical structure consistent with the 1.73 nm repeat observed for macromolecular heparin. The top is a view perpendicular to the chain axis; the bottom, a view parallel to the chain axis. The uronic acid residues are α-L-iduronic acid in the normal C1 chair conformation. There are six sulfate groups per tetrasaccharide denoted by filled circles. The repeat of 1.73 nm could also be obtained with a different backbone involving 1C chairs for the uronic acid residues. Note how the sulfate groups lie on the periphery of the molecule.

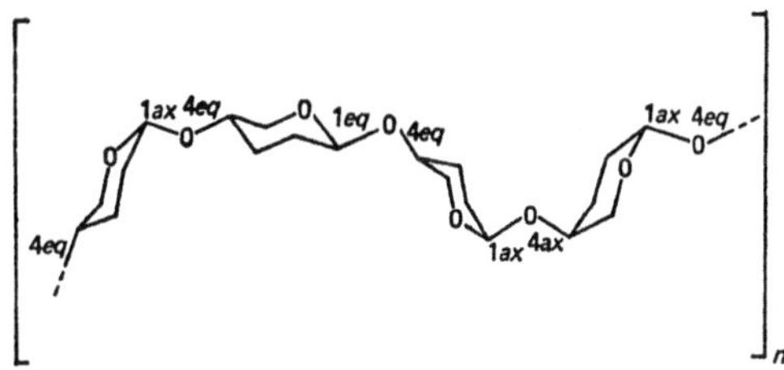

Fig. 14. Schematic diagram of a tetrasaccharide repeat in which one of the uronic acid residues adopts the alternate 1C chair conformation. There are three types of glycosidic linkage.

We would anticipate that the uronic acid residue in the normal C1 chair would be either β-D-glucuronic acid or α-L-iduronic acid, and that the uronic acid residue in the alternate 1C chair conformation would probably be sulfated α-L-iduronic acid. Furthermore, a natural drop in the value of the tetrasaccharide repeat by approximately 0.1 nm in converting a 1-4 diequatorial linked residue to a 1-4 diaxial linked residue would not be surprising and would give rise to a value near the observed repeat of 1.73 nm (1.86 nm observed for heparan sulfate minus 0.1 nm). There is, however, no chemical evidence of a precise alternation in uronic acid residues, and this type of conformation is not thought likely to represent a repeating sequence of tetrasaccharides.

Finally both uronic acid residues for a tetrasaccharide could exist in the alternate 1C chair conformation and result in a pair of 1-4, diaxially linked residues (Fig. 15). The polymer repeat would tend to be reduced again, and model-building trials with space filling models readily give repeats of around 1.65 nm, as illustrated by the computer-generated two-fold helical conformation shown in Fig. 16. In this case, both uronic acid residues would be considered as sulfated α-L-iduronic acid. This model, as drawn, has six sulfate appendages per tetrasaccharide; obviously, if we wish to maintain the average of five sulfates per tetrasaccharide, one could be removed from the glucosamine residues. The results of Linker and Hovingh (30) indicate that more than 30% of the glucosamine residues are disulfated. Indeed, if the model shown in Fig. 16 is roughly correct, we would expect some asymmetry of the tetrasaccharide repeat to account for the observed strength of the 001 reflections (see Fig. 3 and 5).

Conclusion

The structure of heparin has passed through some twists and turns since the pioneering work concerning the covalent repeat beginning with Wolfrom and his collaborators. It is interesting that substituting α-L-iduronic acid residues for the original suggestion of α-D-glucuronic acid residues maintains an α 1-4 linked polymer. The result of this epimerization creates a molecular shape, or more properly shapes, completely different from that originally imagined as formally similar in backbone geometry to amylose. The wealth of evidence is against this latter type of structure for heparin.

At this stage, it is important to keep an open mind regarding the various possibilities for molecular shape. The observation of two distinct shapes for macromolecular heparin could be the result of a readjustment of the polymer by rotation about the glycosidic linkages. Alternatively, the x-ray diffraction method could be monitoring a transition from one chair conformation to another as a

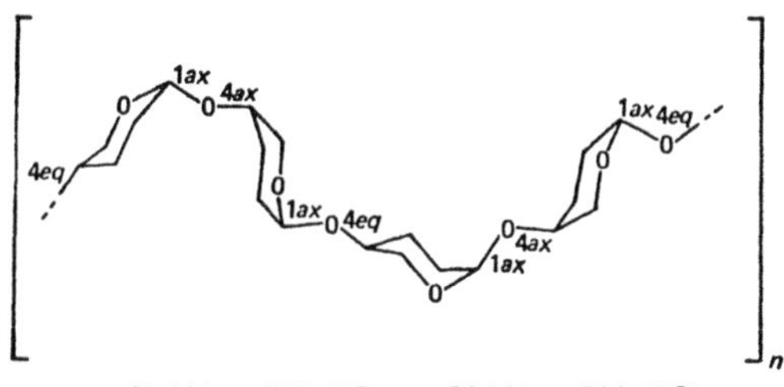

Fig. 15. Diagrammatic representation of a tetrasaccharide repeat in which both uronic acid residues are in the 1C chair form and are 1-4 diaxially linked.

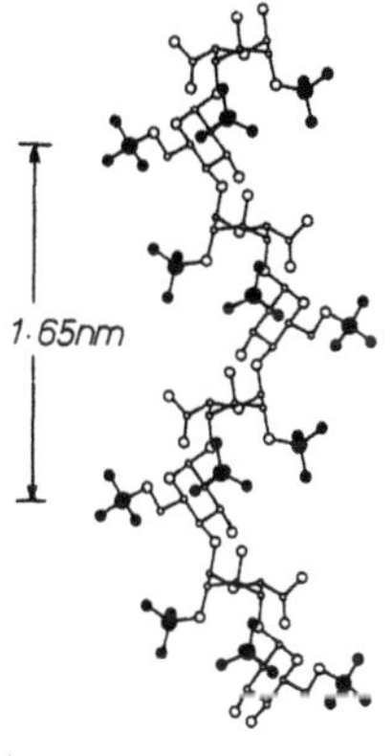

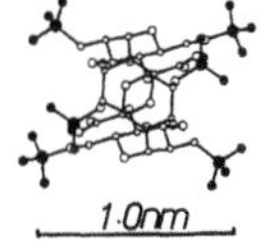

Fig. 16. Trial computer-drawn projections of a two-fold helical structure with both α-L-iduronic acid residues in the alternate 1C chair conformation. The model was designed to fit the repeat of 1.65 nm. Note that the sulfate groups have a distribution different from that shown in Fig. 13.

function of the water environment. Such a change would be exciting and offers scope for this highly charged versatile molecule to perform subtle conformational maneuvers.

The precise molecular shape of heparin is likely to be determined by the amount of α-L-iduronic acid present and by the proportion of the L-iduronic acid that is sulfated. As working models, the two-fold helical structures shown in Fig. 14 and 16 are likely mean conformations for the heparin molecule. In attempts to prepare substances that can mimic heparin's blood anticoagulant activity, backbone shapes similar to those discussed will need to be considered, rather than an amylose backbone.

It is important to remember that the x-ray method, although precise, says nothing about the parts of the molecule, or molecules, that do not crystallize. Any variation in structure (Cifonelli, elsewhere in this volume; Taylor et al.(37)) observed chemically may be accommodated by such noncrystalline regions. What can be said is that molecules that do crystallize exhibit a coherent scattering length accounting for a major part of the molecule. It will be of interest to examine and compare samples from different sources. More detailed analysis of the x-ray patterns given by macromolecular heparin should enable the chair conformations to be elucidated.

REFERENCES

1. ANDERSON, N.S., CAMPBELL, J.W., HARDING, M.M., REES, D.A. and SAMUEL, J.W., J. Mol. Biol., 45 (1969) 85.
2. ARNOTT, S. and SCOTT, W.E., J. Chem. Soc., Perkin Trans II, (1972) 324.
3. ATKINS, E.D.T. and ISAAC, D.H., J. Mol. Biol., 80 (1973) 773.
4. ATKINS, E.D.T., ISAAC, D.H., NIEDUSZYNSKI, I.A., PHELPS, C.F. and SHEEHAN, J.K., Polymer, 15 (1974) 263.
5. ATKINS, E.D.T. and LAURENT, T.C., Biochem. J., 133 (1973) 605.
6. ATKINS, E.D.T., MACKIE, W. and SMOLKO, E.E., Nature, 225 (1970) 626.
7. ATKINS, E.D.T., NIEDUSZYNSKI, I.A. and HORNER, A., Biochem. J., (1974) in press.
8. ATKINS, E.D.T., NIEDUSZYNSKI, I.A., MACKIE, W., PARKER, K.D. and SMOLKO, E.E., Biopolymers, 12 (1973) 1865.
9. ATKINS, E.D.T., NIEDUSZYNSKI, I.A., MACKIE, W., PARKER, K.D. and SMOLKO, E.E., Biopolymers, 12 (1973) 1879.
10. ATKINS, E.D.T., PHELPS, C.F. and SHEEHAN, J.K., Biochem. J., 128 (1972) 1255.
11. BLACKWELL, J., SARKO, A. and MARCHESSAULT, R.H., J. Mol. Biol., 42 (1969) 379.
12. CIFONELLI, J.A. and DORFMAN, A., Biochem. Biophys. Res. Commun., 7 (1962) 41.

13. DANISHEFSKY, I. and STEINER, H., Biochim. Biophys. Acta, 101 (1965) 37.
14. FRANSSON, L-Å, in Proc. Annu. Symp. Soc. Study Inborn Errors Metab. II, (1974) in press.
15. HAUG, A. and LARSEN, B., Carbohyd. Res., 17 (1971) 297.
16. HELTING, T. and LINDAHL, U., J. Biol. Chem., 246 (1971) 5442.
17. HIRANO, S., Int. J. Biochem., 3 (1972) 677.
18. HOOK, M., Abstracts of Uppsala Dissertations from the Faculty of Medicine 197, Univ. of Uppsala, Sweden.
19. HOOK, M., LINDAHL, U. and IVERIUS, P.H., Biochem. J., 137 (1974) 33.
20. HORNER, A., J. Biol. Chem., 246 (1971) 231.
21. HORNER, A., Proc. Nat. Acad. Sci., 69 (1972) 3469.
22. ISAAC, D.H. and ATKINS, E.D.T., Nature New Biol., 244 (1973) 252.
23. JACKOBS, J.J., BUMB, R.R. and ZASLOW, B., Biopolymers, 6 (1968) 1659.
24. LARSEN, B. and HAUG, A., Carbohyd. Res., 17 (1971) 287.
25. LINDAHL, U., in Chemistry and Molecular Biology of the Intercellular Matrix, (Ed. Balazs, E.A.) Academic Press, London and New York, 1970, p. 943.
26. LINDAHL, U. and AXELSSON, O., J. Biol. Chem., 246 (1971) 74.
27. LINDAHL, U., BACKSTROM, G., MALMSTROM, A. and FRANSSON, L-Å, Biochem. Biophys. Res. Commun., 46 (1972) 985.
28. LINDAHL, U., BACKSTROM, G., JANSSON, L. and HALLEN, A., J. Biol. Chem., 248 (1973) 7234.
29. LINDAHL, U., CIFONELLI, J.A., LINDAHL, B. and RODEN, L., J. Biol. Chem., 240 (1965) 2817.
30. LINKER, A. and HOVINGH, P., Biochemistry, 11 (1972) 563.
31. NIEDUSZYNSKI, I.A. and ATKINS, E.D.T., Biochem. J., 135 (1973) 729.
32. NIEDUSZYNSKI, I.A. and ATKINS, E.D.T., in Colston Papers No. 26, Butterworths, London, (1974) in press.
33. PERLIN, A.S., CASU, B., SANDERSON, G.R. and JOHNSON, L.F., Can. J. Chem., 48 (1970) 2260.
34. PERLIN, A.S., MAZUREK, M., JAQUES, L.B. and KAVANAGH, L.W., Carbohyd. Res., 7 (1968) 369.
35. SENTI, F.R. and WITNAUER, L.P., J. Poly. Sci., 9 (1952) 115.
36. SUNDARARAJAN, P.R. and RAO, V.S.R., Biopolymers, 8 (1969) 313.
37. TAYLOR, R.L., SHIVELY, J.E., CONRAD, H.E. and CIFONELLI, J.A., Biochemistry, 12 (1973) 3633.
38. RAMANCHANDRAN, G.N., RAMAKRISHNAN, C. and SASISEKHARAN, V., J. Mol. Biol., 7 (1963) 95.
39. WOLFROM, M.L., HONDA, S. and WANG, P.Y., Carbohyd. Res., 10 (1969) 259.
40. WOLFROM, M.L., WEISBLAT, D.I., KARABINOV, J.V., McNEELY, W.H. and McLEAN, J., J. Amer. Chem. Soc., 65 (1943) 2077.

DISCUSSION OF ATKINS AND NIEDUSZYNSKI PAPER

ROSENBERG

I wonder if Dr. Atkins would comment on when he thinks the end effects in his helix might begin to dominate and alter the conformation of the molecule as one started to reduce the molecular weight of heparin.

ATKINS

It is difficult to really answer that particular question precisely. If you limit the number of repeats to much below 10, then the x-ray diffraction method becomes imprecise and the information that you obtain becomes eventually meaningless. So it depends really whether we are talking about the commercial heparin with about 10 or 12 tetrasaccharide repeats long or the macromolecular heparin. If you mean the low molecular weight heparin, then if the number of coherant repeats decreased by 10% say, then one would expect poor crystallization and less precise x-ray data. This seems to fit in with some of the preparations that we obtained from Tony Cifonelli in which I suspect the region which was described by Tony Cifonelli as the B blocks are quite small, and therefore the end effects are quite large and the material is difficult to crystallize.

JEANLOZ

Dr. Atkins, if Dr. Lindahl and Dr. Horner are correct, and the high molecular weight form is a branched molecule, how do you explain that you can get better x-ray diffraction patterns? How do you arrange those into a micellar form?

ATKINS

Oh yes, that is very interesting! It probably results from the higher molecular weight giving more mechanical strength to the films. Of course, what you have not seen maybe, is that we also crystallized the chondroitin sulfate proteoglycans, which are even more complicated. It is a situation that is giving us, as physicists, some amusement topologically at present. If you take the proteoglycan material bound onto hyaluronic acid, with chondroitin-4-sulfate chains coming off with shorter keratan sulfate chains and there are n of these "bottle brushes" linked to the hyaluronic acid core, the whole system crystallizes and maybe afterwards I can show you some x-ray diffraction patterns (see for example, Atkins *et al*, 1974†). So there are

†Atkins, E.D.T., Hardingham, T.E., Isaac, D.H. and Muir, H. (1974) *Biochem. J.* (in press).

topological difficulties, but certainly these chains, and certainly in the case of macromolecular heparin, the chains manage to aggregate together at the expense of the backbone core. In macromolecular heparin it is somewhat simpler I guess than the aggregated proteoglycan.

JAQUES

My question is simple, you have shown your structure is based on a tetrasaccharide of Pearl and Detrich and myself, and our work used a hexasaccharide as the unit. There are various pieces of evidence supporting this. The most significant I think is the fact that on bacterial enzyme degradation of heparin, one can get a hexasaccharide which has the same or slightly higher metachromatic activity with toluidine blue than with the original heparin, but as soon as you go below that, then the metachromatic activity disappears. Now, is there any significant reason that your structure, your reasoning, wouldn't follow, still be much the same with a hexasaccharide instead of a tetrasaccharide as the repeating unit.

ATKINS

Well, as you saw from the data I presented, there is no evidence at present in this work for a repeating hexasaccharide, but you could imagine one: it would be one and a half tetrasaccharide repeat with a similar backbone conformation. The basic backbone repeat is probably a disaccharide and that by adding appendages, perturbated structures with tetra- or hexasaccharide character could be formed.

THREE-DIMENSIONAL MODEL OF HEPARIN

Leon YUAN[+] and Salvatore S. STIVALA[++]
[+]Department of Biochemistry, Loyola University School of Dentistry, Maywood, Illinois 60153 (USA) and [++]Departments of Chemistry and Chemical Engineering, Stevens Institute of Technology, Hoboken, New Jersey 07030 (USA)

Heparin is best known for its blood anticoagulant activity, and considerable effort has been devoted to determining the structural features responsible for this activity. It has been widely accepted that the anticoagulant activity is related to the degree of sulfation and the molecular conformation (3, 6). The importance of the carboxyl group has recently been suggested by Stivala and Liberti (13). Little is known about what aspects of the three-dimensional structure of heparin that render it a highly active anticoagulant. We would like to present our preliminary thoughts on this subject, which are based on constructed molecular models and other data. In proposing a reasonable three-dimensional model, we have carefully examined the experimental results of previous physical and chemical studies on heparin, which were recently reviewed (4).

Heparin cannot be readily crystallized, so an x-ray diffraction study requires some other appropriate technique for sample preparation. Atkins and Sheehan (2) obtained the x-ray diffraction pattern of a stretched film of heparin, prepared from a deformable gel or putty, which showed crystallinity and orientation of the molecule. They proposed a helical model with a calculated pitch length of 16.5 Å and two disaccharide units per pitch. The calculated value of the pitch length indicates that the heparin backbone is highly extended.

Does heparin exhibit the same type of three-dimensional structure in solution as in the solid state? In answering this question, we should examine various physicochemical studies. Stone (15) studied a heparin-dye complex with optical rotatory dispersion and suggested that heparin behaves like a helical polypeptide. Heparin exhibits typical polyelectrolyte behavior in aqueous solution, owing to its high negative charges (8, 19). According to Lifson's theoretical study of the influence of interactions between neighboring ionizable groups of a polyelectrolyte on the internal rotation around its skeletal bonds (9), the repulsive forces between neighboring groups tend to contract the molecule by twisting the molecular chain, thus acting against other forces that tend to extend the molecular dimension. This theory explains the experimental results of a viscosity study of heparin in different solvent systems (20), in which the shielded intrinsic viscosity (19), $[\eta]_\infty$, decreases as electrostatic repulsive forces between neighboring charge groups increase. These data suggest that strong repulsive forces between charge groups tend to contract the heparin molecule into a more compact conformation. In a self-hydrolysis study of heparin, it was found that the principal change is that the N-sulfate bond is cleaved, which result in an amino group (7, 14). These amino groups can form hydrogen bonds with sulfate groups or can pick up H^+ counterions. This result in a positively charged group on the backbone that can attract the negative sulfate groups (zwitterion formation). The zwitterion formation tends to pull the molecule together, which yields a still more compact conformation. This is in agreement with the viscosity results, which show a decrease in shielded intrinsic viscosity with increasing desulfation (20).

When heparin is dissolved in a urea solution of high concentration, the viscosity changes continuously with time (18, 20). This phenomenon resembles the denaturation of proteins in similar solvents resulting in the helix-coil transition. Therefore, the change in viscosity of heparin-urea solution suggests that there is a transition from an ordered to a less ordered structure. Furthermore, addition of urea to an aqueous heparin solution results in a gradual increase in the shielded intrinsic viscosity, suggesting an unfolding of the chain.

In the acid hydrolysis of the heparin-Cu(II) complex, a sharp decrease in viscosity is observed before desulfation takes place, which may best be explained as follows: the protons from the acid attack the heparin chain, presumably at the sulfamino groups; this results in a change in the charge density and/or charge distribution; a conformational rearrangement of the backbone follows. These data, with those from other solution studies mentioned, suggest that heparin in solution may have some ordered but compact conformation.

Stivala and co-workers (12), using low-angle x-ray scattering, reported that heparin behaves like a Gaussian coil molecule in pure water. They calculated the molecular weight of a fractionated heparin sample to be 12,900, the radius of gyration to be 35.9 Å, and the persistence length to be 21.1 Å. If the molecular shape is known, the radius of gyration can be used to calculate molecular dimensions from the following relations where $\bar{R}$, $\bar{r}$, and L are the

$$(\bar{R}^2) = \frac{(\bar{r}^2)}{6} \quad \text{(coil)}$$

$$(\bar{R}^2) = \frac{L^2}{12} \quad \text{(rod)}$$

radius of gyration, end-to-end distance of a coil, and length of the rod, respectively. The calculated values of $\bar{r}$ and L are 65 Å and 124 Å, respectively. This fractionated heparin sample had an intrinsic viscosity of 0.183 dl/g when measured in 0.1 M NaCl. For the above heparin sample, based on the Flory-Fox model, $(\bar{r}^2)$ was calculated to be 105 Å. If heparin is assumed to behave like a prolate ellipsoid, the axial ratio, a/b, is calculated to be 20 by Simha's equation. Recently, Stivala and Ehrlich (5, 11) calculated $(\bar{r}^2)$ of a series of unhydrolyzed and hydrolyzed heparin fractions in 0.5 M NaCl at a pH of 2.5 from various hydrodynamic models of linear chains, e.g., those of Flory-Fix, Kirkwood-Riseman, Peterlin, Kuhn-Kuhn.

In general, the counterions of polyelectrolytes in solution are not totally dissociated (10). When the sodium salt of heparin is dissolved in water, experiments show that about 63% of the sodium ions are closely associated with the heparin molecule itself (1). From this study, roughly one of three sodium ions of heparin is free; the other two are within the molecular domain. In our proposed molecular model, only O-sulfate groups are assumed to be predominantly charged. Previous dye-binding data showed that two dye molecules react with two disaccharide units. Apparently, the reaction occurs preferentially between two carboxyl or sulfamino groups (16, 17).

The binding of copper ions of heparin was reported by Stivala and Liberti (13) to depend on pH, as shown in Fig. 1. When the pH is changed from neutral to acidic, Cu(II) binding decreases by about 50%. In view of the fact that the pK_a of carboxyl groups in heparin ranged from a pH of 3 to 4, at which a maximal rate of change in binding is observed, one could reasonably conclude that carboxyl groups participate in the binding of Cu(II). Inasmuch as Cu(II) binding of heparin decreases by about 50% when the pH goes from 7 to 3, it is also reasonable to assume that both carboxyl and sulfate groups are involved, in approximately equal numbers. The Cu(II) binding data for the T 79 series of heparin (13) are shown in Table I. For sample T 79-2, only 7.86 moles of copper are bound to one mole of heparin. The molecular weight of this sample is known, and the formula weight of the tetrasaccharide unit is assumed to be

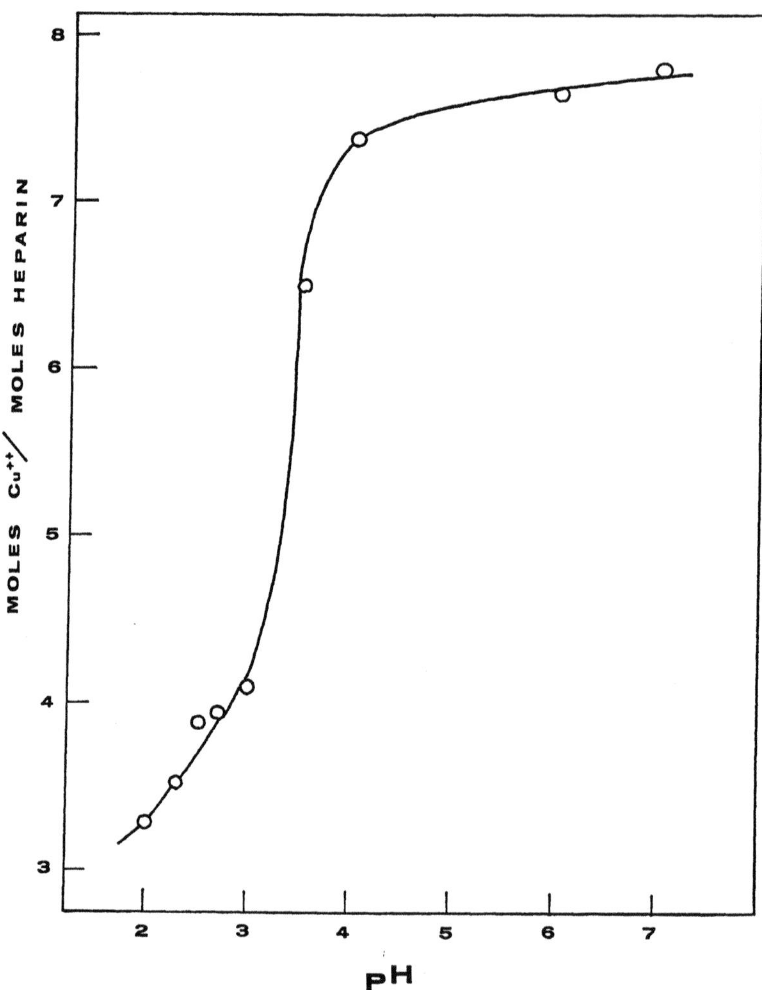

Fig. 1. Cu(II) binding of heparin fraction T 79-2 as a function of pH. Reprinted with permission from Stivala and Liberti (13).

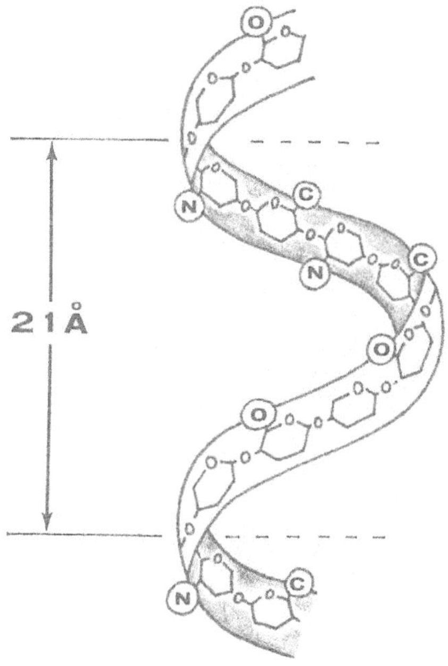

(O): O-SULFATE

(N): N-SULFATE

(C): CARBOXYL

Fig. 2. Schematic representation of the proposed three-dimensional model of heparin.

TABLE I

Cu(II) Binding of Heparin Fractions at a pH of 7

Heparin Fraction	Molecular Weight	Binding per Copper Micromoles of Copper per Milligram of Heparin [a]	Moles of Copper per Mole of Heparin	Sugar Rings per Copper Ion
T 79-2	12,700	0.614	7.86	5.5 ± 0.3
T 79-3	11,400	0.591	6.76	5.8 ± 0.3
T 79-4	7,500	0.542	4.05	6.3 ± 0.3
PR 799 [b]	10,500	0.562	5.90	6.0 ± 0.3

[a] Data from Stivala and Liberti (13).

[b] Original unfractionated sample.

1,177±50, so we can calculate the number of sugar rings associated with each copper ion. This value is shown in the last column of Table I. There are 5.5 monosaccharide units associated with one copper ion for sample T 79-2. Because copper has an "apparent" valence of four, it is reasonable to assert that one copper ion should associate with two disaccharide units, which have two carboxyl, two N-sulfate, and two O-sulfate groups each.* This was not the case for heparin; an average of three disaccharide units per copper ion was observed (Table I). This observation tends to support the premise that in our proposed model the loose secondary helical structure does not hold throughout the entire chain.

Therefore, on the basis of the copper-binding results, including the preceding discussion of the physicochemical data, we propose a molecular model of the heparin-Cu(II) complex in solution. This model is shown schematically in Fig. 2. The molecular chain of heparin is twisted into a regular but loose helical coil with four disaccharide units in each loop. When these four disaccharide units are fully extended, the total length will be about 42 Å. Because the solution studies suggest some ordered and compact conformation for heparin and low-angle x-ray scattering study indicates a persistence length of 21 Å, we are inclined to believe that the segmental length for four disaccharide units is 21 Å. The diameter of this helix is estimated to be 10 Å. Two-thirds of the ionizable groups, including all the N-sulfate and carboxyl groups, are buried inside the groove of the helix, where their counterions are localized. The other one-third of the ionizable groups, the O-sulfate groups, appear on the surface, with their counterions fully dissociated to form an ionic atmosphere that surrounds the whole molecule. This secondary conformation may not necessarily hold throughout the entire chain. When copper ions are introduced into the heparin system, they will be localized inside the heparin molecule to resemble the structure of the starch-iodine complex in solution. This analogue seems to be reasonable, particularly because heparin has α 1-4 linkages, as does the starch molecule.

The model we propose is speculative, in that it would require additional data to substantiate. Under the circumstances, the model appears reasonable. It is to be noted, that two disaccharide units or three disaccharide units per pitch, instead of four disaccharide units, could have been specified.

ACKNOWLEDGEMENTS

We wish to thank Dr. Arthur Veis and Dr. G.W. Rapp for very helpful discussion.

---

* Since we have already assumed that o-sulfate groups are completely ionized, they should have no tendency of binding Cu(II) ions.

REFERENCES

1. ASCOLI, F, BOTRE, C. and LIQUORI, A.M., J. Am. Chem. Soc., 65 (1961) 1991.
2. ATKINS, E.D.T. and SHEEHAN, J.K., Nature New Biol., 235 (1972) 253.
3. BRIMACOMBE, J.S. and WEBBER, J.M., Mucopolysaccharides, Elsevier Press, London, 1964.
4. EHRLICH, J. and STIVALA, S.S., J. Pharmaceutical Sci., 62 (1973) 517.
5. EHRLICH, J. and STIVALA, S.S., Polymer, 15 (1974) 204.
6. FOSTER, A.B. and HUGGARD, A.J., in Advances in Carbohydrate Chemistry (Ed. WOLFROM, M.L.) Academic, New York, 1955, vol. 10.
7. HELBERT, J.R. and MARINI, M.A., Biochemistry, 2 (1962) 1101.
8. LIBERTI, P.A. and STIVALA, S.S., Arch. Biochem. Biophys., 119 (1967) 510.
9. LIFSON, S., J. Chem. Phys., 29 (1958) 79.
10. RICE, S.A. and NAGASAWA, M., Polyelectrolyte Solutions, Academic Press, New York, 1961.
11. STIVALA, S.S. and EHRLICH J., Polymer, 15 (1974) 197.
12. STIVALA, S.S., HERBST, N., KRATKY, O. and PILZ, I., Arch. Biochem. Biophys., 127 (1968) 795.
13. STIVALA, S.S. and LIBERTI, P.A., Arch. Biochem. Biophys., 122 (1967) 40.
14. STIVALA, S.S., YUAN, L., EHRLICH, J. and LIBERTI, P.A., Arch. Biochem. Biophys., 122 (1967) 32.
15. STONE, A.L., Biopolymers, 2 (1964) 315.
16. YEN, F., DAVAR, M. and REMBAUM, A., Biochim. Biophys. Acta, 184 (1969) 646.
17. YOUNG, M.D., PHILIPS, G.O. and BALAZS, E.A., Biochim. Biophys. Acta, 141 (1967) 374.
18. YUAN, L., Polymer Symposia, in press.
19. YUAN, L., DOUGHERTY, T.J. and STIVALA, S.S., J. Polymer Sci., A2 10 (1972) 171.
20. YUAN, L. and STIVALA, S.S., Biopolymers, 11 (1972) 2079.

DISCUSSION OF YUAN AND STIVALA PAPER

FAREED

I was wondering about the dye-binding data. What dye was used in these experiments and what is the association constant for this type of dye?

YUAN

Four basic dyes were used, namely, Azure A, Methylene blue, basic fuchsin, and brilliant cresyl blue. Off hand, I do not have their association constants.

FALB

Our data shows that we get about 6 TDMAC molecules or quaternary ammonium salts binding per tetrasaccharide unit of heparin. With the heparin that we use, this corresponds to a stoichiometric binding of quaternary salt for every negative charge on the heparin molecule. This binding is done by shaking an aqueous solution of heparin together with a toluene solution of TDMAC. A stoichiometric amount of heparin then goes into the toluene solution. Since TDMAC is a large molecule, would your model be able to accomodate a stoichiometric binding of TDMAC or tridecylmethyl ammonium chloride with each negative charge on heparin?

YUAN

This is a very good question. I think the model can accomodate the large tridecylmethyl ammonium chloride molecules, since there is a thick ionic atmosphere surrounding the heparin molecule. In other words, the large TDMAC molecules may be "trapped" within the heparin's ionic atmosphere.

FALB

Then you think that all the charges are on the outside of the helical coil?

YUAN

No, we believe that the charges of heparin are present both inside and outside the helical coil. The point I want to make is that those charges present inside the helical coil are "partially" ionized because their counter-ions have been localized. We consider only those charges appearing on the outside surface of the helical coil fully ionized.

ATKINS

First, the model you are proposing is not stereochemically feasible, that is with all α,1-4 glycosidic linkages and forming a hollow coil like amylose. Secondly, as I thought I tried to show this morning, heparin does not have all α,1-4 glycosidic linkages like the model you suggest. So I do not believe that it is a coil like amylose with an inside surface and an outside surface.

YUAN

Our model was constructed in such a way that it best explains most of the physical properties of heparin in solution. However, no indication is available to suggest that heparin should have the same conformation in solution and in the solid state. It is not surprising to find discrepancies between our model and the findings of x-ray studies. All we want to emphasize is that heparin in solution exhibits ordered conformation. We also suggested in our paper that the helical conformation may not necessarily hold throughout the entire heparin chain.

ATKINS

Well, if I might take your figures, you quoted the value for the mean square distance ($\bar{r}$) of 105 Å and, therefore, the molecule that you are proposing in solution is nearly fully extended. I mean, if you take the simple Flory-type equation, where L over $\bar{r}$ is approximately equal to the square root of the number of units. I have done this calculation and of course your molecule is nearly stretched out. So, again you appear to violate with your own solution data, the model you propose.

YUAN

When a heparin with molecular weight of 12,900 is completely stretched out, the length will be 232 Å. None of our viscosity data suggests that heparin is completely stretched out in solution. According to our model, a heparin molecule (M_w=12,900) will have a length of 116 Å a value in excellent agreement with the value calculated from the low angle x-ray scattering studies (i.e. 124 Å). The calculated volume occupied by this model is 0.46 cc/gm, a value very close to the value of the partial specific volume of heparin, 0.47 cc/gm. We suggest that heparin-copper complex should be examined by x-ray diffraction technique to facilitate further information.

JEANLOZ

I think it is clear from the chemical data that you don't have an α type of linkage like in amylose, because α-L type is structurally equivalent to a β-D type, which means that you have the equivalent of a β-D, α-D, β-D, α-D, and not α-D, α-D, α-D, α-D(1→4) linkages all along the chain.

BROZOVIC

Dr. Derek Calam from the National Institute of Medical Research built a three-dimensional molecular model, and looking at the very tight structure, he suggested that most of the binding is done through calcium bridging or something similar. I don't know how relative it is for this particular discussion, but I think it is something that should be borne in mind if we discuss the binding of large molecules to heparin.

YUAN

We agree with Dr. Brozovic completely that the ideas of "calcium bridging" as well as our "copper ion chelating effect" should be seriously considered for the binding study of large molecules to heparin and for the construction of three-dimensional models. Incidentally, we have studied the binding of fibrinogen to heparin with copper ions and suggested that the copper ions may provide a bridge mechanism to facilitate the formation of protein-polysaccharide linkage. Finally, it must be understood that the proposed model was constructed with the view of accomodating some of the data from solution. It is anticipated that an altered model could have been proposed to accomodate other data.

HEPARIN AND HEPARIN-LIKE SUBSTANCES OF CELLS

Jeremiah E. SILBERT, Hynda K. KLEINMAN and Cynthia K. SILBERT

Boston Veterans Administration Hospital, Boston, Massachusetts 02130 (USA)

The correlation between heparin and mast cells was first recognized by Jorpes (7) in 1937, and the history of heparin has since generally been considered to be the history of mast cells. After this correlation was made, however, heparin-like material with lower anticoagulant activity was found by several investigators as a by-product of heparin isolation from various tissues, such as lung and liver (6, 16). This material, called heparitin sulfate, contained less sulfate per disaccharide unit than heparin and contained a mixture of N-sulfated and N-acetylated glucosamine units (13). Both heparin and heparitin sulfate have been shown to be covalently linked to protein and can therefore by called proteoglycans.

Inasmuch as heparitin sulfate can be isolated from various connective tissues that have few mast cells, does this mean that heparitin sulfate is a separate mucopolysaccharide produced by cells other than the mast cell? Or could this heparitin sulfate come from circulating heparin-like polysaccharides, synthesized by mast cells and discharged into the blood or extracellular spaces to be picked up and stored in a variety of tissues? These questions are further complicated by lack of a clear distinction between the heterogeneous heparin and heparitin sulfate. Do the two polysaccharides constitute two different entities, or are they in a continuum of polysaccharides that includes a range of variations in N-sulfate and total sulfate content?

We hope that some of our work with various tissues and cell sources will help answer these questions. It appears to us that the polysaccharide portions of heparin and heparitin sulfate are synthesized in a similar fashion and that their chemical structures represent a continuum. However, we also believe that heparin and heparitin sulfate do represent two entirely different materials, distinguished by source and location of synthesis and by probable function. It is also possible that the protein portions of the mucopolysaccharide are the source of discrete molecular differences in these compounds, but have no information on this.

Biosynthesis of Heparin by Mast Cells

Our interest in this subject started 15 years ago at Washington University with Dr. David Brown. At that time, it seemed possible that the accumulation of heparitin sulfate in tissues of children with some of the mucopolysaccharidoses syndromes might be related to a defect in synthesis. (The metabolic defect is now known to be a defect in degradation (4)). In particular, we considered that heparitin sulfate might be a precursor of heparin and that a block in conversion to heparin might lead to the accumulation of heparitin sulfate. Because there were no readily available sources of tissues that made heparitin sulfate, and two lines of mouse mast-cell tumors (Furth tumor and Dunn-Potter tumor) that produced large amounts of heparin were available, it seemed reasonable to begin the investigation by looking at the pathway for heparin synthesis.

The role of sugar nucleotides was already well established in the biosynthesis of several different bacterial polysaccharides and of animal hyaluronic acid (5). It therefore seemed probable that sugar nucleotides would be the precursors for the mucopolysaccharide, heparin, but the time and method of forming the N-sulfate and O-sulfate were unclear. The latter were considered to be particularly important, because knowledge of the pathway of sulfation might clarify the distinction between heparin and heparitin sulfate. In particular, the formation of N-sulfate, as opposed to N-acetyl, would be a key to differences or similarities in synthesis. (Formation of this structure still remains elusive, although a partial pathway has been described by us (21) and further defined by Lindahl and his co-workers (12)). The most attractive hypothesis at the time was that sulfation took place at the sugar nucleotide level. There was much precedence for this location for sugar modification, in that many sugar transformations were already known to occur while the sugar was linked to the nucleotide. Furthermore, a sulfated sugar nucleotide, UDP-N-acetylgalactosamine-4-sulfate had already been isolated from an animal source (23).

We thought that a likely candidate for formation of the N-sulfate would be UDP-glucosamine (as opposed to the well-known UDP-N-acetylglucosamine). Our first efforts to show synthesis of this sugar nucleotide from precursor UTP and glucosamine-1-phosphate were successful (22). However, the mast-cell tumor extracts used for this synthesis were able to synthesize UDP-N-acetylglucosamine even more rapidly. Attempts to sulfate the UDP-glucosamine by incubation with a sulfating system (ATP and [^{35}S] sulfate) were unsuccessful. Investigation into this pathway was discontinued before definitive experiments with 3'-phosphoadenosine-5'-phosphosulfate (PAPS) were conducted because of a more promising direction. We found that incubation of the sugar nucleotides, UDP-N-acetylglucosamine and UDP-glucuronic acid, led to formation of a nonsulfated heparin-related "precursor" polysaccharide (19). UDP-glucosamine was not active as a substrate, so its possible role in heparin biosynthesis was discounted. The polymerizing enzyme activity was found to be in a membrane "microsomal" fraction, and the product was linked to protein in this particulate fraction. Figure 1 shows the pattern

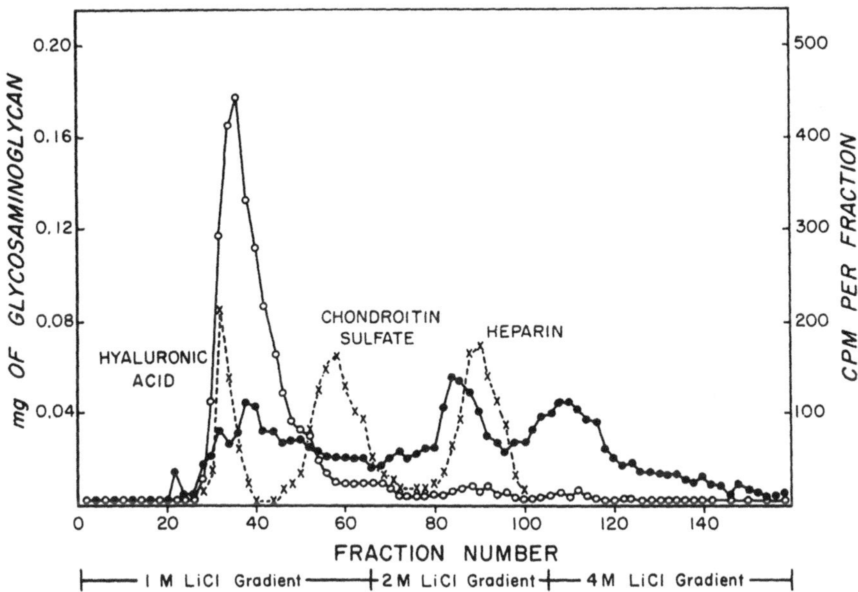

Fig. 1. Gradient elution of [^{14}C]glycosaminoglycan from a DEAE-cellulose column. ^{14}C-labeled glucuronic acid glycosaminoglycan (8300 cpm) - with hyaluronic acid (0.8 mg), chondroitin sulfate (1.6 mg), and heparin (1.6 mg) - was eluted from a column of DEAE-cellulose (1 x 5 cm) with a LiCl logarithmic gradient. There were 125 ml of water in the mixing flask, and 2.5-ml fractions were collected. The reservoir initially contained 1 M LiCl, which was changed to 2 M LiCl and 4 M LiCl at the indicated fractions. Fractions were assayed for radioactivity (o---o) and for the presence of carrier glycosaminoglycan (x---x) (carbazole determination). A separate, identical elution was carried out with the same glycosaminoglycan standards, with [^{14}C]glycosaminoglycan (8300 cpm) formed in the presence of 3'-phosphoadenosine-5'-phosphosulfate. Fractions were assayed for radioactivity (•---•) and for carrier glycosaminoglycan. The results of both columns are superimposed.

of chromatography of the "precursor" polysaccharide on DEAE-cellulose with mucopolysaccharide standards. The polysaccharide was shown to contain glucuronic acid and N-acetylglucosamine units and was similar to hyaluronic acid in its chromatographic behavior. However, this heparin "precursor" was not degradable by bacterial or testicular hyaluronidase and therefore resembled the class of heparin and heparitin sulfate compounds. Further identification revealed that the polysaccharide was larger than commercial heparin as judged by Sephadex G-200 chromatography.

When PAPS was added to the incubation mixture with the sugar nucleotides, considerable sulfation took place, yielding a mixture of products with an average sulfate content approximately half that of commercial heparin (20). Chromatography of this is also shown in Fig. 1. Although the products were not completely identified, they appeared to represent a mixture of polysaccharides with various N-acetyl and N-sulfate contents. Formation of the N-sulfate appeared initially to be through an interchange of a sulfate for an N-acetyl (21). Loss of N-acetyl and substitution of N-sulfate on the preformed polysaccharide is shown in Fig. 2. Recently, Lindahl and co-workers (12) have shown that there is some deacetylation of N-acetylglucosamine before sulfation and that an exchange is not necessary. Thus, it appears that the synthesis of heparin may proceed through acetylated precursors that resemble heparitin sulfate. This lends support to the concept of a continuum in structure between heparin and heparitin sulfate.

Heparitin Sulfate of Cells

Our interests in these compounds have led us away from questions regarding chemical similarities of the polysaccharide chains and toward an examination of possible natural function and tissue localization of the entire class of proteoglycans. We would like to suggest that there is a real difference between heparin and heparitin sulfate. Heparin would be the proteoglycan formed by mast cells, stored in granules, and utilized for as yet unknown purposes. Heparitin sulfate would be the proteoglycan (with a polysaccharide portion similar in structure to some heparin fractions) formed by other cells and probably serving functions related to cell structure and cell-cell or cell-environment interactions.

Commercial heparin is prepared from tissues with a high mast-cell content; but these tissues also contain their own specialized cells and other connective-tissue elements. If these other cells make heparin-like material (heparitin sulfate), this would explain some of the problems of heterogeneity in heparin preparations.

Although the synthesis of heparin-like material by cells other than mast cells was suspected, the work of Kraemer first established

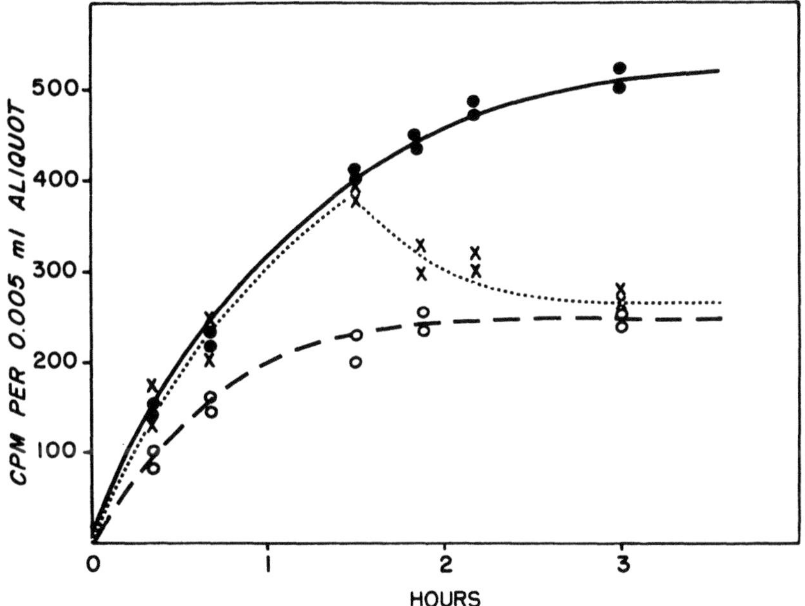

Fig. 2. Presence of [^{3}H]acetyl in glycosaminoglycan with time. Each duplicate reaction mixture contained, in a total volume of 0.040 ml, 0.05 M Tris (pH 7.5), 0.01 M $MgCl_2$, and 0.010 ml of microsomal preparation. In addition, the reaction mixtures contained: (a) UDP-N-[^{3}H]acetylglucosamine (25 millimicromoles, 4 x 10^5 cpm) and UDP-glucuronic acid (25 millimicromoles) (o---o); (b) the same as (a) plus 3'phosphoadenosine-5'-phosphosulfate (175 millimicromoles) (o---o); (c) initially the same as (a), but, after 1.5 hr of incubation, 120 millimicromoles of 3'-phosphoadenosine-5'-phosphosulfate were added (x---x). Aliquots (0.005 ml) of each reaction mixture were removed at the indicated intervals, and the radioactive glycosaminoglycan was isolated and assayed for radioactivity.

it unequivocally (8-10). He and others (1, 3) have used a variety of established lines of cells grown in culture to show production of heparitin sulfate. Furthermore, Kraemer and Smith have localized much of the heparitin sulfate to the external membrane and shown that this material is discharged into the growth medium as the proteoglycan (11).

Heparitin Sulfate Formation by Normal Cultured Fibroblasts

Kraemer and others studied established cell lines that had

undergone multiple transformation in becoming established. We have extended these studies to the production of heparitin sulfate and other mucopolysaccharides by fibroblasts grown from skin biopsies of normal or diabetic patients. Experiments were conducted as follows:

Primary monolayer cultures of human skin fibroblasts were grown from skin obtained by punch biopsy. Skin was cut into small fragments and established in culture using Eagle's minimal essential medium with nonessential amino acids, sodium pyruvate (0.11 mg/ml), penicillin (100 U/ml), streptomycin (100 μg/ml), and 10% fetal calf serum. Good growth of fibroblasts was established after 3-4 weeks. Fibroblasts were maintained with weekly medium changes and incubated at 37° in a humidified atmosphere with of 95% air and 5% CO_2. On the fourth subculture, standard medium was removed and replaced with medium lacking penicillin and streptomycin and containing [^{35}S]sulfate (130 μCi/micromole).

After growth for 3 days (when cell cultures were not yet confluent), the medium was removed and the cells were washed six times with 0.15 M NaCl. Collagenase (Worthington) - 60 units in 0.5 ml of 0.15 M NaCl - was added to each plate; after 10 min at room temperature, the resulting cell suspension was centrifuged at 5000 g for 10 min. All cells were removed by this treatment. Cells were resuspended in collagenase solution, incubated, and centrifuged once more. The cell pellets were then washed by centrifugation two times in 0.15 M NaCl. Cells were then incubated in trypsin - 2.5 mg/ml - for 10 min at room temperature and centrifuged. The pellet was washed once with 0.15 M NaCl and then subjected to five cycles of freezing and thawing. This solubilized the remaining ^{35}S-labeled material.

Aliquots of the growth medium, collagenase supernatants, trypsin supernatants, and lysed pellet supernatants were assayed for radioactivity and when necessary chromatographed on Sephadex G-25 to separate ^{35}S-labeled macromolecular material from [^{35}S]sulfate.

Aliquots of the ^{35}S-labeled macromolecular materials were incubated with:

- Chondroitinase ABC (Miles), 0.75 units; chondroitin-4-sulfate (Miles), 0.5 mg; chondroitin-6-sulfate (Miles), 0.5 mg; and enriched Tris (18), 0.01 ml; total volume, 0.1 ml; duration, 2 hr; temperature, 37°.

- Heparinase (a gift from Dr. A. Linker), 0.5 mg; chondroitin--4-sulfate, 1 mg; and sodium acetate at a pH of 6.0, 10 micromoles; total volume, 0.1 ml, for 21 hr at 20°.

Reaction mixtures were then chromatographed on a column (1 x 20 cm) of Sephadex G-50 with 0.1 M LiCl to separate macromolecular material from degraded material. Fractions were assayed for radioactivity and for the presence of uronic acid by the carbazole method (2).

Aliquots of heparitin [^{35}S]sulfate (macromolecular material not degradable by chondroitinase ABC, but degradable by heparinase) were chromatographed on a column of Sepharose 6B with chondroitin-6-sulfate (Miles; average molecular weight, 40,000) as a standard. Aliquots of heparitin [^{35}S]sulfate were incubated at 20° overnight with 0.5 M NaOH or at 100° for 1 hr with 0.04 M HCl before chromatography on Sepharose 6B.

The total amount of [^{35}S]mucopolysaccharide pooled from 12 culture plates and the heparitin sulfate fraction are shown in Table I. The medium was found to contain only a small percentage of heparitin sulfate, whereas the trypsin fraction (representing external-cell-surface mucopolysaccharide) was almost entirely heparitin sulfate.

Size approximations of the heparitin sulfate from the growth medium, trypsin fraction (external cell surface), and pellet (intracellular) are shown in Fig. 3. The heparitin sulfate of the medium has a pattern indicating a molecular weight greater than 100,000, which was reduced to 50,000 or less by treatment with 0.5 M NaOH. This heparitin sulfate thus would appear to be a proteoglycan with

TABLE I

Sulfated Mucopolysaccharide from Cultured Fibroblasts

Source of [^{35}S]Mucopolysaccharide	Total [^{35}S]Mucopolysaccharide		Heparitin Sulfate, %
	cpm x 10^5	μg	
Medium	14.8	0.93	15
Matrix	6.9	0.41	45
External cell surface	0.53	0.033	90
Intracellular	1.0	0.063	79

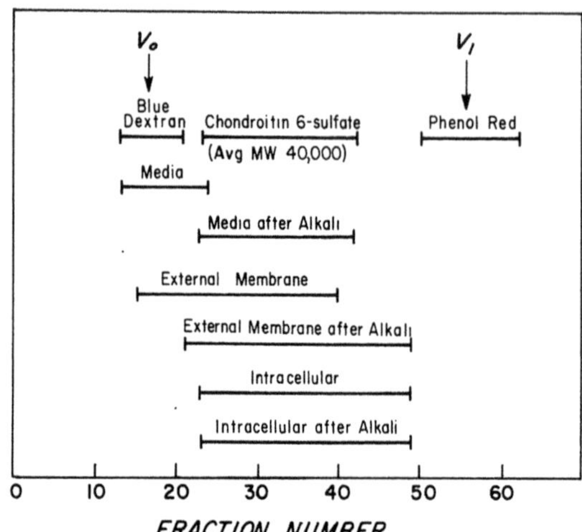

Fig. 3. Chromatography on Sepharose 6B. Aliquots of heparitin [^{35}S]sulfate with blue dextran, phenol red, and chondroitin-6-sulfate were chromatographed on a column (1 x 60 cm) of Sepharose 6B. Columns were eluted with 0.1 M LiCl at a flow rate of 4 ml/hr, and 0.8-ml fractions were collected. Aliquots of each fraction were assayed for radioactivity and for the carrier chondroitin-6-sulfate (carbazole determination).

alkali labile linkages similar to those of other proteoglycans. The collagenase fractions representing extracellular matrix and a mixture of cellular material (collagenase treatment fractured many cells) were similar. The heparitin sulfate of the trypsin fraction (external cell surface) was somewhat smaller than that in the medium, but was further reduced by alkali treatment, suggesting a doublet structure as is found after trypsin treatment of chondromucoprotein (14). The heparitin sulfate within the cells usually did not seem to be proteoglycan, in that this smaller material was not affected in size by alkali treatment. However, with some cell cultures, the intracellular material was a proteoglycan similar to the trypsin fraction. Treatment of the external-cell-surface heparitin sulfate fractions with 0.04 M HCl resulted in hydrolysis of approximately 50% of the [^{35}S]sulfate, confirming the presence of amino sulfate groups.

The heparitin sulfate found in primary cultures of normal skin fibroblasts bears a striking resemblance to the recently described heparitin sulfate from Chinese hamster (CHO) cells grown in suspension culture (11). The CHO cells produced a heparitin sulfate proteoglycan on the external cell surface with a polysaccharide chain size of approximately 25,000-44,000 molecular weight. Material found within the cells was smaller and not linked to protein.

Function of Cellular Heparitin Sulfate

It has previously been suggested that heparitin sulfate may be present on the external surface of all cells (9), but this suggestion was based on findings with established cell lines that have undergone multiple transformations in becoming established. Our results extend the concept of heparitin sulfate as a component of external cell surfaces of normal cells.

We think it is appropriate to speculate on the role of cell-surface heparitin sulfate in the context of this symposium. Heparitin sulfate may simply be a structural component of the cell with no striking activity. But, if this surface proteoglycan proves to have anticoagulant activity or, more directly, to stimulate inhibitors of serine proteases, as suggested by Dr. Rosenberg, then we could consider a primary role in protecting the cell and immediate environment from the coagulation and fibrinolysis mechanisms. Of particular interest would be the part that heparitin sulfate might play in fibrinolysis inhibition as a check in the spread of normal cells, as opposed to fibrinolysis by malignant cells (17). Because "tissue factor" can also be released by trypsinization of skin fibroblast cultures (15), we can speculate on the possibility that release of heparitin sulfate by the same treatment may free the tissue factor from its heparitin sulfate "shield." We can also speculate on the possible role of heparitin sulfate in lipolysis and atherogenesis. We believe that further study in these directions may yield findings of great importance in such major subjects as clotting, atherogenesis, and cancer.

REFERENCES

1. BATES, C.J. and LEVENE, C.I., Biochim. Biophys. Acta, 237 (1971) 214.
2. BITTER, T. and MUIR, H.M., Anal. Biochem., 4 (1962) 330.
3. DIETRICH, C.P. and DeOCA, H.M., Proc. Soc. Exptl. Biol. Med., 134 (1970) 955.
4. FRATANTONI, J.C., HALL, C.W. and NEUFELD, E.F., Proc. Natl. Acad. Sci. US, 64 (1969) 360.
5. GLASER, L. and BROWN, D.H., Proc. Natl. Acad. Sci. US, 41 (1955) 253.
6. JORPES, J.E. and GARDELL, S., J. Biol. Chem., 176 (1948) 271.
7. JORPES, E., HOLMGREN, H. and WILANDER, O., Z. Zellforsch. Mikroskop. Anat., 42 (1937) 279.
8. KRAEMER, P.M., J. Cell Physiol., 71 (1968) 109.
9. KRAEMER, P.M., Biochemistry, 10 (1971) 1437.
10. KRAEMER, P.M., Biochemistry, 10 (1971) 1445.

11. KRAEMER, P.M. and SMITH, D.A., Biochem. Biophys. Res. Commun., 56 (1974) 423.
12. LINDAHL, V., BÄCKSTRÖM, G., JANSSON, L. and HALLÉN, A., J. Biol. Chem., 248 (1973) 7234.
13. LINKER, A., HOFFMAN, P., SAMPSON, P. and MEYER, K., Biochim. Biophys. Acta, 29 (1958) 443.
14. MATHEWS, M.B., Biochem. J., 125 (1971) 37.
15. MAYNARD, J.R., HECKMAN, C.A., PITLICK, F.A. and NEMERSON, Y., Federation Proc., 33 (1974) 243.
16. MEYER, K., DAVIDSON, E., LINKER, A. and HOFFMAN, P., Biochim. Biophys. Acta, 21 (1956) 506.
17. REICH, E., Federation Proc., 32 (1973) 2174.
18. SAITO, H., YAMAGATA, T. and SUZUKI, S., J. Biol. Chem., 243 (1968) 1536.
19. SILBERT, J.E., J. Biol. Chem., 238 (1963) 3542.
20. SILBERT, J.E., J. Biol. Chem., 242 (1967) 5146.
21. SILBERT, J.E., J. Biol. Chem., 242 (1967) 5153.
22. SILBERT, J.E. and BROWN, D.H., Biochim. Biophys. Acta, 54 (1961) 590.
23. SUZUKI, S. and STROMINGER, J.L., J. Biol. Chem., 235 (1960) 2768.

BIOSYNTHESIS OF HEPARIN

Gudrun BÄCKSTRÖM, Anund HALLEN, Magnus HÖÖK, Leif JANSSON and Ulf LINDAHL
Department of Medical Chemistry, The Royal Veterinary College, and Institute of Medical Chemistry, University of Uppsala, The Biomedical Centre, Uppsala (Sweden)

The principal reactions involved in the biosynthesis of glycosaminoglycans are now fairly well characterized. By use of various cell-free systems, the polymerization of sugar units has been shown to occur by stepwise transfer of appropriate monosaccharides from the corresponding UDP-sugar precursors to the nonreducing termini of nascent polysaccharide chains. Sulfation takes place along with, or after, polymerization and requires the presence of 3'-phosphoadenosine-5'-phosphosulfate (PAPS). For recent reviews see Stoolmiller and Dorfman (23), Roden (18), and Phelps (16).

The biosynthesis of heparin and heparan sulfate poses a number of specific problems related to the unique structures of these polysaccharides (Fig. 1). Some important structural features have only recently been elucidated, particularly the identity, anomeric

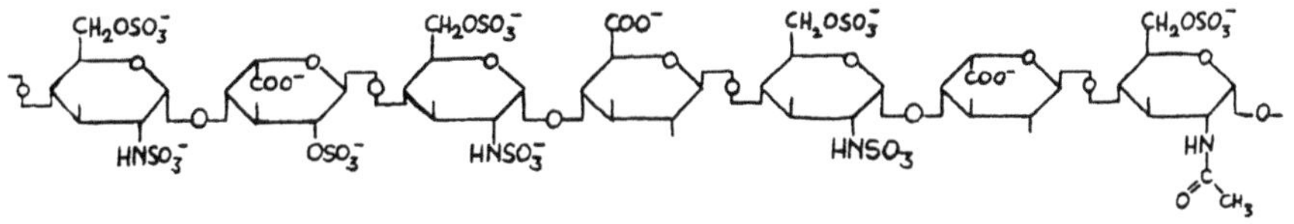

Fig. 1. Structure of heparin. Heptasaccharide shown displays the structural features characteristic of heparin, but does not necessarily indicate strict sequence of variously substituted monosaccharide units. Heparan sulfate is closely related to heparin, but contains generally less sulfate and L-iduronic acid and more N-acetyl and D-glucuronic acid (12, 25).

configuration, and sulfate substitution pattern of the uronic acid constituents (see the review by Jeanloz elsewhere in this volume). Because L-iduronic acid is the predominant uronic acid in heparin (and in dermatan sulfate) and a significant component in heparan sulfate (12, 25), the mechanism of incorporation of this uronic acid into the polysaccharide molecules has been of major interest. Furthermore, the sulfation process, resulting in the formation of sulfamino and various ester sulfate (O-sulfate) groups, has remained incompletely characterized. Finally, it should be noted that, although some individual reactions in the overall biosynthetic process have been fairly well characterized, the concerted action of the corresponding enzymes, cooperating toward formation of the completed polysaccharide molecule, is still poorly understood. In addition to summarizing previous studies, this presentation will deal with recent experiments in our laboratory related to some of these problems.

Previous Investigations

The biosynthesis of heparin has been studied by several workers, all using cell-free systems derived from transplantable mouse mastocytomas. Most of the basic information was provided by the elegant work of Silbert (19) who established a reaction sequence initiated by polymerization of N-acetylglucosamine and glucuronic acid from UDP-N-acetyl-D-glucosamine (UDP-GlcNAc) and UDP-D-glucuronic acid (UDP-GlcUA), respectively. On the addition of PAPS to a microsomal fraction containing the preformed nonsulfated polysaccharide, N-sulfate (sulfamino) and O-sulfate groups were introduced, yielding heparin-like polysaccharides as products (20-22). In later studies by Lindahl *et al.* (15) the formation of N-sulfate groups was found to occur by a two-step mechanism, involving partial N-deacetylation of the nonsulfated microsomal polysaccharide followed by substitution of the exposed free amino groups by sulfate residues. In addition to these studies, based on labeling of endogenous microsomal polysaccharide, the polymerization and sulfation processes have also been investigated by use of exogenous acceptors of glycosyl and sulfate residues. Thus, transfer of galactosyl, glucuronosyl, and N-acetylglucosaminyl residues to appropriate oligosaccharides or carbohydrate-serine compounds was demonstrated, each reaction representing the incorporation of a specific monosaccharide unit into the carbohydrate-protein linkage region (7, 8) or the polysaccharide chain proper (9, 10). The corresponding glycosyltransferases were solubilized from microsomal membranes, and in some cases purified. Sulfotransferases in various microsomal and postmicrosomal fractions were tested with heparan sulfate or N-desulfoheparin as exogenous sulfate acceptors (1, 3, 17, 24). There was evidence of the incorporation of sulfamino, as well as ester sulfate, groups.

Little information is available concerning the cell-free

biosynthesis of heparan sulfate. According to current opinion, heparan sulfate may serve as a precursor of heparin; however, there is no experimental evidence to support this contention.

The mode of incorporation of L-iduronic acid units into glycosaminoglycans has remained unclear. The role of UDP-GlcUA in the formation of glucuronic acid-containing polysaccharides has been amply verified (16, 18, 23), and an analogous function has been postulated for UDP-L-iduronic acid (USP-IdUA). The formation of this nucleotide sugar, by C-5 epimerization of UDP-GlcUA, has in fact been demonstrated in mammalian tissues (4, 13). However, UDP-IdUA has never been isolated, and the precursor properties of this compound have therefore not been subject to straightforward experimental trial. Instead, an indirect approach was tried, using heparin fragments (carbohydrate-serine compounds) with nonreducing terminal uronic acid residues as acceptors in an N-acetylglucosaminyl transfer reaction, catalyzed by a mastocytoma microsomal fraction (10). Transfer of N-acetylglucosamine to acceptor molecules having nonreducing terminal D-glucuronic acid residues was readily demonstrated; in contrast, no such transfer occurred to analogous fragments terminating with L-iduronic acid units. The latter type of fragment would represent an obligatory intermediate in a polymerization reaction involving UDP-IdUA. The results therefore suggested that alternative mechanisms for the incorporation of iduronic acid into heparin should be investigated.

Present Investigation

The biosynthesis of heparin was studied by use of a microsomal fraction from a transplantable mouse mastocytoma, originally described by Furth _et al._ (5). The experiments involved transfer of monosaccharide and sulfate residues to endogenous microsomal primers (see summarized incubation scheme in Table I) and sulfate substitution of exogenous polysaccharide acceptors.

Biosynthesis of L-iduronic acid residues. - In view of previous studies, incorporation of iduronic acid units from UDP-IdUA into heparin appeared unlikely. An alternative mechanism was suggested by the unexpected finding of Haug and Larsen (6, 14) that C-5 epimerization of uronic acid residues may occur on the polymer level. L-Gluronic acid units in alginic acid are thus formed by epimerization of already-polymerized D-mannuronic acid residues. The formation, by a similar mechanism, of iduronic acid units in heparin was demonstrated by the following experiments (11).

A microsomal fraction from mouse mastocytoma was incubated with UDP-[^{14}C]GlcUA and unlabeled UDP-GlcNAc (incubation A, Table I),

TABLE I

Incubations of Mastocytoma Microsomal Fraction[a]

Incubation	Incubation periods 1	2	3
A	UDP-[^{14}C]GlcUA UDP-GlcNAc (60 min)	---	---
B	UDP-[^{14}C]GlcUA UDP-GlcNAc PAPS (60 min)	---	---
C	UDP-[^{14}C]GlcUA UDP-GlcNAc (60 min)	UDP-GlcUA PAPS (0.5, 5, 60 min)	---
D	UDP-[^{14}C]GlcUA UDP-GlcNAc [^{35}S]PAPS (60 min)	---	---
E	UDP-GlcUA UDP-GlcNAc (30 min)	[^{35}S]PAPS (30 min)	PAPS (60 min)

[a]Unlabeled substrates were generally added in large excess over corresponding radioactive substrates; for reactant concentrations, cofactors, and incubation conditions, see Lindahl _et al_. (15) and Höök _et al_. (11). Order of addition of the various substrates was according to numbering of incubation periods, duration of each period was as indicated.

resulting in the formation of nonsulfated polysaccharide. The product was degraded, essentially to the monosaccharide level, by a combination of acid hydrolysis and deaminative cleavage with nitrous acid (12), and the labeled uronic acid was identified by paper chromatography. [^{14}C]Glucuronic acid, but no [^{14}C]iduronic acid, was detected. A similar incubation, carried out in the presence of PAPS (incubation B) yielded sulfated microsomal polysaccharide, in which [^{14}C]iduronic acid constituted approximately one-third of the total labeled uronic acid. These results showed that UDP-GlcUA is a precursor of the iduronic acid in heparin and that the formation of iduronic acid residues depends somehow on the sulfation of the polysaccharide. In a later experiment, the polymerization and sulfation processes were confined to separate incubation periods (incubation C); by adding unlabeled UDP-GlcUA in excess, along with PAPS, the incorporation of radioactivity was restricted to the polymerization period (incubation C1) only. Analysis of the final product showed that sulfation of preformed polysaccharide yielded the same proportions of iduronic acid as did sulfation in conjunction with polymerization. Inasmuch as the preformed polysaccharide lacked iduronic acid, these findings demonstrate that iduronic acid is formed by C-5 epimerization of glucuronic acid residues, on the polymer level. Because the [^{14}C]glucuronic acid units of the nonsulfated polysaccharide were almost exclusively in internal positions (15), it is further concluded that the epimerization process must occur in the interior portions of the molecule.

The relation between sulfation and uronic acid epimerization was studied in more detail, by incubating the microsomal fraction with UDP-[^{14}C]GlcUA and UDP-GlcNAc at various concentrations of [^{35}S]PAPS (incubation D). The doubly labeled polysaccharide preparations were degraded, and the resulting [^{14}C]uronic acid monosaccharides and inorganic [^{35}S]sulfate were separated and quantitated. The results demonstrated that an increased sulfation of microsomal polysaccharide was accompanied by an increase in the ratio of [^{14}C] iduronic acid to [^{14}C]glucuronic acid.

^{14}C-labeled sulfated microsomal polysaccharide was further characterized with regard to the distribution of iduronic acid and sulfate residues. The polysaccharide was treated with nitrous acid under conditions leading to deamination of glucosamine residues having either unsubstituted or sulfated amino groups (N-acetylated glucosamine units remaining intact), with concomitant cleavage of the corresponding glucosaminidic linkages (15). The resulting oligosaccharides were separated with regard to size and sulfate content, using gel chromatography followed by high-voltage paper electrophoresis. Analysis of [^{14}C]uronic acid composition showed that the correlation between sulfate and iduronic acid contents observed with the polymeric preparations was maintained on the oligosaccharide level, thus indicating a preponderance of iduronic acid in heavily sulfated sections of the polysaccharide chain. Because the N-sulfate groups had been lost in the deamination process, the sulfate residues recovered in the oligosaccharides were exclusively ester-bound; it was therefore concluded that uronic acid epimerization was somehow linked to the formation of O-sulfate, rather than N-sulfate, groups. In accordance with this conclusion, oligosaccharides were identified as having iduronic acid residues interspersed between N-acetylated glucosamine units.

The significance of O-sulfation in relation to the biosynthesis of iduronic acid residues was further emphasized by the structural properties of polysaccharide subfractions (M. Höök, U. Lindahl, A. Hallén, and G. Bäckström, unpublished observations). Sulfated, ^{14}C-labeled microsomal polysaccharide was separated, by ion-exchange chromatography on DEAE-cellulose, into two distinct (although incompletely separated) sulfated components (III and IV, respectively, in Fig. 2). Selective deamination (15) showed that both components contained the same relative amounts of N-sulfated and N-acetylated glucosamine residues; no N-unsubstituted glucosamine units were detected. Structural characterization of oligosaccharides revealed that the less anionic polymer (III) lacked O-sulfate, as well as iduronic acid residues, whereas the more anionic component (IV) contained O-sulfate groups (somewhat less than fully sulfated heparin) and [^{14}C]iduronic acid, amounting to approximately 70% of the total [^{14}C]uronic acid (Table II).

TABLE II

Summarized Characterization of ^{14}C-labeled DEAE-Fractions[a]

Fraction	-NAc	$-NH_3^+$	$-NSO_3^-$	$-OSO_3^-$	Iduronic acid
I	+	+	-	-	-
II	+	-	-	-	-
III	+	-	+	-	-
IV	+	-	+	+	+ (70% of total [^{14}C]uronic acid)

[a]Fractions obtained by chromatography, on DEAE-cellulose, of microsomal polysaccharide, isolated after 60-min sulfation period (Fig. 2D). -NAc = acetamido groups, $-NH_3^+$ = unsubstituted amino groups, $-NSO_3^-$ = sulfamido groups, $-OSO_3^-$ = ester sulfate groups.

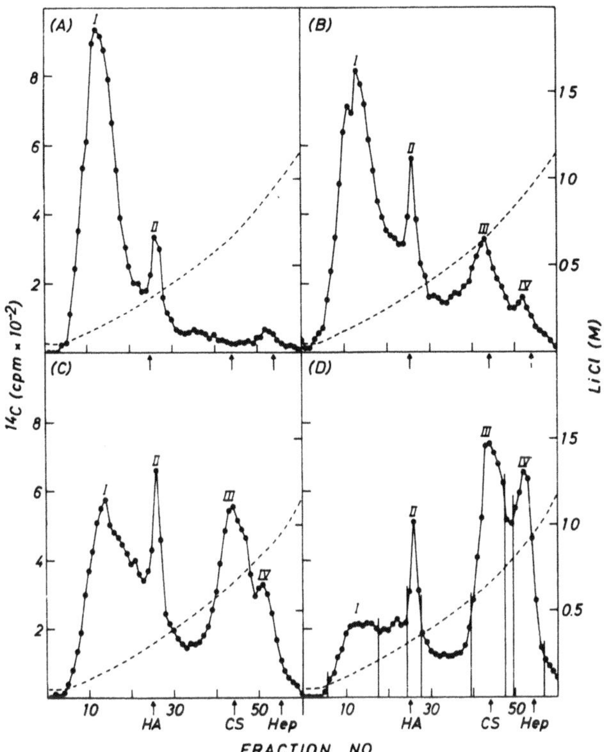

Fig. 2. Chromatography on DEAE-cellulose, of microsomal ^{14}C-labeled polysaccharides isolated (11, 15) from incubation mixtures (A) before, (B) 0.5 min after, (C) 5 min after, and (D) 60 min after introduction of PAPS (for further experimental details, see Table I, incubation C). Effluent fractions were pooled as indicated by vertical lines, yielding Fractions I - IV. Arrows indicate peak elution positions of standard preparations of hyaluronic acid (HA), chondroitin sulfate (CS), and heparin (Hep). Dashed line indicates concentration of LiCl.

The correlation between uronic acid epimerization and the formation of O-sulfate groups is apparently not restricted to the formation of heparin _in vitro_, but appears to be of general significance. Evidence to support this contention was obtained by structural characterization of heparin and heparan sulfate isolated from pig intestinal mucosa and from human aorta, respectively (12). Both polysaccharides were subjected to deamination with nitrous acid, and the products were fractionated with regard to molecular size and charge. The resulting oligosaccharides covered a wide range of O-sulfate: uronic acid molar ratios, which were closely matched by the iduronic acid contents.

These findings also suggest that the L-iduronic acid units of heparin sulfate are formed by a mechanism similar to that operating in the biosynthesis of heparin. Furthermore, preliminary studies on the biosynthesis of dermatan sulfate, using a microsomal fraction from cultured human skin fibroblasts, point to the same conclusion (A. Malmström, L.-Å. Fransson, M. Höök, and U. Lindahl, unpublished observations). It is therefore probable that C-5 epimerization of D-glucuronic acid residues, on the polymer level, is a common route to the formation of L-iduronic acid units in mammalian glycosaminoglycans.

The details of the uronic acid epimerization reaction remain unclear - the mechanism of C-5 inversion and the relation between this process and the formation of O-sulfate groups. Attempts were made to correlate the epimerization of a uronic acid residue with the sulfation of hydroxyl groups at specific positions in the vicinity, by characterizing a large number of oligosaccharides derived from ^{14}C-labeled microsomal heparin (11). The results clearly indicated that completed O-sulfation of both the target uronic acid residue and the adjacent glucosamine units (as indicated in Fig. 3) was not required for the epimerization reaction; for instance, sulfation at C-2 of the epimerizing uronic acid does not seem to be mandatory. However, the epimerization-promoting effect could not with certainty be ascribed to any single sulfate substituent at a specific locus. These considerations bear on the mechanism of the concerted sulfation-epimerization process; it is, thus, not known whether O-sulfation occurs before of after uronic acid epimerization.

—(4)GlcNAc(1)—(4)GlcUA(1)—(4)GlcNAc(1)—

↓

—$GlcNH_3^+$—GlcUA—GlcNAc—

↓

—$GlcNSO_3^-$—GlcUA—GlcNAc—

↓

OSO_3^- (on GlcNSO₃⁻), OSO_3^- (on GlcNAc)

—$GlcNSO_3^-$—IdUA—GlcNAc—

OSO_3^- (on IdUA)

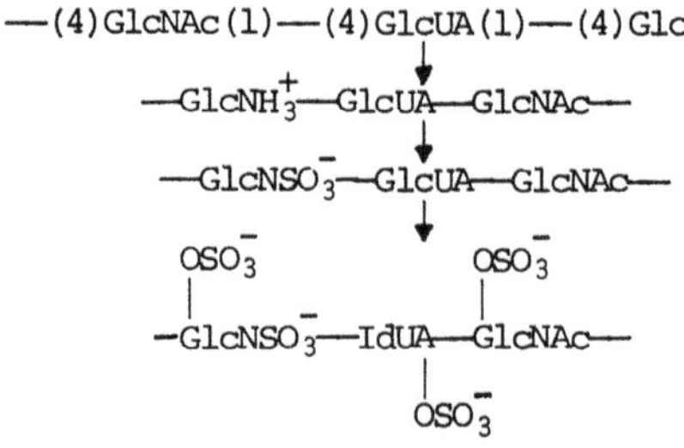

Fig. 3. Tentative scheme of heparin biosynthesis. Fully O-sulfated trisaccharide is shown as representative of end product; however, it should be noted that apparently not all these O-sulfate groups are required for C-5 epimerization of uronic acid unit (11). N-Sulfate and N-acetyl groups of final product have been arbitrarily inserted, because N-substituents of adjacent glucosamine residues do not seem to affect epimerization reaction primarily.

In the latter case, preferential sulfation of iduronic acid-containing polymer sequences would promote the epimerization reaction by withdrawing the product from an equilibrium. But, by sulfation before epimerization (presumably at a specific site in relation to the target uronic acid), the polymer could be modified to satisfy the specificity requirements of an epimerase.

Course of the microsomal sulfation process. - Studies on the microsomal sulfation of heparin were aimed at two specific problems - the relation between the N- and O-sulfation processes, and the rate of sulfation of an individual molecule. As part of this investigation, attempts were made to determine whether one or more catalytic sites were involved in the process.

Assay procedures were developed for the separate determination of N- and O-sulfotransferase activities, by use of N-desulfoheparin and chemically N-acetylated heparan sulfate, respectively, as exogenous sulfate acceptors (L. Jansson, M. Höök, Å. Wasteson, and U. Lindahl, unpublished observations). Structural characterization of the products obtained on incubating polysaccharides with solubilized (8) microsomal enzymes, in the presence of [^{35}S]PAPS, showed selective incorporation of [^{35}S]sulfate into N-sulfate groups of the N-desulfoheparin and into O-sulfate groups of the heparan sulfate preparation. The assay procedures were optimized with regard to pH, divalent metal ions, and incubation time and were then used to study the effects of increased temperature and ionic strength on the sulfate transfer reactions. The O-sulfotransferase was found to be more sensitive to heat inactivation, 60% of the activity being lost after 1 min at 50°, whereas 85% of the N-sulfotransferase activity was retained. In addition, the N-sulfotransferase was selectively inhibited (or inactivated) by sodium chloride - with 0.125 M NaCl, the O-sulfotransferase activity was essentially unaffected, whereas the N-sulfotransferase activity was decreased by 80%. The enzymes also differed with regard to affinity for PAPS, K_m for the N-sulfotransferase and the O-sulfotransferase being approximately 2 x 10^{-5}M and 1 x 10^{-4}M, respectively. These results strongly indicate that the N- and O-sulfate transfer reactions should be ascribed to different enzymes or to separate and independent active sites on the same enzyme molecule. Inasmuch as O-sulfate groups occur in more than one position in heparin, it seems likely that at least three different enzymes (active sites) would be involved in the entire sulfation process.

The concerted action of N- and O-sulfotransferases was studied with particulate enzymes, acting on endogenous microsomal polysaccharide (M. Höök *et al.*, unpublished observations). Polymerization and sulfation of such polysaccharide were conducted during separate incubation periods (incubation C, Table I), the latter of which thus involved sulfation (by introduction of unlabeled sulfate groups) of preformed partially N-deacetylated, nonsulfated [^{14}C]polysaccharide

(10). After various periods of sulfation, polysaccharide was isolated and fractionated by chromatography on DEAE-cellulose, yielding four distinct peaks of labeled material (Fig. 2). Structural characterization of these components - by selective deamination (15), in conjunction with gel chromatography and paper electrophoresis of the resulting oligosaccharides - indicated (Table II) in order of elution: (I) partially N-deacetylated nonsulfated polysaccharide, (II) fully N-acetylated nonsulfated polysaccharide, (III) partially N-acetylated N-sulfated polysaccharide lacking O-sulfate groups, and (IV) partially N-acetylated N- and O-sulfated polysaccharide. In the course of sulfation, components III and IV were formed while component I was eliminated, each type of component retaining the same elution position throughout the entire sulfation period (60 min). Component II, which is of unknown significance, remained unchanged and did not seem to be involved in the sulfation process. Completed heparin molecules (component IV) were detected after only 0.5 min of sulfation. These results show that sulfation is a rapid and highly organized process. Sulfation of preformed nonsulfated polysaccharide does not simultaneously engage the total pool of molecules available. Instead, a limited number of molecules are rapidly passed through the sulfation process; this procedure is successively repeated until the nonsulfated polysaccharide pool is depleted. Independent recent studies by DeLuca _et al_. (2) showed that sulfation of chondroitin sulfate in embryonic chick cartilage occurs by mechanisms similar to those inferred from the present investigation.

In addition to heparin-like molecules, containing both N- and O-sulfate groups (component IV), the products obtained on sulfation of preformed microsomal polysaccharide included an N-sulfated polymer, apparently devoid of O-sulfate substituents (component III). The formation of this novel polysaccharide was unexpected and suggested that sulfation of heparin may occur in a step-wise fashion, the formation of exclusively N-sulfated intermediates preceding the introduction of O-sulfate groups (Fig. 3). In testing this hypothesis, the preferential formation of component III at low PAPS concentration (M. Höök _et al_., unpublished observations) was turned to advantage. Nonsulfated, preformed, unlabeled microsomal polysaccharide (incubation E1, Table I) was pulse-labeled with [^{35}S]PAPS at low concentration (incubation E2). A portion of the resulting [^{35}S]polysaccharide was isolated, and the remainder was subjected to chase-sulfation at high concentration of unlabeled PAPS (incubation E3). Analysis of the resulting labeled polysaccharide preparations, by chromatography on DEAE-cellulose, showed that a considerable portion of the N-sulfated component III had become converted into the N- and O-sulfated component IV during incubation with unlabeled PAPS (Fig. 4). These results demonstrate that O-sulfate groups may be introducted after completed N-sulfation of the polysaccharide, as illustrated in Fig. 3, but do not, of course, exclude the formation of heparin by concomitant N- and O-sulfation of nonsulfated precursors. The differentiation between these two alternatives is of interest in

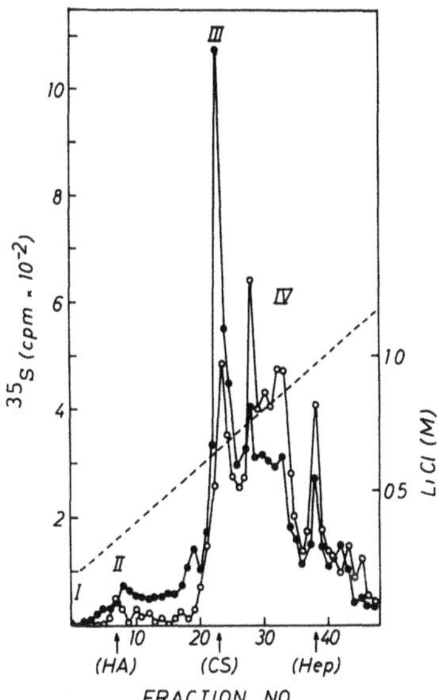

Fig. 4. Chromatography, on DEAE-cellulose, of microsomal pulse-labeled [^{35}S]polysaccharide (incubation E, Table I), before (•---•) and after (o---o) chase-sulfation with unlabeled PAPS. Column was eluted with LiCl (linear gradient from 0.2 M to 1.5 M, - - -), by use of Ultragrad gradient mixer. Improved resolution of components in fraction IV, compared with chromatograms in Fig. 2, is probably due to more shallow gradient used. Complex elution pattern of fraction IV was highly reproducible; furthermore, similar resolution was obtained on separation, under identical conditions, of ^{14}C-labeled sulfated polysaccharide (same material as in Fig. 2D).

relation to the biosynthesis of L-iduronic acid residues, in that this process apparently depends on the formation of O-sulfate groups. Furthermore, a general biosynthetic pathway involving consecutive, rather than concomitant, incorporation of N- and O-sulfate groups would have important implications for the biosynthetic relation between heparan sulfate and heparin. If all N-sulfate groups must be incorporated before O-sulfation can be initiated, the postulated role of heparan sulfate as a biosynthetic precursor of heparin appears to be ruled out. Although speculative at present, these considerations certainly warrant further experimental work.

ACKNOWLEDGEMENTS

This work was supported by grants from the Swedish Medical Research Council (13X-2309, 13X-4, 13P-3431), the Swedish Cancer Society (53), Konung Gustaf V:s 80-årsfond, and the Faculty of Medicine, University of Uppsala.

REFERENCES

1. BALASUBRAMANIAN, A.S., JOUN, N.S. and MARX, W., Arch. Biochem. Biophys., 128 (1968) 623.
2. DeLUCA, S., RICHMOND, M.E. and SILBERT, J.E., Biochemistry, 20 (1973) 3911.
3. FOLEY, T. and BAKER, J.R., Biochem. J., 135 (1973) 187.
4. FRANSSON, L.-Å., in The Chemistry and Molecular Biology of the Intercellular Matrix, vol. II, (Ed. Balazs, E.A.) Academic Press, New York, 1970, p.823.
5. FURTH, J., HAGEN, P. and HIRSCH, E.I., Proc. Soc. Exp. Biol. Med., 95 (1957) 824.
6. HAUG, A. and LARSEN, B., Carbohyd. Res., 17 (1971) 297.
7. HELTING, T., J. Biol. Chem., 246 (1971) 429.
8. HELTING, T., J. Biol. Chem., 247 (1972) 4325.
9. HELTING, T. and LINDAHL, U., J. Biol. Chem., 246 (1971) 5442.
10. HELTING, T. and LINDAHL, U., Acta Chem. Scand., 26 (1972) 3515.
11. HÖÖK, M., LINDAHL, U., BÄCKSTRÖM, G., MALMSTRÖM, A. and FRANSSON, L.-Å., J. Biol. Chem., (1974) in press.
12. HÖÖK, M., LINDAHL, U. and IVERIUS, P.-H., Biochem. J., 137 (1974) 33.
13. JACOBSON, B. and DAVIDSON, E.A., J. Biol. Chem., 237 (1962) 638.
14. LARSEN, B. and HAUG, A., Carbohyd. Res., 17 (1971) 287.
15. LINDAHL, U., BÄCKSTRÖM, G., JANSSON, L. and HALLÉN, A., J. Biol. Chem., 248 (1973) 7234.
16. PHELPS, C.F., Biochemical Society Transactions, 1 (1973) 814.
17. RINGERTZ, N.R., Ann. N. Y. Acad. Sci., 103 (1963) 209.
18. RODÉN, L., in Metabolic Conjugation and Metabolic Hydrolysis, vol. II, (Ed. Fishman, W.H.) Academic Press, New York, 1971, p. 345.
19. SILBERT, J.E., J. Biol. Chem., 238 (1963) 3542.
20. SILBERT, J.E., J. Biol. Chem., 242 (1967) 2301.
21. SILBERT, J.E., J. Biol. Chem., 242 (1967) 5146.
22. SILBERT, J.E., J. Biol. Chem., 242 (1967) 5153.
23. STOOLMILLER, A.C. and DORFMAN, A., in Comprehensive Biochemistry, vol. 117, (Eds. Florkin, M. and Stotz, E.H.) Elsevier, Amsterdam, 1969, p. 241.
24. SUZUKI, S. and STROMINGER, J., J. Biol. Chem., 235 (1960) 257.
25. TAYLOR, R.L., SHIVELY, J.E., CONRAD, H.E. and CIFONELLI, J.A., Biochemistry, 12 (1973) 3633.

DISCUSSION OF SILBERT *ET AL* AND LINDAHL *ET AL* PAPERS

SILBERT

In our work on chondroitin sulfate biosynthesis, we found that our products were rather heterogeneous in size. I wonder if you found similar heterogeneity, since the pattern that you see on the DEAE-cellulose chromatogram could be a reflection of size differences in the polysaccharide chain as well as a reflection of different amounts of sulfate.

LINDAHL

Both the nonsulfated and the sulfated fractions obtained on DEAE-chromatography were heterogeneous with regard to molecular size. The average molecular weight of either fraction was in the order of 30 - 70,000 as estimated by gel chromatography; at such high molecular weights the effect of chain size on the chromatographic behavior is minor. We therefore believe that the separation into distinct peaks on DEAE-cellulose essentially reflects charge differences.

HORNER

I'd like to point out that we have some evidence which I now realize completely supports Dr. Lindahl's thesis that the N-sulfation does precede the O-sulfation, although these were not experiments done to test biosynthetic pathways. For completely different reasons we have given normal rats ^{35}S sulfate, killed them 16 hours later, and isolated the skin macromolecular heparin. Although N-sulfate obviously does not account for two-thirds of the sulfate in the heparin molecule, it does account for two-thirds of the label, the ^{35}S sulfate in the molecule, after 16 hours. So that is indirect support for Dr. Lindahl's thesis

LINDAHL

How long did you wait after injecting the ^{35}S sulfate before recovering the labeled polysaccharide?

HORNER

Sixteen hours. I should point out that there is some work by other Swedish workers who worked with isolated mast cells from the peritoneal cavity, that is not mastocytoma mast cells, normal mast cells. They have found that the ^{35}S-sulfated heparin in these has a very long half life, something like 30 or 40 days.

LINDAHL

One might argue, maybe, that in our microsomal system the final N-sulfate/O-sulfate ratio seemed to be established within a few minutes or less, that is in much shorter time than you allowed for your *in vivo* experiment. At any rate your data show that N- and O-sulfation may occur as distinctly separate processes.

JEANLOZ

The epimerization at C-4 is quite a complex mechanism, as you know, in monosaccharides, and to see it applied directly to a high molecular compound is a very revolutionary concept. This is why I'd like to see a clearer demonstration of this mechanism. What is the stoichiometric transformation between these two fractions?

LINDAHL

So far we have only been able to study the epimerization of uronic acid residues in conjunction with sulfation of ^{14}C-labeled, microsomal polysaccharide. However, we know that during sulfation no further labeled uronic acid is incorporated into the polymer, and it's therefore quite clear that some of the glucuronic acid residues present in the nonsulfated precursor are ultimately recovered as iduronic acid residues in the final, sulfated product.

The mechanism of the epimerization reaction is still unknown. I guess one would have to assume the abstraction of a hydrogen at C-5, one way or another, which is then reintroduced from the other side, thus causing inversion of the configuration. Whether this occurs by formation of an intermediate carbanion or by removal of an additional hydrogen at C-4, forming an intermediate unsaturated uronic acid, we don't know. We are testing these possibilities at present by carrying out similar incubations with UDP-glucuronic acid which is specifically tritiated in various positions.

WALDMAN

First, you find intracellular heparitin sulfate; what do you think that might be doing or is it on its way to the membrane. Secondly, how much of the heparitin sulfate might possibly be heparin by some of the classic definitions?

SILBERT

Some of Elizabeth Neufeld's work pertains to this first question. She has shown that there is some heparitin sulfate in normal cells and suggests that this is material which has been put out by the cell into the media and then taken up and degraded intracellularly with lysosomal enzymes. Consequently, this material is not linked to protein, and represents smaller pieces. We found similar results with some of our cultures, but in others the intracellular material appeared to be mainly proteoglycan. Since the heparitin sulfate is synthesized by the cells, there has to be some heparitin sulfate as proteoglycan in the cell during the time of synthesis, but this may represent a small fraction of the total.

In regard to the second question, the answer depends upon the definition of heparin. I define heparin as the sulfated glucosamine-containing mucopolysaccharide that is produced by mast cells. I define sulfated glucosamine-containing mucopolysaccharide (other than keratan sulfate) produced by other cells to be heparitin sulfate. I believe there are real differences. The material that we have looked at is essentially homogeneous as far as its sulfate content is concerned, and chromatographs on DEAE-cellulose like a standard of chondroitin sulfate. We conclude from this that it has approximately one sulfate per disaccharide repeating unit, and we find no material that has the high degree of sulfate content that one associates with heparin. If we can ever get at the nature of the protein portion of the heparitin sulfate and compare it with the protein portion of heparin, perhaps we might find that these molecules may be totally different. We may be looking at one small aspect of the molecule, the polysaccharide portion, which may be similar for heparin and heparitin sulfate, but the entire molecules might be very different in structure.

HORNER

Dr. Silbert, at the end of your talk, you were throwing in some provocative remarks about the biological activity of heparitin sulfate. I'd like to know how strong the evidence is, in your opinion, for heparitin sulfate activating lipoprotein lipase, - and I think you also mentioned the fibrinolytic system.

SILBERT

We know of no evidence, but we are going to start looking for it.

HORNER

That was my feeling.

DAVIS

I should like to ask some practical questions. At what pH are the sulfated polysaccharide chains appreciably hydrolyzed into shorter fragments? It is well known that heparin administered directly (intravenous, intrafat, intraperitoneal, intramuscular, or sublingual) is effective in producing anticoagulant effects, but swallowed oral heparin is without effect. Is this because the sulfated polysaccharide chains are hydrolyzed by stomach acidity into ineffective low molecular fragments?

The next question relates to recognized hypercoagulability in older people, and the beneficial effects of heparin in senile manifestations. Do older people have a deficiency of endogenous heparin, and is this related to a deficiency of available sulfate for conversion to anticoagulant compounds?

The final question concerns the separations achieved on the cationic columns. Are these separations related to the increasing charge and hydration which result when the hydroxyl and amino groups are sulfated - thus markedly increasing both the negative charge and hydration on these polyanions? In other words, is heparin so effective as an anticoagulant and in other biochemical reactions because it carries both a maximum negative charge and a maximum hydration?

SILBERT

I think you are referring to the degradation of heparin. The N-sulfate is quite labile to acid (0.04 N HCl) but further degradation requires stronger conditions. Biological degradation is mainly an unanswered questions, since the only well described heparinases are of bacterial origin, while mammalian enzymes for heparin degradation have not really been described.

I don't think there is any evidence for a sulfate deficiency state. Sulfate is quite plentiful in the diet, and I think it would be very unlikely that this would be a factor. I think that the relatively low sulfate content of heparitin sulfate is intrinsic to the structure and is not related to the amount of PAPS that happens to be present.

LINDAHL

The third question referred to the mechanism of separation on the ion exchange column. I am no expert at these matters; we have carried out the separations without bothering really about the mechanism. Analysis of the separated polysaccharides showed that they differed with regard to the number of sulfate groups per disaccharide unit. I would therefore explain the separation simply in terms of electrostatic interaction between the ion-exchanger and polyanions of different charge density.

SILBERT

I'd like to add something to that. In the use of DEAE-cellulose, one is actually looking at two parameters. One is looking at the charge density per disaccharide repeating unit and also looking at the size of the molecule. Even though a larger molecule may have the same charge density per disaccharide unit as a smaller molecule, it will be more polyanionic and will be eluted from the column later. I brought this up in my initial question to Dr. Lindahl. We find that the heparitin sulfate of cell surface is quite polydisperse as far as its chromatography on DEAE-cellulose, but still appears to be relatively homogeneous as far as sulfate content. We conclude from this (and confirmed by chromatography on Sepharose) that it consists of polysaccharide chains with a great deal of variation in size, but little in sulfate content.

LINDHAL

I think I'll expand somewhat on the size matter, which is a rather dissatisfying point at present. We know that the final biosynthetic product resembles the macromolecular heparin discovered in rat skin by Dr. Horner; such heparin molecules apparently consist of several polysaccharide chains joined by a polysaccharide core. Since heparin generally contains covalently bound residual peptides it seems likely that the initial polymerization product is more like a conventional proteoglycan, in which the various polysaccharide chains are joined by a protein core. If so, we don't know at what stage during biosynthesis the "proteoglycan type" heparin is converted to "macromolecular type" heparin.

MARX

Do you have any quantitative data regarding the fractions in your pulse-chase experiment, particularly in regard to the position of the N-sulfate, and whether or not the N-sulfate once attached was transferred to give some degree of O-sulfation. I believe that that would be relatively easily tested in a nitrous acid deamination.

LINDAHL

This is what we are doing at present, by electrophoretic analysis of ^{35}S-labeled DEAE-fractions after deamination with nitrous acid. After such treatment 95% of the label incorporated into component 3 occurs as inorganic ^{35}S-sulfate, in agreement with the contention that this component is N- but not O-sulfated. We will analyze component 4 in the same way, after direct sulfation with ^{35}S-PAPS and after pulse-chase sulfation, but such data are not yet available.

METABOLISM OF MACROMOLECULAR HEPARIN IN MURINE NEOPLASTIC MAST CELLS

Sören ÖGREN[+] and Ulf LINDAHL[++]
[+]Department of Medical Chemistry, Royal Veterinary College, Biomedical Centre and [++]Institute of Medical Chemistry, University of Uppsala, S-751 23 Uppsala (Sweden)

Most glycosaminoglycans occur in the native state as proteoglycans, composed of several polysaccharide chains covalently linked to a common polypeptide core (9). Attempts to isolate a heparin proteoglycan have been unsuccessful, yielding essentially single polysaccharide chains, with molecular weights ranging from 7,000 to about 25,000 (6, 10). However, in rodent tissues heparin has been shown to occur as a multichain, macromolecular species (molecular weight up to 10^6) distinctly different from a conventional proteoglycan (4, 10); in this molecule, the individual polysaccharide chains are joined in an unknown manner, possibly by a polysaccharide core (4). The results of Horner suggest that the macromolecular heparin represents a precursor form of the polysaccharide, which must be depolymerized to acquire biological activity (5). The present investigation was undertaken in order to obtain experimental evidence for the postulated conversion *in vivo* of the macromolecular precursor (in the following denoted "macromolecular heparin") to the low-molecular heparin product ("conventional heparin"). In addition, the enzyme responsible for this degradation was isolated and characterized in some detail.

Polysaccharide in murine Furth mastocytoma was pulse-labeled *in vivo* by means of a single intraperitoneal injection of inorganic [^{35}S]sulfate (carrier-free, 75 μCi/mouse), and was then isolated after varying periods of time from subcellular fractions. Such fractions were recovered from tissue homogenates, by consecutive centrifugations at 1,000g for 10 min, 20,000g for 20 min and 100,000g for 1 hr, respectively, yielding three particulate preparations from each sample. Initially, the [^{35}S]polysaccharide formed occurred largely in the fraction sedimenting at 1,000 - 20,000g, with smaller amounts in the 0 - 1,000g (granule) and 20,000 -

100,000g (microsomal) fractions; the amounts of labeled polysaccharide in the 100,000g supernatant were negligible. The molecular size of the newly-synthesized polysaccharide (about 90% of which was identified as heparin) considerably exceeded that of conventional heparin, as indicated by gel chromatography on Sepharose 4B.

Within the first hour after injection of [^{35}S]sulfate, most of the labeled polysaccharide was redistributed from the 1,000 - 20,000g to the 0 - 1,000g fraction. During, and possibly after this shift, the macromolecular heparin was degraded (Fig. 1), ultimately to the size of conventional heparin. The degradation process appeared to be completed 6 hr after injection of [^{35}S]sulfate.

Particulate subcellular fractions were analyzed for heparinase activity, by incubation with macromolecular [^{35}S]heparin, followed by gel chromatography of the resulting labeled products. Significant degradation of the substrate occurred in the 1,000 - 20,000g fraction only. Further characterization of this fraction, by density-gradient centrifugation in iso-osmotic colloidal silica (12) revealed a single visible band of particles, with the same density as lysosomes. These particles contained all the heparinase, β-glucuronidase and ^{35}S-labeled, endogenous polysaccharide of the 1,000 - 20,000g fraction.

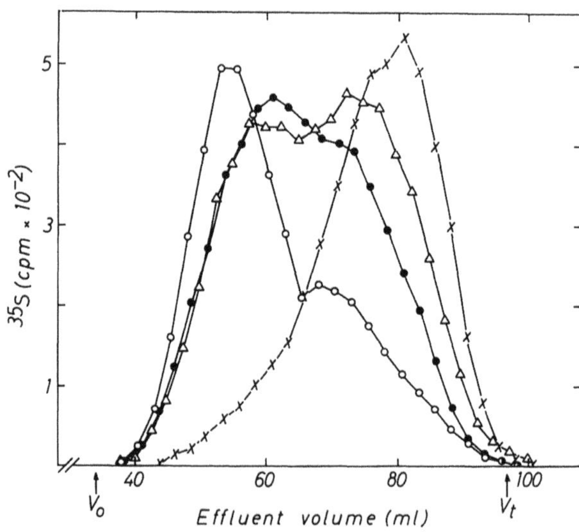

Fig. 1. Gel chromatography on Sepharose 4B (1.2 x 95 cm column, eluted with 1 M NaCl in 0.05 M Tris-HCl, pH 8.0) of [^{35}S]polysaccharide isolated from 1,000 - 20,000g fraction, obtained (O) 30 min; (●) 1 hr; (Δ) 2 hr; (x) 6 hr after injecting tumour-bearing mice with [^{35}S]sulfate.

Heparinase was isolated from the 1,000 - 20,000g particulate fraction by salt extraction followed by precipitation from saturated ammonium sulfate. By use of gel chromatography on Sepharose 4B, the enzyme was shown to degrade macromolecular, ^{35}S-labeled, mastocytomal heparin (K_{av} about 0.25) to products similar in size to conventional heparin (K_{av} about 0.85), apparently by nonrandom cleavage of a limited number of glycosidic linkages per molecule. Prolonged incubation times (up to 5 days, with repeated addition of enzyme) did not result in further degradation of the product. No significant depolymerizing activity was observed with any other glycosaminoglycan tested, including chondroitin sulfate, dermatan sulfate, hyaluronic acid, heparin sulfate and commercial heparin. The pH-optimum for degradation of macromolecular heparin was around pH 5.

The nature of the linkage cleaved by the heparinase was investigated by reduction of the polysaccharide degradation products with sodium [^{3}H]borohydride. The degraded molecules (but not the macromolecular substrate) incorporated significant amounts of tritium (Fig. 2). An essentially monodisperse fraction of the labeled,

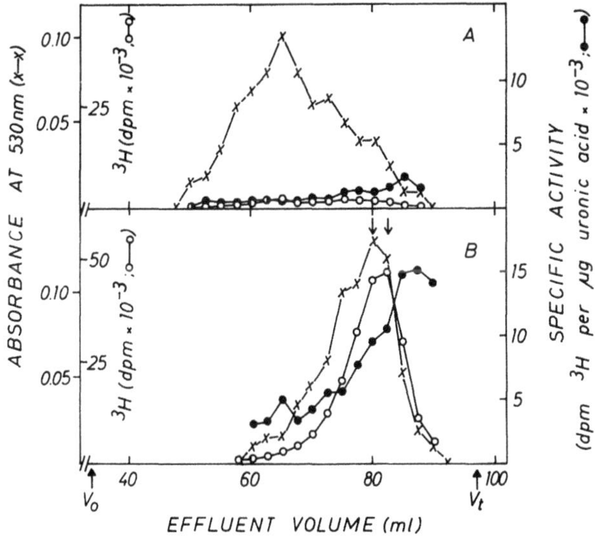

Fig. 2. Gel chromatography on Sepharose 4B of macromolecular mouse mastocytoma heparin after treatment with (A) heat-inactivated; (B) active heparinase, followed by reduction with [^{3}H]borohydride. The gel column was eluted as described in the legend to Fig. 1. Effluent fractions were analyzed for uronic acid (x) and for radioactivity (0), and the specific activity (0) of each fraction was calculated. The two fractions indicated by arrows in the lower panel were combined and subjected to molecular weight determination as described in the text.

degraded heparin was subjected to sedimentation equilibrium ultracentrifugation, indicating a molecular weight of 14,500. By relating the molecular weight to the specific activity of the preparation, the amount of reducible groups was calculated to be approximately one per molecule.

The ^{3}H-labeled heparin was degraded to monosaccharides by a combination of acid hydrolysis and cleavage by deamination with nitrous acid (2). Analysis of the degradation products, by paper electrophoresis and paper chromatography, showed a major radioactive component which behaved like L-gulonic acid. Since [^{3}H]gulonic acid would be the expected labeled reduction product of a polysaccharide molecule containing a glucuronic acid residue in reducible terminal position, these results tentatively suggest that the heparinase is an endo-glucuronidase and that the macromolecular, branched heparin is degraded to single-chain molecules of conventional heparin by the action of this enzyme.

A number of observations support the contention that the formation and degradation of macromolecular heparin is not restricted to rodents, but may be of general significance: (a) macromolecular heparin has been isolated from monkey and human tissues (3); (b) heparin preparations, isolated from various mammalian species by mild methods, generally contain polysaccharide chains lacking the polysaccharide-protein linkage region (1, 6, 8 and L. Jansson, S. Ögren, and U. Lindahl, submitted for publication); (c) uronic acid was the only reducible terminal residue detected in heparin isolated from bovine liver capsule (L. Jansson *et al*., submitted for publication), and could be released as free monosaccharide on treatment of a number of different commercial heparin preparations with nitrous acid (F and M. Höök , unpublished observations); (d) mastocytomal, macromolecular [^{35}S]heparin was depolymerized during incubation with homogenates of human intestine and spleen, respectively (S. Ögren, unpublished observations); (e) basophilic leucocytes from a patient with chronic myeloid leukemia incorporated [^{14}C]glucosamine into a heparin-like polysaccharide of high molecular weight, while a pool of smaller molecules (representing the bulk of polysaccharide) remained unlabeled (11). Further work should be devoted to the isolation and characterization of the various postulated mammalian heparinases as well as their physiological substrates.

ACKNOWLEDGEMENTS

This work was supported by grants from the Swedish Medical Research Council (13P-3431; 13X-2309), the Swedish Cancer Society (53), Konung Gustav V:s 80-årsfond and the Faculty of Medicine, University of Uppsala.

REFERENCES

1. CIFONELLI, J.A. and RODÉN, L., Biochim. Biophys. Acta, 165 (1968) 553.
2. HÖÖK, M., LINDAHL, U. and IVERIUS, P.H., Biochem. J., 137 (1974) 33.
3. HORNER, A.A., Proc. Can. Fed. Biol. Soc., 13 (1970) 109.
4. HORNER, A.A., J. Biol. Chem., 246 (1971) 231.
5. HORNER, A.A., Proc. Nat. Acad. Sci. US, 69 (1972) 3469.
6. LINDAHL, U., Biochem. J., 116 (1970) 27.
7. LINDAHL, U. and AXELSSON, O., J. Biol. Chem., 246 (1971) 74.
8. LINDAHL, U., CIFONELLI, J.A., LINDAHL, B. and RODÉN, L., J. Biol. Chem., 240 (1965) 2817.
9. LINDAHL, U. and RODÉN, L., in Glycoproteins, (Ed. GOTTSCHALK, A.) Elsevier Publishing Co., Amsterdam, Vol. A, (1972) 491.
10. ÖGREN, S. and LINDAHL, U., Biochem. J., 125 (1971) 1119.
11. OLSSON, I., BERG, B., FRANSSON, L.A. and NORDÉN, Å., Scand. J. Haematol., 7 (1970) 440.
12. WOLFF, D. and PERTOFT, H., Biochim. Biophys. Acta, 286 (1972) 197.

DEMONSTRATION OF ENDOGENOUS HEPARIN IN RAT BLOOD

Alan A. HORNER

Department of Physiology, University of Toronto,
Toronto, Ontario (Canada)

Dr. Lindahl has referred to my concept of macromolecular heparin as a multichain, whose structural integrity depends on a polysaccharide core. In 1971, I postulated that macromolecular heparin must be depolymerized to become biologically active (6). Later, I found that macromolecular heparin is a potent inhibitor of heart lipoprotein lipase (LPL) *in vitro* (7). Injected macromolecular heparin has a relatively low potency as an initiator of LPL activity in the blood. This activity appears in the blood rather slowly, compared with that of low-molecular-weight heparin (i.e., commercial heparin or enzymatically depolymerized macromolecular heparin) (7). These facts support the concept that depolymerization precedes the release of biologically active heparin from the mast cells. Endogenous heparin in the blood should therefore be of low molecular weight, and I would now like to demonstrate that this is so.

Anaphylactic shock in dogs causes the release of heparin into the blood, with concomitant degranulation of mast cells in the liver (8). It has therefore been assumed that mast cells normally release heparin into the blood. However, Riley has pointed out that the dog is exceptional in releasing heparin from its mast cells so readily (13). In fact, the normal presence of heparin in the blood cannot be called an accepted fact. A direct demonstration by chemical means was made by Eiber and Danishefsky in 1957 (4). They injected [^{35}S]sodium sulfate into dogs, collected blood 26 hr later, added commercial heparin to the plasma as carrier, and obtained a barium salt of heparin with constant radioactivity after repeated recrystallization. By comparison with carrier-free [^{35}S]heparin prepared from the livers of the same dogs, they calculated that the blood contained at least 500 µg of heparin per 100 ml.

The work has been criticized (2) on the grounds that this concentration of heparin is well above the limit of detection by normal isolation procedures. The data may also have received insufficient attention because of Riley's contentions about the atypical nature of dog mast cells (13).

In the present work, a similar experimental approach was made with rats, but with the additional aim of determining the molecular size of the heparin and its distribution among the blood components.

Each of eight male Wistar rats (body weights, 241 $\pm$ 8.0 g) received 2 mCi of carrier-free [^{35}S]sodium sulfate by intraperitoneal injection and was exsanguinated by cardiac puncture under ether anesthesia 18 hr later. Blood (total volume, 82 ml) was collected into one-ninth volume of anticoagulant (disodium EDTA, 1.5 g/100 ml of 0.15 M NaCl) with plastic syringes and transferred to plastic centrifuge tubes. The blood was centrifuged (300 _g_ for 20 min) at room temperature, and platelet-rich plasma was removed with a siliconed Pasteur pipet. The remaining blood was respun under the same conditions, and more platelet-rich plasma was removed. Platelets were obtained by centrifuging at 2,000 _g_ for 20 min, and platelet-poor plasma was pipetted off from the plug of platelets. The red blood cells and buffy layer were gently resuspended in an equal volume of a modified Tyrode solution containing EDTA but no calcium or magnesium ions (7) and centrifuged at 150 _g_ for 20 min. The buffy layer was carefully removed with a siliconed Pasteur pipet.

The platelet and buffy layer preparations were resuspended in ethanol. The red blood cells and the plasma were each mixed with three volumes of ethanol. After standing overnight at 4° the four products were centrifuged (1000 _g_ for 30 min) at 4°. The supernatants were discarded. The sediments were washed with ethanol and hexane at room temperature and dried under reduced pressure. Each dry product was homogenized in 0.10 M Tris-HCl and 0.02 M $CaCl_2$ at a pH of 8.0 and incubated with shaking at 40°. Pronase (10 mg) was added to each. The addition of Pronase was repeated after 24 hr; at this time, commercial pig mucosal heparin was added (5 mg to the red blood cells and 2 mg each to the plasma, platelet, and buffy layer digests). After 48 hr, heparin was recovered by precipitation with cetylpyridinium chloride in the presence of 1.2 M NaCl, using general procedures described previously (6). Each partially purified heparin sample was applied to a column of microgranular DEAE-cellulose (Whatman DE 32) and eluted in a NaCl gradient at a pH of 2.5 (6). The elution of platelet [^{35}S]heparin and carrier pig mucosal heparin from such a column is illustrated in Fig. 1. The carrier heparin was measured by a carbazole method for uronic acid (1) and endogenous heparin by determining sulfur-35 radioactivity. The endogenous and carrier heparin samples were combined and reprecipitated with ethanol. Their relative molecular sizes were estimated by gel filtration on a column of Bio-Gel A-15m agarose gel

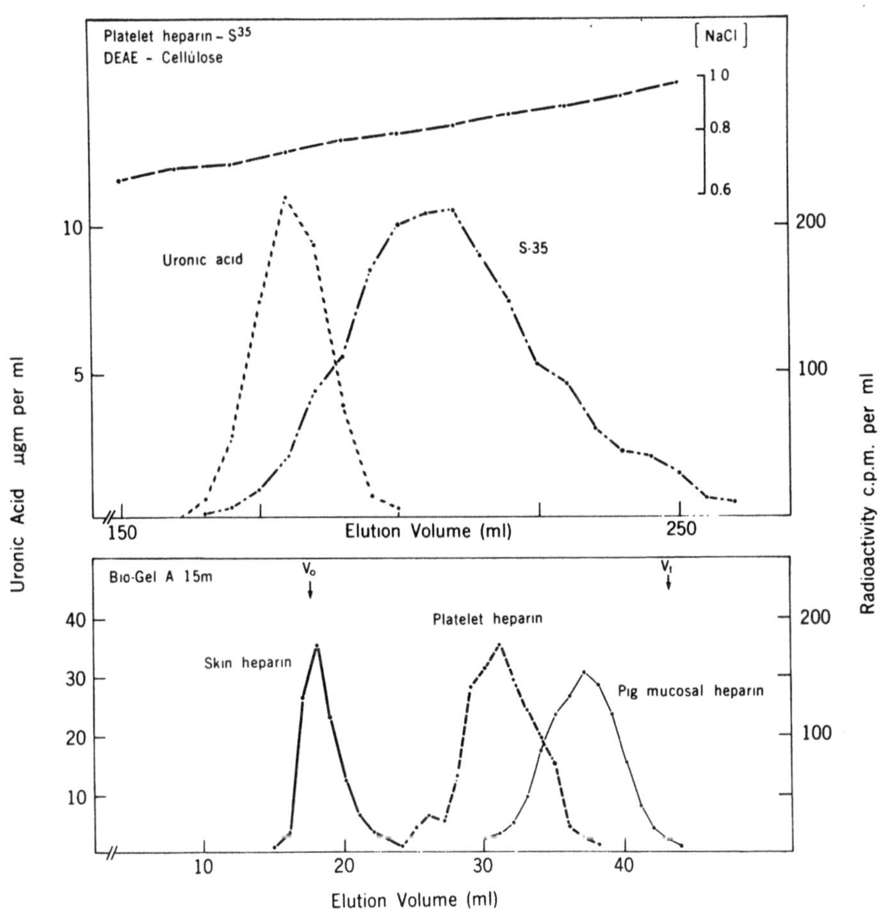

Fig. 1. Top elution of carrier pig mucosal heparin and endogenous [^{35}S]heparin, recovered from rat platelets after digestion with Pronase and precipitation with cetylpyridinium chloride, from a column of microgranular DEAE-cellulose (Whatman DE 32; 1.5 cm x 20 cm; volume, 35 ml) with a linear NaCl gradient at a pH of 2.5 (6). NaCl concentration was determined with a conductivity meter. Bottom gel filtration of samples of macromolecular [^{35}S]heparin from rat skin, [^{35}S]heparin from rat platelets, and pig mucosal heparin on a column of Bio-Gel A 15m agarose gel granules (100-200 mesh; 1.5 cm x 24 cm; volume, 43 ml) equilibrated with 1.0 M NaCl. V_0 = void volume, determined with tobacco mosaic virus. The pig mucosal heparin was quantitated by uronic acid determinations, and the rat heparins by measuring sulfur-35 radioactivity.

granules, through which a sample of macromolecular rat skin heparin was also run for comparison. The elution patterns (Fig. 1) show that the endogenous platelet heparin is larger than pig mucosal heparin, but much smaller than macromolecular heparin. Thus, its size is similar to that of the lower-molecular-weight fraction obtained by gel filtration of enzymatically degraded macromolecular heparin (7).

Further tests were made to confirm that the endogenous polyanion containing sulfur-35 was heparin. These tests could not be anticoagulant assays or chemical analyses, because of the presence of carrier heparin. Therefore, the behavior of the sulfur-35 label was studied in two well-characterized methods of degrading heparin by chemical means.

Hydrolysis of heparin with 0.04 N HCl yields N-desulfated heparin and inorganic sulfate (5). These products are readily separated on a column of Sephadex G-25. The gel filtration of the hydrolysis products from platelet [^{35}S]heparin is shown in Fig. 2. For comparison, [^{35}S]heparin and [^{35}S]dermatan sulfate prepared from the skin and [^{35}S]chondroitin sulfate from the sternums of the rats from which the blood was obtained were treated in the same way. The elution patterns of skin and platelet heparin were the same, the material eluting in the void volume being N-desulfated heparin. Dermatan sulfate and chondroitin sulfate were completely degraded to lower-molecular-weight material.

Lagunoff and Warren (9) used nitrous acid to degrade heparin. Under their conditions, glucosamine-N-sulfate residues are degraded to anhydromannose. Lindahl and co-workers have used this reaction, with other nitrous acid methods, to characterize heparin from mouse mastocytomas (10). N-Desulfation with nitrous acid is accompanied by cleavage of the adjacent glycosidic bond. Therefore, more low-molecular-weight products are obtained by degradation with nitrous acid than by hydrolysis with 0.04 N HCl. Glycosaminoglycans that have no N-sulfate groups are resistant to nitrous acid under these conditions. Figure 3 shows gel filtration patterns on Sephadex G-25 after nitrous acid treatment. Again, the patterns for platelet heparin and skin heparin are the same, both having been extensively degraded. In contrast, dermatan sulfate and chondroitin sulfate were not broken down.

Thus, the presence of N-sulfate in platelet heparin in the same proportion found in authentic heparin from the skin was established by two different methods of degradation. The platelet product therefore meets chemical criteria for a heparin. The reason for its elution from DEAE-cellulose at a higher salt concentration than the carrier pig mucosal heparin is not known, but the skin heparin shows exactly the same behavior.

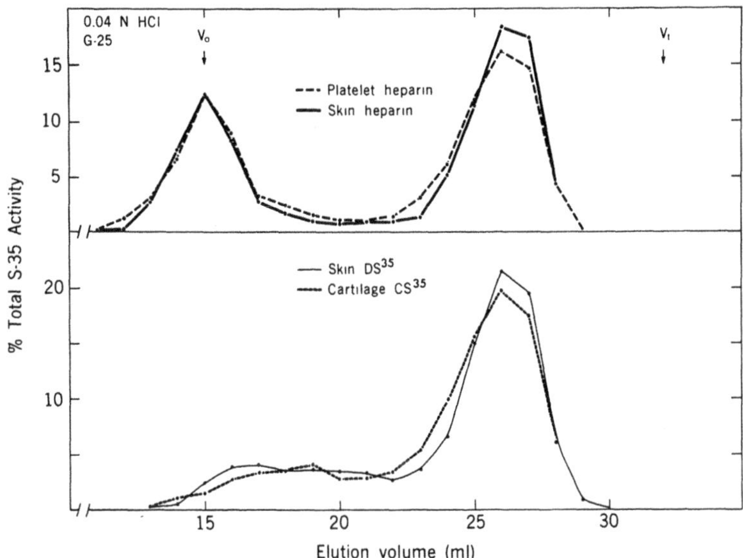

Fig. 2. Gel filtration of samples of platelet [^{35}S]heparin, skin [^{35}S]heparin, skin [^{35}S]dermatan sulfate, and cartilage [^{35}S] chondroitin sulfate after hydrolysis with 0.04 N HCl at 98° for 2 hr (5). The column of Sephadex G-25 (0.9 cm x 50 cm, volume, 32 ml) was equilibrated with 1.0 M NaCl. The sulfur-35 content of each fraction is expressed as a percentage of the total counts in the sample. V_o = void volume, determined with blue dextran 2,000.

I have used the platelet product to illustrate the methods for isolating and characterizing the endogenous heparin in blood. The reason for this is that 50% (in terms of sulfur-35 radioactivity) of the heparin recovered was in the platelet product, 30% was recovered from the red cells, 10% from the buffy layer, and 10% from the plasma. Thus, 90% of the blood heparin was associated with formed elements. The high proportion associated with platelets is particularly interesting, inasmuch as they contained a protein with very high binding affinity for heparin, platelet factor 4 (11).

The [^{35}S]heparin radioactivity in blood was 143 counts/min/milliliter. Carrier-free [^{35}S]heparin was recovered from the skins and peritoneal mast cells of the same rats (3, 6). The specific/activities of these products were: skin heparin, 1,100 counts/min/microgram of uronic acid; and peritoneal heparin, 8,400 counts/min/microgram of uronic acid. On the possibly unwarranted assumption that the specific activity of the heparin in blood is within this range, the concentration of heparin in blood can be estimated. By these criteria, it is in the range of 1.7 - 13 μg of

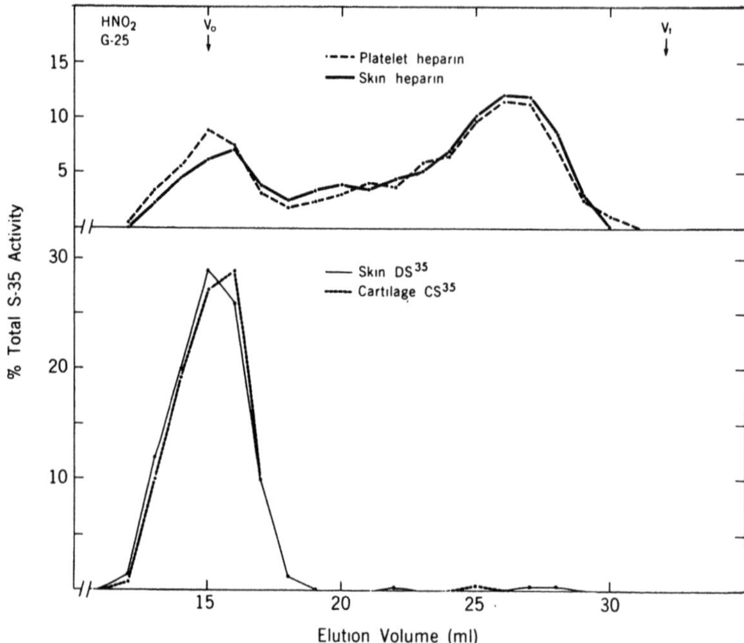

Fig. 3. Gel filtration of samples of the same four glycosaminoglycan preparations labeled with sulfur-35 as in Fig. 2 on the same Sephadex G-25 column, after treatment with nitrous acid at room temperature for 90 min (9).

uronic acid per 100 ml, 5.5 - 42 μg of heparin per 100 ml, or, assuming an anticoagulant activity of 160 units/mg, 0.88 - 6.7 units/100 ml.

Obviously, a more direct method of determining the concentration of heparin in blood is needed. However, this study has demonstrated the presence of heparin in blood and its association with the formed elements, particularly platelets. Furthermore, the heparin in blood is of relatively low-molecular-weight, supporting the concept that depolymerization of the macromolecular form probably precedes its release into the blood.

ACKNOWLEDGEMENTS

This work was supported by the Ontario Heart Foundation and the Medical Research Council of Canada

REFERENCES

1. BITTER, T. and MUIR, H.M., Anal. Biochem., 4 (1962) 330.
2. BRIMACOMBE, J.S. and WEBBER, J.M., Mucopolysaccharides, Elsevier Publishing Co., New York, 1964, p. 103.
3. CHAKRAVARTY, N. and ZEUTHEN, E., J. Cell Biol., 25 (1965) 113.
4. EIBER, H.B. and DANISHEFSKY, I., Proc. Soc. Exp. Biol. Med., 94 (1957) 801.
5. FOSTER, A.B., MARTLEW, E.F. and STACEY, M., Chem. Ind., London, 1953, p. 899.
6. HORNER, A.A., J. Biol. Chem., 246 (1971) 231.
7. HORNER, A.A., Proc. Nat. Acad. Sci. US, 69 (1972) 3469.
8. JAQUES, L.B. and WATERS, E.T., J. Physiol., 99 (1941) 454.
9. LAGUNOFF, D. and WARREN, G., Arch. Biochem. Biophys., 99 (1962) 396.
10. LINDAHL, U., BÄCKSTRÖM, G. and JANSSON, L. and HALLÉN, A., J. Biol. Chem., 248 (1973) 7234.
11. NATH, N., NIEWIAROWSKI, S. and JOIST, J.H., J. Lab. Clin. Med., 82 (1973) 754.
12. RENAUD, S., KINLOUGH, R.L. and MUSTARD, J.F., Lab. Invest., 22 (1970) 339.
13. RILEY, J.F., Ann. New York Acad. Sci., 103 (1963) 151.

DISCUSSION OF HORNER PAPER

SILBERT

I don't remember whether you showed that this was more anionic than commercial heparin on DEAE-cellulose chromatography.

HORNER

Yes, I did. This was the subject that we were getting involved in earlier this morning and I avoided it then, because I think there are some other factors involved that we do not understand yet. Certainly it appears on a DEAE-cellulose column to be more anionic. In other systems we have looked at, we have no evidence that it may be conformationally different. You would think it would move faster on electrophoresis than commercial heparin. It does not. In our pH 3 agarose gel system it moves at about the same speed as the slower component in pig mucosal heparin. Then there is the Wessler technique for doing electrophoresis in dilute HCl in which mobility is supposed to be directly proportional to sulfate content. In this system it is slower moving than commercial heparin or the low molecular weight heparin that we can recover from rat skin. (There is a trace of low molecular weight heparin in rat skin, and we have used that as a control, too.) However, it is faster moving than chondroitin sulfate or dermatan sulfate. So I don't want to commit myself, it's different.

SILBERT

The reason I asked is that you have shown that your material has a preponderance of N-sulfation. This means that you couldn't have more than two sulfates per disaccharide repeating unit in this molecule, and it would have to be of lower sulfate content than the usual heparin.

HORNER

I don't think you can say that, because we have only shown counts, not specific activity. I suggested this morning that the reason for that is that the incorporation of counts into the N-sulfate is more rapid than into the O-sulfates.

SILBERT

This is how long after the injection of the animal?

HORNER

This is overnight, 16 hours after injection of the animals.

SILBERT

I would be surprised to find differences in specific activity between N-sulfate and O-sulfate after such a long time interval.

HORNER

All you can say is that the distribution of the label is almost exactly the same in the skin heparin at that time.

GLUECK

It may be a little naive from someone working in a coagulation laboratory, but it is interesting that when you aggregate the platelets with an aggregating agent, the heparin will interfere with this effect; the so-called platelet factor 4 activity and it is particularly intriguing to me that in your studies platelets themselves contain a heparin-like material. Do you believe there are two kinds of heparin?

HORNER

I think you may very well be right that this is a physiological neutralization system. There are mucopolysaccharides on the surface of platelets, that has been shown. We would like to go on and see if the heparin is on the surface or if it is inside, for the platelet factor 4 is known to be in the granules inside the platelets. It might be that the platelet factor 4 and the heparin don't get together until the platelets disrupt. That is a possibility, but we have no evidence for or against it.

FAREED

I'm not sure if you mentioned the amount of carrier heparin you added in your preparation.

HORNER

No, I did not. I did not think it was too relevant. It is a purely empirical sort of thing related to the size of the fractions. In the case of the red blood cells it was 5 mg, in the case of all other blood fractions, it was 2 mg. We recovered quite a small percentage of that, which is why I am suggesting that there are heparin binding factors in blood. This was what gave us all our trouble at first and why we couldn't find any heparin without this carrier. You can get endogenous heparin into solution without having carrier heparin, but you can only do this with strong salt, and this appears to bring out the binding factors too. As soon as you dilute the strong salt so that you can put the extract on a DEAE-cellulose column, everything comes out of solution again. The ^{35}S-heparin will stay at the top of the column *ad infinitum*, and this was a frustration in our preliminary experiments.

FAREED

Well then, this brings out another interesting question of exchange between the ^{35}S you used, between the carrier heparin and the ^{35}S heparin.

HORNER

No, I don't think there is any exchange because the carrier heparin is added *in vitro*. The blood fractions are isolated and precipitated with ethanol. The solid material recovered is then subjected to proteolytic digestion and it is only at the proteolysis stage that the carrier heparin is added.

FAREED

Did you look at the urine macromolecular fractions of these animals.

HORNER

The urine? No we did not.

RELATION OF CHEMICAL STRUCTURE OF HEPARIN TO ITS ANTICOAGULANT ACTIVITY

J. A. CIFONELLI

Departments of Pediatrics and Biochemistry, Joseph P. Kennedy, Jr. Mental Retardation Research Center, University of Chicago, Chicago, Illinois 60637 (USA)

The relation of the chemical structure of heparin to its anticoagulant activity has been of interest for many years (4, 8, 11), and much information has accumulated on it. However, study of this has been complicated by the variability in composition of heparin (1) and by its chemical similarity to some other mucopolysaccharides with which it is often associated in tissues. It is now possible to isolate heparin that is nearly free of contaminating polysaccharides and has high biologic activity, but this was not always so. Thus, some of the early reports of the effects of molecular size or sulfate concentration on biologic activity are difficult to assess. Because earlier preparations generally had lower and more variable anticoagulant activity than current ones, conflicting reports of the relation of various characteristics to biologic activity have appeared, and the problem is made more difficult by the lack of full knowledge of the chemical structure of heparin and by the complexity of its possible structural arrangements.

Nevertheless, the recent strides made in clarifying structural aspects of heparin, as noted earlier by Jeanloz, allow a more comprehensive and definitive study of the effects of chemical characteristics on biologic activity. Numerous reports have shown an apparent effect of heparin on different biologic systems. This discussion, however, will be concerned solely with the influence of the variations in heparin on its anticoagulant activity. Although the difficulties inherent in the biologic methods used for the assay of heparin, as well as their unreliability as emphasized by Jaques and co-workers (12, 18), are well recognized, that aspect will not be considered here, inasmuch as the data to be presented are of value in a comparative way (10).

If we use the simplified version of the chemical structure of heparin in Fig. 1, we can see the effects of some variations. Thus, when the section labeled "c" (uronic acid and N-acetylglucosamine) represents an increasingly large proportion of the molecule, it approaches heparan sulfate; when the section labeled "b" (sulfated iduronic acid and N-sulfated glucosamine) enlarges, through various biosynthetic steps, it progresses from heparan sulfate to heparin, at which stage section b becomes a major proportion of the polysaccharide. Ester sulfate on the N-sulfated glucosamine residues, because of its variable nature, is not illustrated on the model.

Using the structural illustration is not intended to imply the sequences of the sections in the molecule, but only to suggest the variabilities possible in both the proportions and the arrangements of the components. It is based on several considerations: N-acetylglucosamine residues are bound mostly to glucuronic acid; iduronic acid may be sulfated, but not glucuronic acid (22); nonsulfated uronic acid associated with N-sulfated glucosamine may be either glucuronic or iduronic acid; N-sulfated glucosamine is ester-sulfated in variable proportions, ranging up to 75% (or more) in beef lung heparin (7, 17); and ester sulfate on uronic acid units may also vary, perhaps in a range similar to that of the hexosamine (21). We may note further that variations in molecular weight of heparin fractions from the same source have been reported in a number of instances (13, 16, 20).

It is apparent that several characteristics of heparin may be studied in regard to their effect on bioactivity. In addition, one would like to know the effect of the distribution of ester sulfate in the iduronic acid and N-sulfated glucosamine on anticoagulant activity. Although other characteristics - those presumed to be constant, such as molecular shape and number and arrangement of carboxyl groups may also influence the anticoagulant activity, only the variable functions discussed earlier will be considered here. (An illustration of the specific effect of the carboxyl moiety on anticoagulant activity will be presented by Danishefsky later in this volume.)

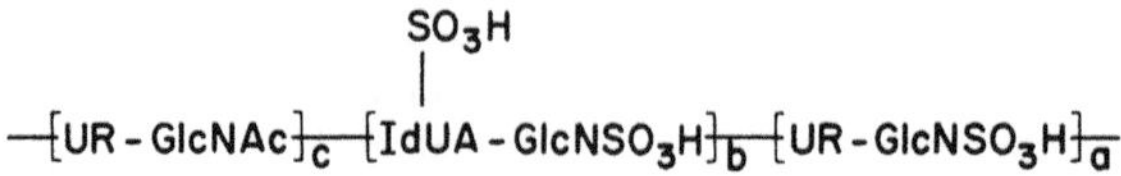

Fig. 1. Structural representation of heparin. UR, uronic acid; GlcNAc, N-acetylglucosamine; SO_3H-IdUA, sulfated iduronic acid; $GlcNSO_3H$, N-sulfated glucosamine.

Effect of molecular weight. - Conflicting reports are found in the literature on the effect of molecular size of heparin on its anticoagulant activity (2, 3, 13, 14, 16, 20). Unfortunately, in some instances, results are based on compounds with meager analytic data, as well as relatively low biologic activity; it is therefore difficult to compare the results properly.

In the present study, fractionation of a highly purified heparin preparation has supplied samples that, by various criteria (19), appear to be free of contaminating polysaccharides. The starting material was commercial beef lung heparin with a stated activity of 159 I.U./mg. This was first subjected to purification with cetylpyridinium chloride in the presence of 1.5 M sodium chloride by established procedures (19) and was then separated into four fractions by gel filtration with Sephadex G-75, as described previously (6). Analytic and bioactivity data for these fractions are listed in Table I.

It is apparent that all the fractions are similar with respect to N-sulfate and ester sulfate contents. None of the fractions showed the presence of free amino groups in measurable amounts; this is of some importance, in that such groups themselves influence anticoagulation and would therefore constitute an interference, with respect to the interpretation of molecular-size effects. The bioactivity of preparations is found to be highest for the heaviest,

TABLE I

Effect of Molecular Weight of Beef Lung Heparin on Anticoagulant Activity

Fraction	Hexosamine Content, %	N-Sulfate Content[a]	Ester Sulfate Content[a]	Molecular Weight X 10	Activity, I.U./mg
I	22.2	0.96:1	1.42:1	15.5	149
II	21.6	0.98:1	1.48:1	11.8	171
III	21.6	0.99:1	1.60:1	8.1	123
IV	20.4	0.98:1	1.47:1	6.0	45

[a] Given as molar ratios to hexosamine.

it decreased by about 75%, as compared to heparin standard of about 180 units, when the molecular weight got down to 6×10^3. (Similar data have been obtained with heparin fractions isolated from hog mucosa (5)). It appears that beef lung heparin with a molecular weight below approximately 5×10^3 ought to be inactive. This could not be determined in this instance, because a fraction of this molecular weight was not obtained. However, heparin fractions with molecular weights estimated at approximately 5×10^3 had been isolated previously from beef lung heparin byproducts, and they showed an activity similar to that of fraction IV or slightly lower. Comparison of these results with other reported data shows general agreement with the conclusion that anticoagulant activity varies with chain size. However, differences are apparent with regard to the chain size at which such a relation is observed, previous reports (13, 14) indicating that it occurs at higher molecular weights than shown here. However, other data (5) showing high bioactivity with preparations of molecular weights approximating 7.5×10^3 support the present results and indicates that chains containing 10-11 repeating units suffice for establishing maximal anticoagulant activity.

Effect of N-desulfation and chemical constituents. - Although it has been stated that as little as 7% free amino groups on hexosamine units are sufficient to remove the anticoagulant activity of heparin (23), other studies have indicated that somewhat more of the hexosamines could be N-sulfated before complete inactivation occurred. Furthermore, the presence of 25 - 30% glucosamine with N-acetyl groups in heparin preparations from whale and other mammalian tissues (6) suggests that appreciable N-desulfation must occur before heparin is completely inactivated.

Table II lists samples obtained after gel filtration of a beef lung heparin preparation* that had free amino groups on nearly 30% of its hexosamine residues. It is observed that reasonable activities are found in preparations of molecular weights of 11×10^3 or more, even when these have nearly one-third of their hexosamine units with unsubstituted amino groups. N-sulfation increases the anticoagulant activity most markedly in the fractions of lower molecular weight (J. A. Cifonelli and J. King, unpublished data). The results support the suggestion that extensive N-desulfation must occur before major decreases in anticoagulant activity are produced in samples of normal

*The sample used for obtaining these fractions was a gift from Dr. D. D. Dziewiatkowski, who obtained it from the laboratory of Dr. P. A. Levene. Discussion with Dr. L. B. Jaques at this symposium indicated that the heparin preparation, crystallized as the barium salt, was given to Dr. Levene over 30 years ago, The derivation of the free amino groups is therefore believed to stem, at least partially, from the use of acetic acid during crystallization.

TABLE II

Effect of N-Desulfation and Molecular Weight of Beef Lung Heparin on Anticoagulant Activity

Fraction	Hexosamine Content, %	Free Amino Groups[a]	N-Sulfate Groups[a]	Ester Sulfate Groups[a]	Molecular Weight, $\times 10^{-3}$	Activity, I.U./mg
I	23.7	0.32:1	-	1.46:1	14.5	133
I-S[b]	22.3	0.01:1	0.98:1	-	14.0	153
II	23.2	0.26:1	-	1.57:1	11.0	132
II-S[b]	22.0	0.02:1	0.96:1	-	10.5	156
III	23.0	0.28:1	-	1.62:1	8.0	92
III-S[b]	21.7	0.01:1	0.93:1	-	7.9	162
IV	24.0	0.26:1	-	1.53:1	6.7	44
IV-S[b]	22.2	0.01:1	0.97:1	-	6.5	92

[a]Given as molar ratios to hexosamine.
[b]After N-sulfation.

molecular weight. The finding of an increased inhibitory effect on anticoagulant activity from the presence of free amino groups in fractions of lower molecular weight suggests that interruption in sequences of N-sulfated hexosamine units cannot be as easily tolerated in chains of minimal molecular weight without affecting bioactivity. This would be in harmony with the assumption that at least about 10 repeating units with N-sulfated glucosamine residues are necessary for maximal manifestation of anticoagulant activity of heparin.

The effects of ester sulfate on biologic activity are not as well defined as those of N-sulfate. However, ratios of this group to hexosamine ranging from less than 1.0:1 to over 1.5:1 have been found in heparin preparations of comparable anticoagulant activity (Table III), indicating that this group may vary by 50% or more without influencing such activity. With the finding of iduronic acid in heparin, another component was added to the list of possible influences on biologic activity. Because iduronic acid may be sulfated, whereas sulfated glucuronic acid has not been found, variation in proportion of iduronic acid in heparin samples may reflect

TABLE III

Variation of Iduronic Acid in Heparin

Heparin Sample[a]	Hexosamine Content, %	Sulfate Content[b] N-	Sulfate Content[b] O-	Molecular Weight, X 10^{-3}	Activity, I.U./mg	Iduronic Acid Content[c]
BL	23	0.98:1	1.48:1	11.0	180	75
HM	21	0.89:1	1.41:1	10.5	177	73
BM	24	0.86:1	1.56:1	11.0	152	66
W	23	0.74:1	1.06:1	11.0	176	60
M	26	0.83:1	0.86:1	17.0	132	47

[a]Sources: BL, beef lung; HM, hog mucosa; BM, bovine mucosa; W, whale; M, clam (quahog).
[b]Given as molar ratio to hexosamine.
[c]As percent of total uronic acid.

differences in distribution and local density of sulfate groups. In any event, as shown in Table III, pronounced variability of iduronic acid is possible without apparent effect on bioactivity.

Effect of Distribution of Heparin Components

The existence of heparin preparations of similar composition but different anticoagulant activity suggests that the distribution of the various constituents may affect the degree of biologic activity shown by individual fractions. This has been indicated for N-acetyl group distribution (5). Thus, a hog mucosal heparin that was partially N-desulfated by acid hydrolysis, followed by N-acetylation, to provide a sample in which about one-fifth of the hexosamine units contained N-acetyl groups showed only about half of the activity of a whale heparin that had an even higher concentration of N-acetyl groups. Chemical studies showed that the former compound contained an appreciable content of repeating units with N-acetylhexosamine residues in multiple sequences. In contrast, the whale heparin had such units in single sequences, indicating that optimal anticoagulant activity will tolerate 25% or more singly placed N-acetyl groups along the molecule. And multiple sequential units of this constituent, although tolerated when at the protein-linkage end, may affect bioactivity when they are in the interior of the molecule. In agreement with this, fractions isolated from heparin byproducts (5) contained such multiple sequences of N-acetylglucosamine units in the interior of the molecule and had decreased anticoagulant activity.

Information on the distribution of ester sulfate groups is more complex than that of N-acetylhexosamine units, because such sulfate moieties may reside on both the hexosamine and the uronic acid residues. Chemical degradative techniques (15) indicate that heparin preparations with high proportions of N-acetylglucosamine, as found in whale and other mammalian heparin, generally have sulfated uronic acid residues in chains of three or more repeating units, which are separated by nonsulfated uronic acid units in single sequences. Similar results have been found with preparations that have relatively low concentrations of ester sulfate, such as mactins and some mammalian heparin fractions. This is in harmony with the results, stated earlier, indicating that N-acetylhexosamine units in multiple sequences in the interior of the heparin molecule inhibit anticoagulant activity.

The data accumulated from study of heparin preparations of high purity suggest that the relation of chemical structure to anticoagulant activity may be summarized as follows:

(1) Free amino groups appear not to influence anticoagulant activity to the extent reported by some investigators (9, 23). However, in assessing the effect of such groups on bioactivity, both the distribution of the amino groups and the molecular weight of the preparation must be considered.

(2) Similar anticoagulant activity has been observed for fractions, from a single source, that varied in molecular weight from 8.0×10^3 or less to 14.5×10^3. Very low activity occurs only below a molecular weight of approximately 6×10^3, suggesting that a chain with 10-12 repeating units suffices for maintenance of optimal anticoagulant activity.

(3) Introduction of N-acetyl groups in multiple sequences within the chain may affect bioactivity, although single disbursement of such residues, even to the extent of one in three or four hexosamines, does not appear to decrease anticoagulant activity.

(4) Specific effects due to distribution of ester sulfate groups on hexosamine or iduronic acid cannot yet be stated, but chains containing sulfated iduronic acid generally are found to be separated from each other by single units of nonsulfated uronic acid, suggesting that this arrangement is important.

(5) The various characteristics may act singly or in concert

to influence the anticoagulant activity.

ACKNOWLEDGEMENTS

We wish to thank Dr. D. D. Dziewiatkowski for a generous supply of heparin in this study and are grateful to Dr. L. L. Coleman, The Upjohn Company, for supplying purified beef lung heparin and for anticoagulant assays.

REFERENCES

1. BARLOW, G.H., COEN, L.J. and MOZEN, M.M., Biochim. Biophys. Acta, 83 (1964) 272.
2. BARLOW, G.H., SANDERSON, N.D. and McNEILL, P.G., Arch. Biochem. Biophys., 115 (1961) 360.
3. BRASWELL, E., Biochim. Biophys. Acta, 158 (1968) 103.
4. BRIMACOMBE, J.S. and WEBBER, J.M., Mucopolysaccharides, (1964) Elsevier Publishing Co., Amsterdam, p. 92.
5. CIFONELLI, J.A., Carbohyd. Res., in press.
6. CIFONELLI, J.A. and KING, J., Biochim. Biophys. Acta, 320 (1973) 331.
7. DANISHEFSKY, I., STEINER, H., BELLA, A. and FRIEDLANDER, A., J. Biol. Chem., 224 (1969) 1741.
8. FOSTER, A.B. and HUGGARD, A.J., in Advances in Carbohydrate Chemistry, (Eds. Wolfrom, M.L. and Tipson, R.S.), Academic Press, New York, 1955, vol. 10, p. 335.
9. FOSTER, A.B., MARTLEW, E.F. and STACEY, M., Chem. Ind., 899 (1953).
10. JAQUES, L.B. and BELL, H.J., in Methods of Biochemical Analysis, (Ed. Glick, D.), 1959, vol. 7, p. 253.
11. JEANLOZ, R.W., in The Carbohydrates (Eds. Pigman, W. and Horton, D.), Academic Press, New York, vol. IIB, p. 609.
12. KUO, S.H., JAQUES, L.B. and MILLAR, G.J., J. Pharm. Pharmac., 24 (1972) 858.
13. LAURENT, T.C., Arch. Biochem. Biophys., 92 (1961) 224.
14. LIBERTI, P.A. and STIVALA, S.S., Arch. Biochem. Biophys., 119 (1967) 510.
15. LLOYD, A.G., EMBERY, G. and FOWLER, L.J., Biochem. Pharmacol., 20 (1971) 637.
16. PATAT, F. and ELIAS, H.-G., Naturwissenschaften, 46 (1959) 322.
17. PERLIN, A.S., MACKIE, D.M. and DIETRICH, C.P., Carbohyd. Res., 18 (1971) 185.
18. REZANSOFF, W.E. and JAQUES, L.B., Thromb. Diath. Haemorrh., 18 (1967) 150.
19. RODÉN, L., BAKER, J., CIFONELLI, J.A. and MATHEWS, M.B., in Methods of Enzymology (Ed. Ginsburg, V), Academic Press, New York, 1972, vol. 28, p. 73.

20. STIVALA, S.S., YUAN, L., EHLICH, J. and LIBERTI, P.A., Biochem. Biophys., 122 (1967) 32.
21. TAYLOR, R.L., SHIVELY, J.E., CONRAD, H.E. and CIFONELLI, J.A., Biochem., 12 (1973) 3633.
22. WOLFROM, M.L., WANG, P.Y. and HONDA, S., Carbohyd. Res., 11 (1969) 179.
23. FOSTER, A.B., MARTLEW, E.F. and STACEY, M., Chem. Ind., 899 (1953).

SYNTHESIS AND PROPERTIES OF HEPARIN DERIVATIVES

I. DANISHEFSKY

Department of Biochemistry, New York Medical College, Valhalla, New York (USA)

The most pronounced actions of heparin are its anticoagulant effect and its promotion of the clearance of chylomicrons and various lipoproteins from the bloodstream. In both activities, heparin is far more efficient than any available synthetic materials. Two restricting factors in its use are that it has to be administered intravenously and that its effect is comparatively short-lived (5, 10-12). The aims of current studies on chemical modification of heparin therefore include the synthesis of products that would be cleared more slowly from the circulation and the preparation of materials that might be absorbed from the intestine. Another aim is to make it possible to use heparin that is modified at specific sites for investigating the mechanism of its action. It would be of interest to obtain derivatives with only one of the biologic activities - lipid clearance or anticoagulation. The other purpose in studying heparin modification is to elucidate the mechansim of its actions. This is complicated by the lack of definition of the processes involved in the release and activation of lipoprotein lipase and in the individual steps of the blood coagulation sequence. However, some aspects of the mechanisms are amenable to study with heparin derivatives.

The findings on the chemistry of heparin modification may also be used for linking heparin to various adsorbants and polymers. These, in turn, can be used to probe the mechanism of heparin or to prepare nonthrombogenic materials for vascular prostheses. The procedures to be described here also may be adapted for the synthesis of heparin-peptides for immunochemical investigations. Such derivatives may elicit antibodies directed at heparin, which is otherwise nonantigenic. This is reasonable, in view of the findings in experiments with various galactose-peptide conjugates (23).

Structural Modifications

Heparin is composed of glucosamine, uronic acids (glucuronic and iduronic), O-sulfate or sulfate ester groups, and N-sulfate or sulfamino groups (see R. Jeanloz elsewhere in this volume). Although some aspects of the structure remain to be defined, the overall structure can be indicated, as in Fig. 1. The most striking feature of the heparin macromolecule is its large number of negatively charged groups, of which the most prominent are the highly acidic sulfate groups.

If heparin is modified to yield a product in which the N-sulfates are lost (Fig. 1), both biologic activities are also lost. This is generally done with dilute acid (4). Acylation of the free amino groups (Fig. 1) does not regenerate the anticoagulant activity.

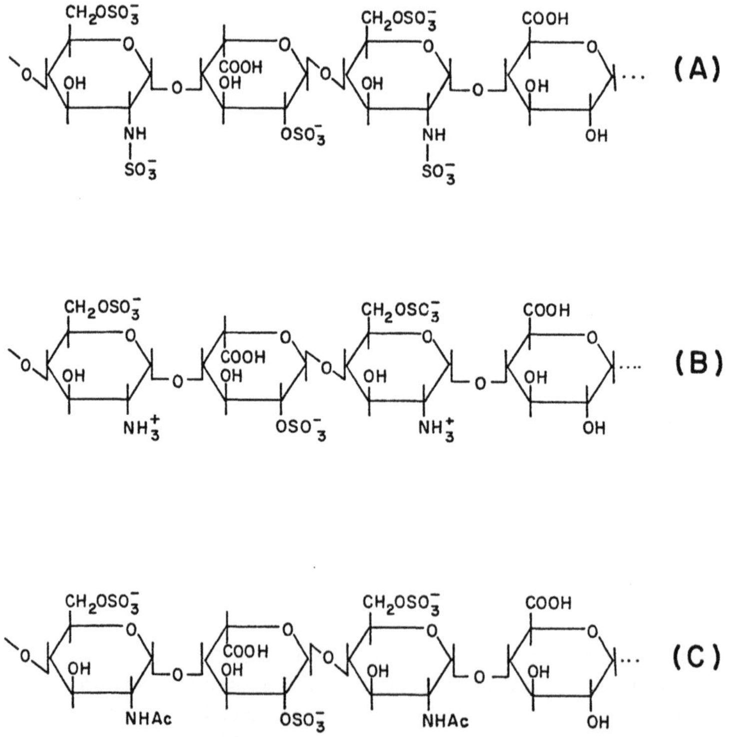

Fig. 1. Structural representations showing hexosamine, iduronic acid, glucuronic acid, O-sulfates, and N-sulfates. A, heparin; b, N-desulfated heparin; c, N-acetylated heparin.

Various derivatives of this type have been synthesized (26). One series of such acylated compounds are shown in Fig. 2. None of these has significant anticoagulant activity, but several have been shown to promote lipoprotein lipase activity or clearing action (25).

It has been concluded from a number of studies (13, 27) that the N-sulfate group of heparin is critical for its anticoagulant activity. Nonetheless, a number of nonnitrogenous compounds, containing sulfate groups, possess substantial anticoagulant action. Prominent among these are dextran sulfate and sulfated alginic acid (18). It appears that dextran sulfate and other polysaccharide sulfates exert their effect by different mechanisms from that of heparin. Alternatively, the activity of heparin may result from the effects of more than one factor.

Assuming that N-sulfate is an essential requirement for anticoagulant action, we have prepared a number of derivatives wherein the sulfate groups are undisturbed (6). The basic reaction is shown in Fig. 3. The mechanism involves the interaction of the carboxyl group of the uronic acid with a water-soluble carbodiimide - in this case, 1-ethyl-3-(dimethylaminopropyl)carbodiimide - to form an O-acylurea. This is followed by a nucleophilic displacement by the comparatively activated amino group of glycine methyl ester to yield the amide of the heparin and glycine methyl ester. Concomitantly, the urea derivative is released into solution.

From a practical viewpoint, the conditions are very mild. Heparin is mixed with the carbodiimide at a pH of 4.75, glycine methyl ester is added, and the solution is stirred in an ice bath for 1-2 hr. The product isolated by precipitating with ethanol is essentially uncontaminated. It can be purified further by dialysis and reprecipitation

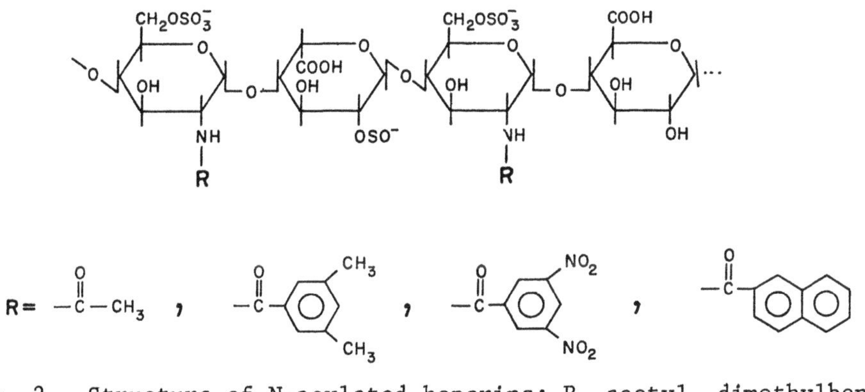

Fig. 2. Structure of N-acylated heparins; R, acetyl, dimethylbenzoyl, dinitrobenzoyl, or naphthoyl.

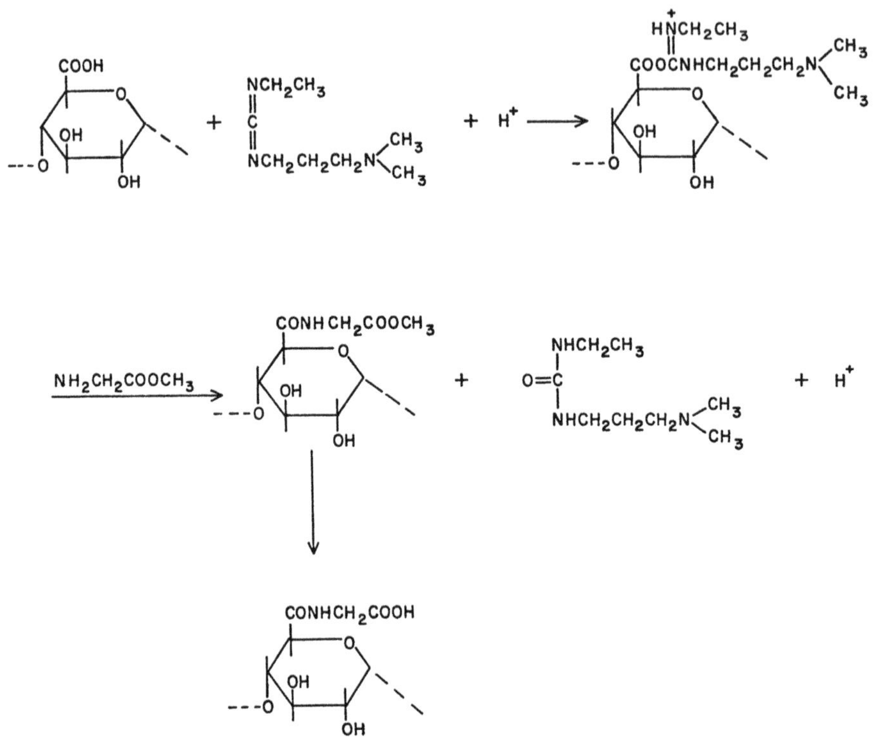

Fig. 3. Equations for the synthesis of heparinylglycine methyl ester and heparinylglycine.

(13). The glycine ester moiety is saponified by treatment with dilute alkali at 0° for a few hours. In the product of this reaction, the carboxyl groups of the glycine are free.

The titration curves of the derivatives substantiate the proposed structure of the products (Fig. 4). The mucopolysaccharides were converted to the acid form with Dowex 50(H) and titrated with sodium hydroxide. Heparin is titrated at a pH of 1.8-2.5, at which the sulfates are neutralized, and a pH of 4-5, at which the carboxyls are neutralized. In contradistinction, the glycine methyl ester derivative is titrated only at the lower pH, indicating the absence of free carboxyl groups. The product obtained after saponification has free carboxyl groups and thus exhibits titration characteristics similar to those of heparin.

Other heparin derivatives, represented in Fig. 5, were prepared in a manner analogous to that described for the glycine methyl ester

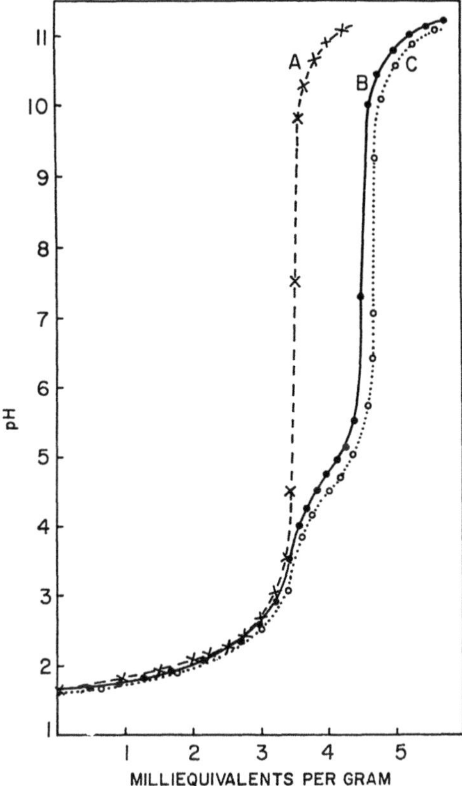

Fig. 4. Titration curves for heparin and derivatives: A, heparinylglycine methyl ester; B, heparinylglycine; C, heparin. Results are in milliequivalents of sodium hydroxide per gram of each compound.

derivative. These included the compound formed with phenylalanine methyl ester, its saponification product, and derivatives with glycylphenylalanine amide, aminomethane sulfonic acid, and taurine.

Analytic data for some of the derivatives are shown in Table I. The molar ratio of sulfate to hexosamine for all of these is similar to that of heparin, indicating that the reactions did not incur any loss of sulfate groups. The amount of amino acid linked to the heparin is about the same as the hexosamine value, as would be expected for a complete reaction.

The anticoagulant activities of the heparin derivatives are indicated in Table II. Although the heparinylglycine methyl ester is inactive, the product obtained after de-esterification

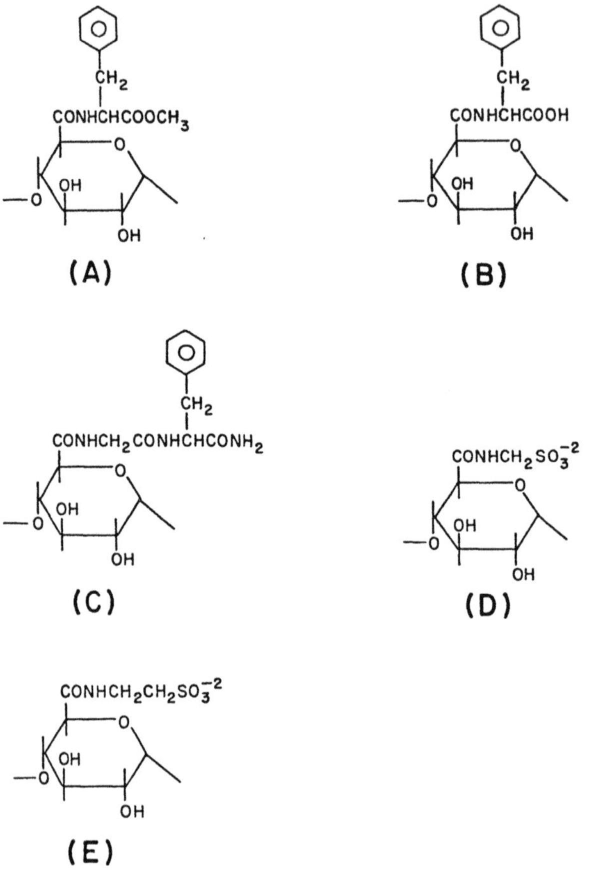

Fig. 5. Structure at the modified uronic acid for the derivatives of heparin with: A, phenylalanine methyl ester; B, phenylalanine; C, glycylphenylalanine amide; D, aminomethane sulfonic acid; E, taurine.

(heparinylglycine) has over 35% of the activity of heparin. Similarly, heparinylphenylalanine methyl ester has practically no anticoagulant activity, but heparinylphenylalanine has significant activity. It thus appears that a free carboxyl group is necessary for the anticoagulant action of heparin; however, considerable activity is maintained even if the carboxyl group is not directly on the uronic acid moiety. The finding that the derivative obtained with aminomethanesulfonic acid is almost as potent as heparin suggests that the primary requirement is that there be an acidic group in the specified position, but it need not be a carboxyl group. The taurine derivative, which differs from the aminomethanesulfonic acid

TABLE I

Composition of Heparin-Amino Acid Derivatives, Millimoles/g

Sample	Constituent Content, millimoles/g			
	Glucosamine	Sulfate	Glycine	Phenylalanine
Untreated heparin	1.39	3.48		
Derivative with:				
glycine	1.22	3.16	1.17	
L-phenylalanine	1.21	3.13		0.90
glycyl-L-phenylalanine	1.13	2.89	0.92	0.85

TABLE II

Properties of Heparin Derivatives

Derivative	Anticoagulant Activity, U.S.P. units/mg
Untreated heparin	135
Derivative with:	
glycine methyl ester	3
glycine	51
phenylalanine methyl ester	14
phenylalanine	26
glycyl-L-phenylalanine amide	9
aminomethanesulfonic acid	121
taurine	10

compound by an additional methylene unit, is inactive.

To substantiate the conclusion that the carboxyl in heparin is critical for its anticoagulant activity, we prepared the methyl ester of heparin itself, by treating it with ethereal diazomethane. Analysis of the product showed that over 95% of the carboxyl groups were esterified and that sulfate was not lost in the process. Assay of this product revealed that it was inactive.

Reduction of heparin with sodium borohydride and a water-soluble carbodiimide brings about the conversion of the carboxyl groups to alcohol units (24). Assay of this product also showed that it had no activity.

The reaction between the carboxyl of heparin and amino groups of various compounds to yield the respective amides was extended to the preparation of heparin-linked agarose. Although agarose does not contain the required amino groups, derivatives of agarose containing amino groups can be prepared (3). A preparation of this type - aminoethyl-agarose - when treated with heparin and a water-soluble carbodiimide yielded a product in which heparin was linked to agarose (8). In another modification, heparin bound to agarose through a six-carbon bridge was prepared (I. Danishefsky and F. Tzeng, unpublished data). The general structures are shown in Fig. 6. Inasmuch as the reactants for each of these preparations are polymeric materials, it is assumed that a minimal number of the heparin carboxyl groups are involved in the linkage. Steric factors would not allow for a considerable fraction of the uronic acid to react with the agarose derivatives. Thus, it is expected that most of the carboxyl groups in these products are free.

Heparin has also been linked directly to agarose by treatment with cyanogen bromide (16). This procedure depends on the availability of free amino groups on the heparin macromolecule. The source of such groups is either an unsulfated hexosamine unit or some N-terminal amino acid retained from the native heparin proteoglycan.

Biochemical Applications

The available experimental evidence indicates that, when heparin functions as an anticoagulant or as an activator of lipoprotein lipase, it interacts with one or more proteins or enzyme components. It is therefore reasonable to expect such substances, when applied to heparin-linked agarose, to react at the heparinized sites and thus be preferentially absorbed.

Investigations in our laboratory dealt with the interaction of plasma with heparinized aminoethyl agarose (8). When plasma was

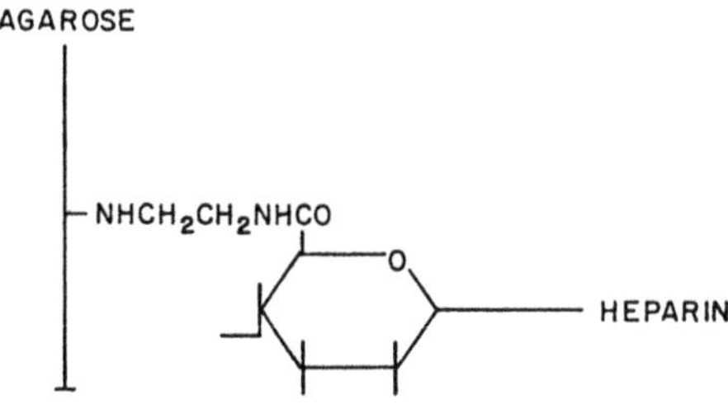

AGAROSE

$NH(CH_2)_6$ NHCO

O

HEPARIN

Fig. 6. Structures of heparin linked to agarose via a two-carbon bridge and a six-carbon bridge.

treated with this adsorbant, the coagulation time was increased considerably. Assays of the treated plasma revealed that it was deficient in factors IX and XI. These factors, in addition to heparin cofactor, could be eluted in separate fractions from the adsorbant with aqueous NaCl of different concentrations. Such experiments suggest that the anticoagulant action of heparin involves an interaction and binding of factors IX and XI, thus rendering the blood incoagulable.

Studies on the binding of coagulation factors from plasma to heparin-agarose prepared according to the procedure of Iverius (16) were reported by Gentry and Alexander (15). In addition to the adsorption of factors IX and XI and heparin cofactor from plasma, they showed that thrombin is also bound to the heparin-agarose. The heparinized adsorbants prepared by the two procedures may differ in their specific binding capacities for individual factors. These characteristics could be exploited for various purification methods. Indeed, heparin-agarose linked through an amino group of heparin has been used in studies on interactions with lipoproteins (17) and for final steps in the purification of lipoprotein lipase (1, 9), heparin cofactor (22), and factor IX (14).

The amino acid derivatives of heparin described earlier were also used for various biochemical studies. Preliminary results indicate that heparinylglycine methyl ester (Fig. 3) is almost as active as heparin in eliciting lipoprotein lipase when injected into dogs. Heparinylmethyl ester also has considerable effectiveness in this action. It therefore appears that, by blocking the carboxyl groups of heparin, it is possible to separate the lipoprotein-clearing activity from the anticoagulant effects, and quantitative studies on this activity are in progress. Because heparin has been shown to release lipases with specific characteristics from different tissues (19), it will be interesting to determine whether the modified heparins can distinguish between the lipases from various sources.

The overall effect of heparin on blood coagulation is the result of its actions at a number of sites in the coagulation sequence. Inasmuch as different enzymes and protein factors are involved at each stage, the molecular specificity of heparin is probably not the same for all its inhibitory effects. For example, a free carboxyl group may be critical for heparin's interference with one of the early steps in coagulation, but not for a later reaction, or vice versa. As a first approach to investigate this problem, we studied the effects of modified heparin on the clotting of plasma, as measured by different systems, such as activated thromboplastin time, prothrombin time, and thrombin time.

The inhibitory action of heparin and several derivatives on coagulation of plasma, on the basis of activated partial thromboplastin time, is shown in Fig. 7. These experiments show the total effect of the inhibitor on the complete "intrinsic" system. Heparinylglycine methyl ester and heparinylmethyl ester that is completely esterified have minimal activity. But there is considerable inhibition with heparinylglycine (which has free carboxyl groups) and heparinylmethyl ester in which only 82% of the carboxyl groups are esterified (I. Danishefsky and E. Perricone, unpublished data). The results can be expressed semiquantitatively by comparing the amounts of the different materials required to produce a given increase in clotting time. Compared with heparin, the activities for heparinylglycine, partially esterified heparinylmethyl ester, completely esterified heparinylmethyl ester, and heparinylglycine methyl ester are 28%, 13%, 7%, and 4%, respectively.

The anticoagulant effects of some of these materials when clotting was assayed by the one-stage prothrombin time are plotted in Fig. 8. These studies indicate the action of the agents on the "extrinsic" system and the later steps of the clotting sequence. Heparinylglycine methyl ester and completely esterified heparinylmethyl ester (not shown on the graph) showed no inhibitory effect,

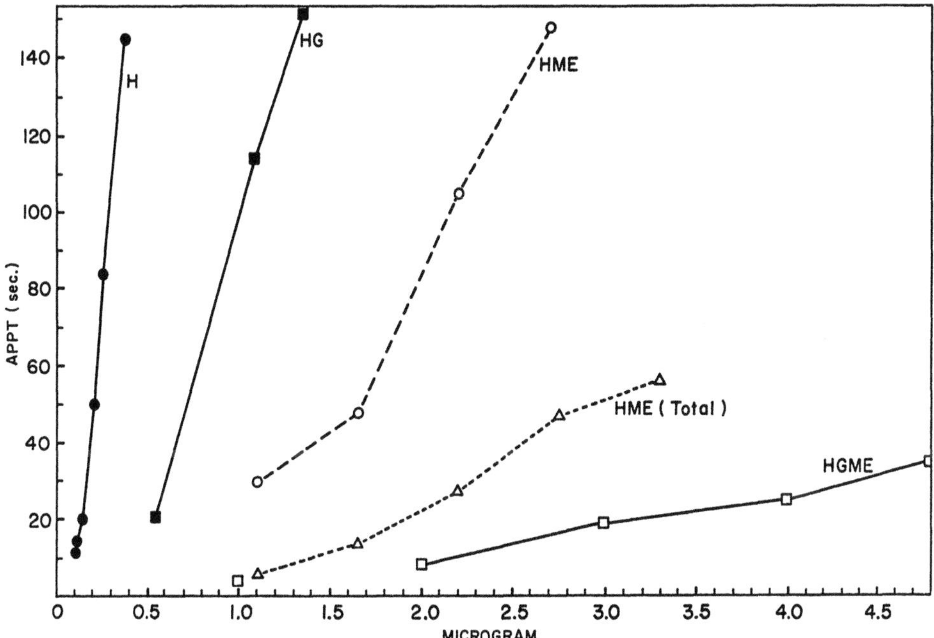

Fig. 7. Activated partial thromboplastin time on addition of different amounts of heparin and derivatives. H, heparin; HG, heparinylglycine; HGME, Heparinylglycine methyl ester; HME (Total), completely esterified heparin methyl ester, HME, heparin methyl ester in which 82% of the carboxyl groups are esterified. The assay was performed in a total volume of 0.4 ml, of which 0.1 ml was citrated plasma. The micrograms of heparin or derivative in the assay mixture are shown on the abscissa. The ordinate shows the increase in time, compared with that of the blank, which did not contain inhibitor.

even at high concentrations. Both heparinylglycine and the partially esterified heparinylmethyl ester had about 20% of the activity of heparin in this system. Similar results were obtained with these inhibitors when clotting of plasma was initiated with Russell's viper venom. The latter assay system also bypasses the early stages of coagulation by activating factor X.

Experiments of this type were also performed when clotting of plasma was affected by addition of thrombin. In this system, which measures inhibition on terminal clotting reactions, the relative inhibitory effects of the heparin derivatives were significantly different from those of the other systems. In terms of percent of heparin activity, the potencies for heparinylglycine, partially esterified heparinylmethyl ester, completely esterified heparinyl-

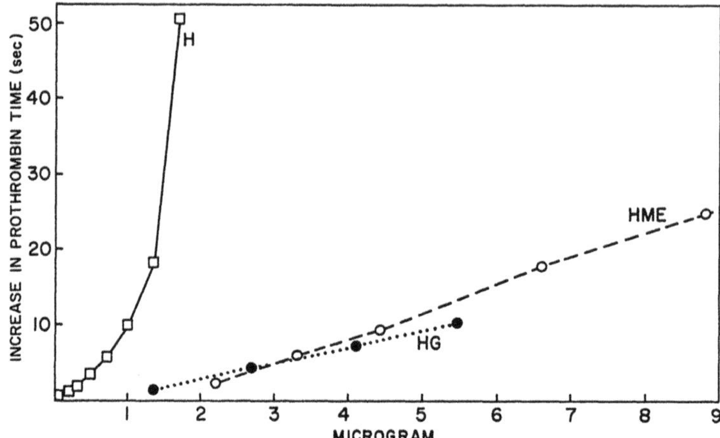

Fig. 8. Prothrombin time on addition of different amounts of heparin and derivatives. H, heparin; HME, partially esterified (82%) heparin methyl ester; HG, heparinylglycine. The assay was performed in a total volume of 0.4 ml, of which 0.1 ml was citrated plasma. The micrograms of heparin or derivative are shown on the abscissa. The ordinate shows the increase in time, compared with that of the blank, which did not contain inhibitor.

methyl ester, and heparinylglycine methyl ester were 12%, 20%, 5%, and 5%, respectively.

Inhibition of blood coagulation by heparin is due to interference in at least three sites of the coagulation sequence. Heparin antagonizes the action of thrombin on fibrinogen. This requires an antithrombin generally termed "heparin cofactor" (2). Heparin in conjunction with the antithrombin also neutralizes activated factor X (28). Another known inhibitory site that does not seem to require a cofactor is a reaction in which factor IX is involved (20, 21). The studies described above suggest that the free carboxyl groups of heparin are equally critical for all the inhibitory sites. However, modification of heparin by insertion of a glycine unit appears to decrease the inhibitory action at the terminal steps to a greater degree than in the earlier processes. It is realized that these conclusions are based on experiments with whole plasma, in which various complex interactions are conceivable. Further investigations are therefore being carried out with purified factors in isolated systems.

These studies are examples of a general approach wherein heparin derivatives may help to elucidate the mechanism of action of heparin and to probe some of the biologic processes that heparin inhibits.

In all these experiments, it is important that only a specific functional groups is altered and that other portions of the heparin macromolecule remain intact. It was demonstrated that various heparin-linked adsorbants can be used to identify the blood components that interact with heparin. These materials were found to be useful in purifying various enzymes and coagulation factors.

Modified heparin can also be used to identify the relation between specific functional groups and inhibitory action at specific sites. It is expected that some modifications will provide products with greater specificity both with respect to overall lipoprotein lipase and anticoagulant effects and with respect to individual steps in these processes. Such materials with well-defined molecular specificities could serve as agents for elucidating the relevant biologic processes. It is hoped that they will also have the necessary pharmacologic properties to be useful in the treatment of various disorders.

REFERENCES

1. ASSMANN, G., KRAUSS, R.M., FREDRICKSON, D.S. and LEVY, R.I., J. Biol. Chem., 248 (1973) 1992.
2. BRINKHOUS, K., SMITH, H.P., WARNER, E.D. and SEEGERS, W.H., Am. J. Physiol., 125 (1939) 627.
3. CAUTRECASAS, P., J. Biol. Chem., 245 (1972) 3059.
4. DANISHEFSKY, I., in Methods in Carbohydrate Chemistry, (Ed. Whistler, R.), Academic Press, New York, 1965, vol. 5, p. 407.
5. DANISHEFSKY, I. and EIBER, H.B., Arch. Biochem. Biophys., 85 (1959) 53.
6. DANISHEFSKY, I. and SISKOVIC, E., Carbohydrate Res., 16 (1971) 199.
7. DANISHEFSKY, I. and SISKOVIC, E., Thrombosis Res., 1 (1972) 173.
8. DANISHEFSKY, I. and TZENG, F., Thrombosis Res., 4 (1974) 237.
9. EGELRUD, T. and OLIVECRONA, T., J. Biol. Chem., 247 (1972) 6212.
10. EIBER, H.B. and DANISHEFSKY, I., Nature, 180 (1957) 1359.
11. EIBER, H.B., DANISHEFSKY, I. and BORRELLI, F.J., Angiology, 11 (1960) 40.
12. EIBER, H.B., DANISHEFSKY, I. and CARR, J.J., Journ. Am. Med. Assoc., 176 (1961) 871.
13. FOSTER, A.B., MARTLEW, E.F. and STACEY, M., Chem. and Ind., 899 (1953).
14. FUJIKAWA, K., THOMPSON, A.R., LEGAZ, M.E., MEYER, R.G. and DAVIE, E.G., Biochem., 12 (1973) 4938.
15. GENTRY, P.W. and ALEXANDER, B., Biochem. Biophys. Res. Comm., 50 (1973) 500.
16. IVERIUS, P.H., Biochem. J., 124 (1971) 677.
17. IVERIUS, P.H., J. Biol. Chem., 247 (1972) 2607.

18. JAQUES, L.B., in Progress in Medicinal Chemistry, (Eds. Ellis, G.P. and West, G.B.), Plenum Press, New York, 1967, vol. 5, p. 139.
19. LaROSA, J.C., LEVY, R.I., WINDMUELLER, H.G. and FREDRICKSON, D.S., J. Lipid Res., 13 (1972) 356.
20. LUNDBLAD, R.L. and DAVIE, E.W., Biochem., 3 (1964) 1720.
21. PITLICK, F.A., LUNDBLAD, R.L. and DAVIE, E.W., Journ. Biomed. Mater. Res., 3 (1969) 95.
22. ROSENBERG, R.D. and DAMUS, P.S., J. Biol. Chem., 248 (1973) 6490.
23. RUDE, E., WESTPHAL, O., HURWITZ, E., FUCHS, S. and SELA, M., Immunochemistry, 3 (1966) 137.
24. TAYLOR, R.L. and CONRAD, H.E., Biochem., 11 (1972) 1383.
25. VELLUZ, L., Bull. Soc. Chem. Biol., (France) 61 (1959) 415.
26. VELLUZ, L., NOMINÉ, G. and PIERDET, A., Comptes Rendus des Séances de l'Acemic des Sciences, 247 (1958) 1521.
27. WOLFROM, M.L. and McNEELY, W.H., Journ. Am. Chem. Soc., 67 (1945) 748.
28. YIN, E.T., WESSLER, S. and STOLL, P.J., J. Biol. Chem., 246 (1971) 3712.

LOW-MOLECULAR-WEIGHT DERIVATIVE OF HEPARIN THAT IS ORALLY ACTIVE IN MICE

Sigmund E. LASKER

New York Medical College, New York, New York 10029 (USA)

An important characteristic of heparin as a therapeutic anticoagulant or a prophylactic agent for deep venous thrombosis is its rapid onset of action. The need for parenteral injection, however, severely limits its use as a long-term therapeutic agent. An orally active heparin-like compound could find wide application.

Claims have been made for the gastrointestinal or sublingual absorption of commercial heparin and synthetic heparinoids (2, 11). These claims have been challenged by Wright (13) and others (8, 9) and have not survived a critical investigation by Windsor and Freeman (12). An extensive investigation of the routes of administration of commercial heparin - including inhalation, instillation into small intestines, and sublingual administration in dogs, rabbits, and man - failed to show evidence of absorption.

Drug absorption is influenced by a number of physiochemical properties, including solubility, charge, and molecular weight. Increasing absorption of high doses of poorly absorbed high-molecular-weight polydisperse substances, such as heparin, could be rationalized as the availability of a larger mass of the lower-molecular-weight fraction (3). The molecular-weight range of fractionated commercial preparations is 4,000 - 16,000 Daltons when fractionation is accomplished by elution from cellulose, gel filtration on Sephadex (G), or differential solubility in alcohol-water mixtures (6). The biologic activity ranges from 70 to 176 I.U. for these fractions. When they are depolymerized with heparinase, the molecular weight is 3,000 - 6,000 Daltons, and the biologic activity, 45 - 95 I.U. (5).

Materials and Methods

Commercial heparin preparations, derived from gut or lung and supplied by several manufacturers, were used during this study, with similar results. An unbleached lung preparation was also used. Heparinase was prepared by adapting flavobacterium (4) to heparin. The bacteria were grown in a 500-liter aerated trypticase medium for 24 hr.* When 60 of heparin was used in the medium, 30 of acetone powder of adapted cells was recovered. Ascorbic acid depolymerization was accomplished when 2.0 mg/ml of heparin was incubated in a mixture containing 4 mM ascorbic acid and 0.5 mM copper sulfate in 0.5 M sodium chloride and 0.1 M sodium phosphate at a pH of 7.8 for 24 hr at 40°. A 3% hydrogen peroxide solution could be used in place of copper sulfate. Weight-average molecular weights were estimated by equilibrium ultracentrifugation in 0.5 M sodium chloride. A single centifuge speed of 44,770 rpm was used to generate a plot of r^2 versus lnC that was generally a straight line. Proton magnetic spectra were measured with the 220-mHz Varian superconducting spectrometer with a fast Fourier transform computer.

Bleeding time was measured by a micro-Lee-White method. Melting-point capillaries were used to bleed mice from the ocular sinus. Lack of free movement of the blood plug in the capillary was taken as a first sign of clotting. The presence of a fibrin thread in the broken capillary was used as the endpoint. *In vitro* anticoagulant activity was measured by the U.S.P. method.

Results and Discussion

The dispersion of molecular weights of a preparation of commercial porcine gut heparin that was also depolymerized and fractionated is shown in Table I. Fractionation was accomplished with Sephadex G-150. Although the molecular-weight dispersion of the commercial preparation is wide - 6,680 - 15,800 Daltons - there is no overlap in molecular weights with the heparinase depolymerized preparation. The sulfur, nitrogen, and uronic acid contents of fractions of heparin are similar, as are infrared, proton, and carbon-13 magnetic-resonance spectra (5). The high-resolution nuclear-magnetic-resonance spectra of commercial and depolymerized heparin are shown in Fig. 1. The water peak is 180° out of phase. At the degree of depolymerization that is represented by these fractions-$(\text{tetramer})_1$ to $(\text{tetramer})_4$-there is no evidence of an unsaturated elimination product (7, 10). The spectra are similar, with a possible quantitative reduction in the N-acetyl content of the depolymerized products suggesting a

* The adaped cells were prepared at the New England Enzyme Center of Tufts University, Boston, Mass.

TABLE I

Distribution of Molecular Weights and Anticoagulant Activity in Heparin and Depolymerized Heparin Fractions

Preparation	Molecular Weight[a]	Anticoagulant Activity, Units[b]
Heparin[c]	11,600	139
Fraction A	15,800	170
B	13,200	150
C	9,100	100
D	6,680	70
Depolymerized	5,300	70
Fraction A	5,900	95
B	4,880	80
C	4,920	60
D	3,260	45

[a]Weight average molecular weight estimated by equilibrium ultracentrifugation in 0.5 M sodium chloride.
[b]Anticoagulant activity measured by the U.S.P. method.
[c]Unbleached.

nonrandom distribution of N-acetyl groups along the polymer chain.

The analytic data for enzyme-treated heparin are presented in Table II. The elemental analysis, uronic acid content, and optical rotation are not significantly altered after depolymerization, but molecular weight is reduced by one-half to two-thirds and the anticoagulant activity is reduced by one-half. The electrophoretic mobility is increased by 25%. Our analytic data reveal no loss in sulfur content. This observation appears to be consistent with the

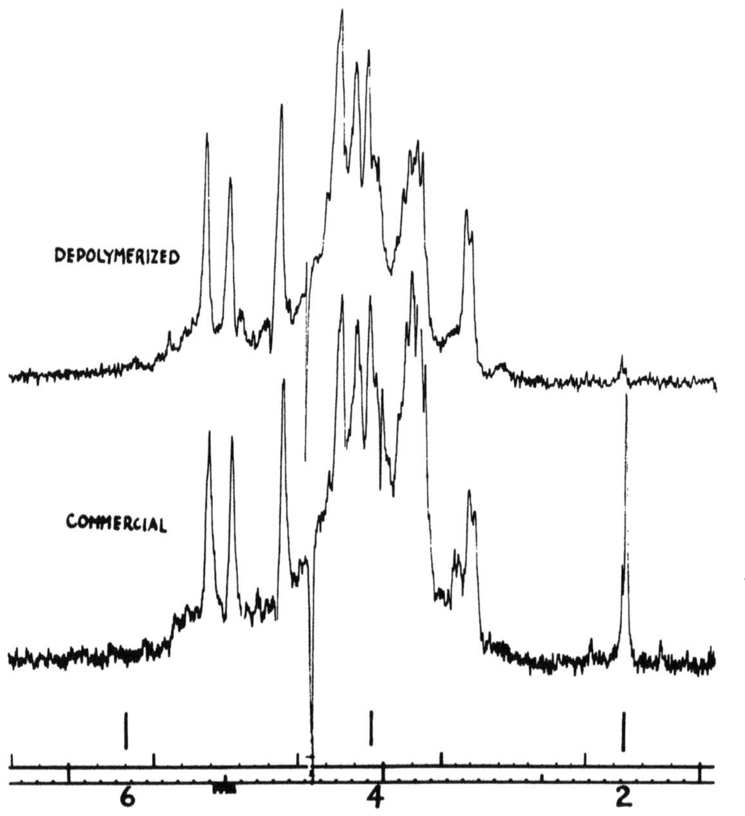

Fig. 1. Proton magnetic resonance spectra at 220 mhz of commercial and depolymerized heparin were measured by fast Fourier transform spectroscopy. The water peak is presented 180° out of phase. The similarity of the major proton resonances is shown.

apparent sensitivity of the desulfation process to the substrate molecular weight (1). Lower-molecular-weight sulfur compounds ranging from dimers to a molecular weight of 2,000 were, however, isolated.

As a potentially practical approach, the depolymerization of heparin was also accomplished with ascorbic acid catalyzed by copper ions. Because it was difficult to remove the trace amounts of copper complexed to heparin, peroxide was used in combination with ascorbate to depolymerize heparin. Data on the fractionated depolymerization product are summarized in Table III.

TABLE II

Typical Analytic Data for Enzyme-Treated Heparin

	Elemental Analysis, %				Molecular Weight	Uronic Acid (%)		Optical Rotation, $(\alpha)^{20}_{D}$	Activity, units
	C	H	S	N		Carbazole test	Orcinol test		
Porcine gut heparin	23.1	5.1	10.9	2.24	10.130	34.4	13.5	+53.6	150
Bovine lung heparin	22.9	3.4	10.3	1.94	12.740	43.2	11.9	+47.6	143
(Enzyme-treated) porcine gut heparin	21.1	4.36	11.0	2.35	5,300	38.0	12.5	+52.7	70
(Enzyme-treated) bovin lung heparin	22.0	4.11	11.0	2.13	4,500	40.8	13.3	+45.5	68

Animal Studies

Pentothal-anesthetized mice were used to test the anticogulant activity of selected low-molecular-weight fractions. Control clotting measurements were made by bleeding from the ocular sinus with a capillary before administration of the compound. Solutions containing 2.5 - 18 mg in 0.5 - 0.75 ml of water or isotonic salt solution were slowly applied to the buccal cavity of anesthetized mice, which were immobilized on a small-animal board. Care was taken to minimize losses by applying the solutions slowly over a period of several minutes. When a sample was lost, the data on it were excluded from the tabulation.

The results of tests in mice are shown in Table IV. The clotting time, when measured 1-1.5 hr after oral administration of commercial heparin, did not appreciably increase above the control range of 0.5-3 min. In less than 10% of the tests, a small effect was demonstrated with commercial heparin. In contrast, the low-molecular-weight product increased the Lee-White clotting time significantly.

TABLE III

Alcohol-Fractionated, Ascorbic Acid-

Hydrogen Peroxide-Depolymerized Heparin

Alcohol Content, %	mg Recovered	Anticogulant Activity, Units	Molecular Weight
63	57.5	---	
60	71	12	4000
55	121	45	5280
50	175	80	6420
45	230	100	7100
	654.5		

TABLE IV

Oral Efficacy of Low-Molecular-Weight Heparin in Individual Tests in Mice

Preparation	Weight, g	Dose, mg	Clotting Time, min	Time after Administration, hr
Commercial heparin	32.5	9.5	2	1
Lot #(111Rts)	22.5	7.5	5	1.5
molecular weight, 12,740	34	4.9	2.5	1
	24	4.5	1	1
	26	5	14	2
	25	2.5	3	1
Depolymerized heparin	27	18	30	1.5
Prep #(1A(D12))	24	13	210	2
molecular weight, 4,700	25	10	30	5
	30.5	7.2	90	2
	36	5	120	1.5
	23	2.5	15	1.5

TABLE V

Oral Efficacy of Low-Molecular-Weight Heparin in Mice

Preparation	No. Mice	Dose, mg	Clotting Time, min
Control	105	---	0.5-3
Commercial heparin	18	2.4-9.5	1-3
Commercial heparin + NaCl	13	2.4-10	1-6
Prep #1A(D1) (molecular weight, 5,300) + NaCl	15	4-7	16-45
Prep #1A(D12) (molecular weight, 4,700) + NaCl	18	2.5-18	15-120
Prep #1A(D12) (molecular weight, 4,700) ---	14	2.4-9.8	10-30

There were 15% failures with the low-molecular-weight product. Table V presents individual mouse data. When two anesthetized rhesus monkeys were given, at 90 mg/kg, a degraded unfractionated heparin with an average molecular weight of 5,500, no significant increase in Lee-White whole-blood clotting time was observed 1 hr after administration. Too few animals were involved in this primate study to warrant any conclusions.

ACKNOWLEDGEMENTS

Mrs. Marie Chiu performed the anticogulant testing with meticulous care. Dr. Alex Silverglade of Riker Laboratories and Dr. L. Coleman of the Upjohn Company provided generous supplies of various heparin preparations. This work was supported in part by grants from the National Institutes of Health, the National Science Foundation, Research Corporation, and Sloan Foundation (to the Nuclear Magnetic Resonance Facility at the Rockefeller University).

REFERENCES

1. DIETRICH, C.P., Biochem. J., 111 (1969) 91.
2. ITWINS, J., VORZIMER, J.J., SUSSMAN, L.N., APPLEZWEIG, N. and ETESS, A.D., Proc. Soc. Exper. Biol. and Med., 77 (1951) 325.
3. KOH, T.Y., Canad. Jour. Biochem., 47 (1969) 951.
4. KORN, E.D. and PAYZA, N., J. Biol. Chem., 223 (1956) 859.
5. LASKER, S.E. and CHIU, M.L., Annals N.Y. Acad. Sci., (Eds. Lasker, S.E. and Milvy, P.) 222 (1973) 971.
6. LASKER, S.E. and STIVALA, S.S., Arch. Biochem. Biophys., 115 (1966) 360.
7. LINKER, A. and HOVING, P., J. Biol. Chem., 240 (1965) 3724.
8. LOOMIS, T.A., Proc. Soc. Exper. Biol. and Med., 101 (1959) 447.
9. McDEVITT, E., HEUBNER, R.D. and WRIGHT I., J. Amer. Med. Assoc., 148 (1952) 1123.
10. PERLIN, A.S., MACKIE, D.M. and DIETRICH, I.P., Carbohydrate Res., 18 (1971) 185.
11. WINDSOR, E. and CRONHEIM, G.E., Nature, London, 190 (1961) 263.
12. WINDSOR, E. and FREEMAN, L., Amer. Jour. Med., 37 (1964) 408.
13. WRIGHT, I.S., Amer. J. Med., 14 (1953) 720.

DISCUSSION OF LASKER PAPER

ESTES

You talked about a heparinase preparation. How do you define your heparinase, where does it come from, and what are its specifications?

LASKER

We adapted flavobacteria on a phytone media that contained heparin. These organisms were harvested after maximal growth and freeze dried. We isolated an active enzyme from this crude preparation. It was also possible to use the crude preparation to depolymerize the heparin.

ESTES

Does this have any analog in man or other mammals?

LASKER

I really don't know of an analog isolated from mammals.

JAQUES

The first point I want to make to Dr. Lasker is really a response to one remark by him. He mentioned substances which enhance the absorption of heparin. I would point out that a few years ago Jarrett in my laboratory, in an investigation of one of the heparinoids included one of these chelating agents, namely NTA, since these have been reported to enhance absorption, I think the significance of this paper (Thromb. Diath. Haem. 25, 187-200, 1971) has been overlooked. Jarrett demonstrated that NTA and, presumably then, EDTA and similar agents when given by mouth, can be absorbed and can mobilize sulfated mucopolysaccharide, presumably dermatan sulfate or chondroitan sulfate, in sufficient quantities to make significant increases in the clotting time and this is responsible for the reports in the literature. This brings out the point that in all these experiments, one must be very careful about the controls to be carried out. Now the more significant question I'd like to ask is regarding Dr. Lasker's reported values for molecular weights. I was particularly interested to note this morning that Dr. Cifonelli expressly stated that the values he reported for molecular weights were not to be taken as absolute values. The reason I raise this point is that I can remember in 1939, when ultracentrifuges first became available, Dr. Tom Waters going down to Harvard and asking Dr. Oncley, who was the expert at that time in molecular weight determinations on polymers, if he would run a sample of heparin for us and give us his opinion on it. Dr. Oncley even refused to put the heparin sample in the ultracentrifuge. Simply because, he said, with such a highly charged molecule as heparin it was not possible to use the formulae which had been developed originally in terms of essentially spherical particles with essentially a balance of the charges, so balanced that they are essentially neutral. This has bothered me ever since and I therefore ask Dr. Lasker what his molecular weight values mean.

LASKER

I think there are two important points here. One is that it is difficult to accept a random number for molecular weight for a highly charged polymer system when the conditions of measurement are not defined. But when you examine a homologous series of polymers under proper solution conditions, one can place greater confidence in the individual values of molecular weight as part of a dispersion of molecular weights. The values of molecular weight approach the true value as a limit, but the dispersion gives you an added picture of what is happening. Now as far as actually determining the molecular weight in a highly charged system is concerned, we have explored this rather thoroughly, both with a mathematical model and in real systems. Suppose one starts with a molecular weight of say 10,000 and simply plugs this molecular weight into a formula for molecular weight for a charged system varying say from 10 charge units to 7 charge units per mole, one can get a value of 1,000 Daltons for the molecular weight that one knows is 10,000 Daltons depending on the salt concentration. So the numbers one gets can be preposterous if one is not cautious about swamping out charge. I think the simplest analogy is the measurement of viscosity in charged polymer systems. If one doesn't swamp out the charge, one can get a very high viscosity. This was demonstrated many years ago as a classical work on the viscosity of polyelectrolyte systems in water and in salt solutions of varying concentrations. So if one swamps out the charged system with salt and also reduces the pH in the case of an anionic polymer as I indicated on the slide, one is approaching a system where one can have some confidence in the numbers. One can have faith in these numbers if one is cautious about swamping out the charges with high salt concentration and makes measurements at as close to the isoelectric point as possible without altering the polymer. If one has a homologous series of charged systems with a dispersion of molecular weights and makes individual measurements of the fractionated polymers, this also lends considerable confidence to the data.

ENGELBERG

Dr. Lasker, do you have any preliminary observations about whether this degraded heparin as absorbed in man?

LASKER

I've done only one experiment on myself but we really don't have enough material to do any extensive experiments. That is one of the reasons we became interested in Drs. Wessler and Yin's work on low levels of heparin and its analytical measurement. We hope that instead of measuring anticoagulant activities, we can measure anti-Xa. But we have no data on that. We have a couple of failures of absorbtion in monkeys and one failure in one dog. Because we don't have enough material we have not yet done a sufficient number of experiments in large animals or man to really make any statement about that. We expect to do this.

HEPARIN IS AN ANIONIC HYDRATED ANTICOAGULANT

Herbert L. DAVIS

Departments of Surgery and Biochemistry, University of Nebraska College of Medicine, Omaha, Nebraska 68105 (USA)

The major objective of this symposium is to seek substances and procedures by which blood can be kept flowing. The major determinant of whether blood continues to circulate or grows sluggish and forms an intravascular gel is the colloidal stability of fibrinogen (7, 8). Although fibrinogen constitutes only about 3-6% of total plasma protein, it is the dominant one in viscosity and gel formation (3). Fibrinogen continues to circulate normally as long as it carries a moderate negative charge and minimal degree of hydration on both its negative and positive polar sites. Reduction in either charge or hydration causes progressive shifts toward hypercoagulability, with increased surfactant property of the fibrinogen as it is driven from the plasma onto formed elements, vascular walls (4), and any other surface in or exposed to plasma. Progressive destabilization of fibrinogen causes marked increase in blood viscosity; aggregation and segregation of platelets, then of leukocytes, and then of red cells (Knisely sludging) (13); and finally disseminated intravascular coagulation. Dintenfass and Forbes (10) have shown increased viscosity of blood and plasma in many conditions in which gross hypercoagulability is later evident.

Figure 1 represents a modification of a figure of Kruyt's (14). All proteins owe a portion of their characteristics to a balance of positive and negative charges and to hydration of all charged sites. Destabilization of the protein from the sol state may be achieved by neutralizing the charge or by removing or displacing the water. Loss of these two stability factors leads to precipitation, and proteins are defined in terms of the reduction in charge and hydration necessary to remove them from the dispersed sol. We have added the diagonal line to the Kruyt figure to represent the finding that partial reduction in charge and/or hydration can destabilize the

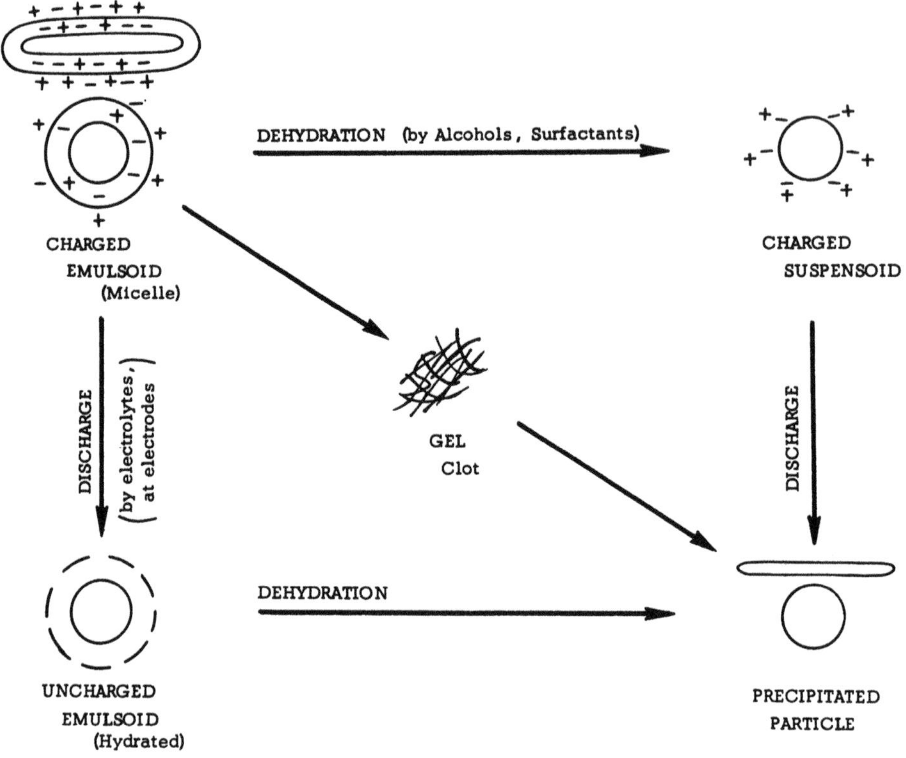

CONCEPT OF GEL FORMATION

Modified from Kruyt - Emulsoid Sol Stability
Hydrophilic Colloids
Proteins - FIBRINOGEN, Gelatin
Carbohydrates - Starch

Fig. 1. Schematic representation of the factors affecting gel formation by alteration in charge and/or hydration on proteins.

protein fibrinogen (5, 7, 8) to an intermediate gel state, and that further additions of reagent(s) from the cationic or dehydrant classes will then precipitate fibrinogen. It is thus clear why any procoagulant will show minimal clot times at some critical concentration, and why excess destabilizing agent retards or prevents gel formation.

Figure 2 emphasizes the interdependence of charge and hydration. Fibrinogen sols or plasma may be converted to the gel state by

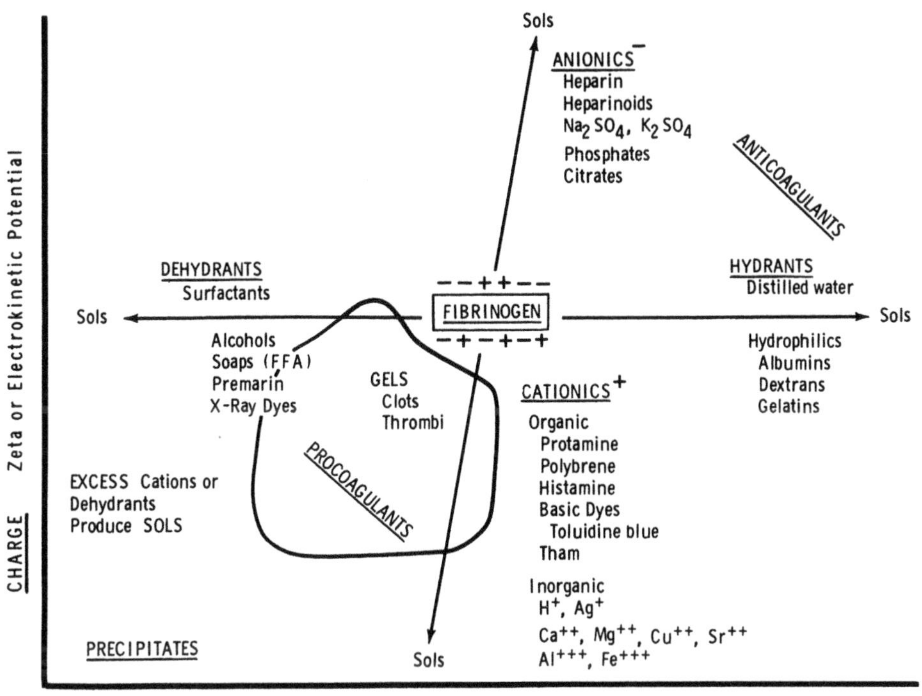

Fig. 2. Charge and hydration are interdependent determinants of fibrinogen dispersion stability. Gel structures result from moderate additions of cationics and/or dehydrants. Anionics and/or hydrants have anticoagulant stabilizing effects. Note that excess procoagulants redisperse fibrinogen into the sol state (outside the heavy line), and are thus anticoagulant in effect.

critical additions of cationics and/or dehydrants; gels are generally formed by a conjunction of these classes in quadrant 3. However, anionics or hydrants tend to stabilize fibrinogen and thus act as anticoagulants. Clot formation is achieved only within a moderate range of cationics and/or dehydrants, and beyond this range, gel formation is prevented. It is thus no paradox that excess procoagulants have anticoagulant effects, and this is especially evident in such compounds as aspirin and steroids. Thrombin splits from fibrinogen two small highly negative fragments, and this is equivalent to adding cationics, as Niewiarowski (17) showed with protamines.

Table I shows some of the many members of the four classes that can modify the colloidal stability of fibrinogen and thus alter the fundamental flow characteristics of blood. It is unfortunate that the anionic and hydrant classes are so small, because these are the substances that keep people alive, if used appropriately. Industry is making synthetic polyanionic reagents (2) that may be found effective and tolerable, and this symposium has included discussion of

TABLE I

Some Representatives of the Four Classes of Substances Which Modify Charge and Hydration on Fibrinogen - and on Other Living Proteins, Such as Those of Nerves, Membranes, Enzymes.

PROCOAGULANTS

CATIONIC$^+$	DEHYDRANT (Surfactants)
CALCIUM Magnesium Strontium Barium Zinc ACIDS (H^+) Lactic Phosphoric Citric H Cl Carbonic Trichloracetic Organics Protamine Polybrene HISTAMINE Histidine Antihistamines Basic Dyes Toluidine blue Glutamine Glucosamine Zephiran Tris buffer Imidazole	Lipids generally Glycerides Cholesterol Phospholipids Lecithins Thromboplastins Tachostyptan SOAPS C_8-C_{18} ("FFA") Saturated Unsaturated Steroids Premarin Adrenosem Experimental cpds. Protein Precipitants Tungstate Phosphotungstate Phosphomolybdate Trichloracetic (H^+) X-Ray Opaque Dyes Hypaque Angio-Conray Diodrast Urokon Cholegrafin Methyl glucamine Ditriokon

Styptic Agents

$FeCl_3$ $AgNO_3$ $AlCl_3$ $Al_2(SO_4)_3$ $CuSO_4$	Tannic Acid Sodium Tannate

ANESTHETICS

Local	General
Procaine Xylocaine Blockain Carbocaine	Nembutal Ether Chloroform Fluothane Penthrane Cyclopropane Nitrous oxide

ANTICOAGULANTS

ANIONIC$^-$	HYDRANT
HEPARIN Heparinoids Mepesulfate Dextran Sulf. Alginic Sulf.	WATER - Dilution Hydrophilic Colloids Albumins Gelatins Dextrans Polyvinyl- pyrrolidone Urea

Many More Dehydrating Agents

ALCOHOLS	Miscellaneous
Methyl to Octyl Polyhydroxy alcs. Glycols Glucose ENZYMES Proteolytic THROMBIN Trypsin Bapain Ficin Lipolytic Stypven Stypturon High Ionic Strength Solvents Acetone Dioxane Dimethyl Sulfoxide (DMSO)	Phenol Gluconate Glucuronate Glucuronolactone Salicylate Acetylsalicylate Acid Dyes Congo Red Ellagic Acid Aminocaproic acid Pantothenyl alc. Viadril Tetracyn Naphthionin Butazolidin

many heparinoids that may prove valuable. The two larger classes, however, include substances that kill people every day, and it is time that clinicians recognize and correct the hypercoagulability that is the common mechanism.

Anyone who reads the literature thoughtfully will see many examples of these phenomena. The first two lines in Fig. 3 include the major recognized killers of Americans. When these cases are discussed, "ischemia" is a common explanation and sometimes people will suggest "thrombosis". But only recently have these discussions begun to include the question of what caused the thrombosis. Virchow (18) suggested three mechanisms of thrombosis. Two of these are somewhat incidental, but hypercoagulability is a continuously variable property of flowing blood, and we are now gaining some information as to the controlling factors (12). Increased concentration or activity of cationics and/or of dehydrants will initiate or accelerate the gradual transformation of the blood sol to a gel. These changes are involved in all the many forms of stress and become manifest in susceptible tissues and organs (11).

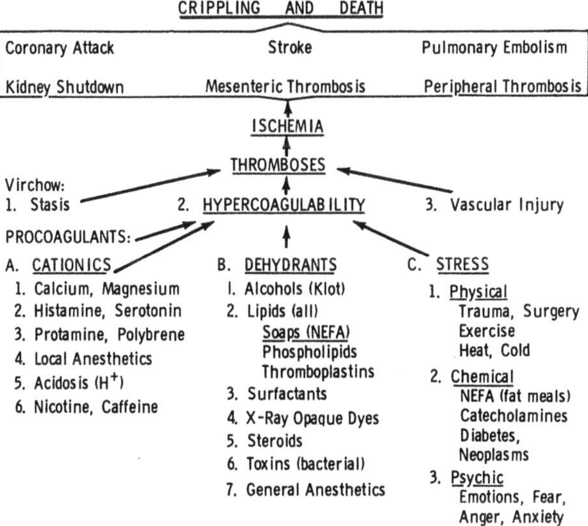

Fig. 3. Clinical applications and mechanisms of hypercoagulability. Effects of the two procoagulant classes are cumulative, and may be counteracted by anticoagulant substances. Stress and many disease syndromes involve these changes in coagulability.

There is growing appreciation that hypercoagulability is an imminent hazard in nearly all forms of stress and disease (15). If the blood continues to perfuse the tissues, the body can overcome almost any challenge (13). But if blood distribution fails, no drug or procedure can save the patient. The major need in medicine today is for a reliable, convenient index of hypercoagulability. It appears unlikely that any of the orthodox clotting factor procedures will prove useful; they are designed to recognize hypocoagulability, and they depend on grossly excessive amounts of procoagulants, which obscure the factors killing the patient. It seems likely that a heparin tolerance test will prove useful (9). Heparin functions because of its high negative charge and hydration, so that it is able to counteract either or both of the procoagulant classes. An intermediate amount of heparin will prevent clotting in normal blood and will permit clotting in hypercoagulable blood. This is analogous to the BaSon test (1) coming into use for control of heparinization itself; celite and thromboplastin accelerate clotting about eightfold.

Another important need is for improvement in the technique of using heparin or other anticoagulants. Intermittent slugs of intravenous heparin cause a patient to swing from hemorrhagic to thrombotic states with great hazard in both directions. Bolus injections intramuscularly or intraperitoneally are equivalent to intravenous injection. Intrafat injections of heparin best simulate and supplement endogenous release, without the wide swings in coagulability. This technique of "minidose" heparin promises to be a most important advance in surgery and medicine generally (11, 19). The French use of calcium heparin is claimed to minimize hemorrhagic possibilities in the intrafat injection. The objective is to prevent or reverse hypercoagulability not to produce hypocoagulability.

It seems likely that recognition of the value of heparin, adequately and appropriately used, will lead to markedly increased demands. Some signs of shortage have already been suggested, and it may be desirable to validate the effectiveness and safety of heparinoids, such as the Hercules Paran (2). This polyanion has been extensively studied by Regelson (16) who has demonstrated the ability of strongly negative substances generally to inhibit cell division, as of neoplastic cells in culture and *in vivo*. All proteins depend on charge and hydration for optimal form and function, so that representatives of the four classes (Table I) modify nerve function (6), membrane permeability, and enzyme activity. Most people clot to death, and heparin or heparinoids can be valuable adjuncts in the fight against disease and death.

REFERENCES

1. BADEN, J.P., SONNENFIELD, M., FERLIC, R.M. and SELLERS, R.D., Amer. J. Surg., 124 (1972) 772.
2. BRESLOW, D.S., EDWARDS, E.L. and NEWBURG, N.R.L., Nature, 246 (1971) 160.
3. CHIEN, S., USAMI, S., DELLENBACK, R.S. and GREGERSON, M., Amer. J. Physiol., 219 (1970) 143.
4. COPLEY, A.L., Proc. 5th Internatl. Congr. Rheology, Univ. Park Press, 2 (1970) 25.
5. DAVIS, H.L. and MUSSELMAN, M.M., Int. Rec. Med. Gen. Pract. Clin., 167 (1954) 439.
6. DAVIS, H.L., Stroke, 2 (1967) 3.
7. DAVIS, H.L. and DAVIS, N.L., Abstr. Third Congr. Internatl. Society on Thrombosis and Haemostasis, 76 (1972).
8. DAVIS, H.L. and DAVIS, N.L., Myocardial Biology, Univ. Park Press, Baltimore, 1974, vol. 4, p. 591.
9. DE TAKATS, G., Surgery, 70 (1971) 318.
10. DINTENFASS, L. and FORBES, C.D., Biorheology, 10 (1973) 457.
11. GALLUS, A.S., HIRSH, J., New Eng. J. Med., 288 (1973) 545.
12. KAKKAR, V.V., FIELD, E.S., NICOLAIDES, A.N., FLUTE, P.T., WESSLER, S. and YIN, E.T., Lancet, 2 (1971) 669.
13. KNISELY, M.H., Blood Must Circulate, (Ed. Malinin, T.L.), Thomas (1971).
14. KRUYT, H.R., Colloids, A Textbook, (Translated by van Klooster, H.S.), John Wiley & Sons (1930) 200.
15. MC KAY, D.G., Disseminated Intravascular Coagulation, Harper and Row, Philadelphia, 1965.
16. REGELSON, W., Advances in Cancer Res., 11 (1968) 223.
17. STEWART, G. and NIEWIAROWSKI, S., Biochim. et Biophys. Acta, 104 (1969) 462.
18. VIRCHOW, R., in Vigran, I.M., Clinical Anticoagulant Therapy, Lea Febiger, Phila. (1965) 47.
19. WESSLER, S. and YIN, E.T., Circulation, 47 (1973) 671.

WHAT IS "HEPARIN"?

Louis B. JAQUES

Department of Physiology, University of Saskatchewan, Saskatoon, Saskatchewan (Canada)

"Heparin" has been and is still used with four different meanings, depending on the author: (1) any heat-stable naturally occurring anticoagulant activity (neutralized by protamine, etc.); (2) a sulfated mucopolysaccharide present in tissues, highly metachromatic, with a high critical electrolyte concentration, with mast cells; (3) a commercial drug of varying composition and activities; and (4) a sulfated mucopolysaccharide with distinctive chemical and biologic properties. I have discussed the problem of a term having different meanings in different disciplines (9). The four definitions of "heparin" are those of the hematologist, the histologist, the clinician-pharmacologist, and the biochemist. To our regret, the first definition is still found in the literature. The second definition will be applicable later in these proceedings. This portion of the conference is directed to the third and fourth definitions, and the third shades imperceptibly into the fourth. The basic difficulty for Drs. Jeanloz, Cifonelli, and Brozovic has been the relation in operational terms of these definitions. One must work with the "heparin" of the third definition. Does the heparin in the fourth definition exist? This question caused the controversy between Jorpes and Charles and Scott in 1937-1939, and it must still trouble any serious workers.

Table I shows the variations in chemical analytic data on commerical heparin preparations since 1939. The variability in values for optical rotation, uronic acid, and hexosamine has increased.

The sources of this variability are differences in material prepared by different extraction procedures, differences in material from different sources, and problems in standardizing heparin by tests for inhibition of blood coagulation systems.

TABLE I

Values for Some Physical Parameters and

Chemical Components of Commercial Heparin-Na

Received	1936 - 1939[a]	1955 - 1965[b]	1968 - 1970[c]
Sulfur, %	9.6 - 11.0	6.2 - 12.1	8.0 - 10.8
Nitrogen, %	2.0 - 2.3	1.9 - 4.5	1.8 - 3.7
$\alpha_D^{20^\circ}$, deg	+42 - + 44	+33 - + 54	+29 - + 52
Amino-sugar, %	29.6[d]	15.9- 21.6	23.6- 31.4
Uronic acid, %	32.2[d]	32.0- 39.6	32 - 40
Volatile, % material	6.2 - 15.2	6.9- 19.5	6.4 - 24.5
Sodium, %	5.5 - 6.6	8.7 - 13.2	10.0-12.4
No. mfrs.	2-4	13	5

[a]Data from Jaques (6).

[b]Data from Jaques *et al.* (12).

[c]Data from Kavanagh and Jaques (14).

[d]Single preparation.

Differences Related to Extraction Procedure

The early preparations by the procedure of Charles and Scott (11) used alkaline ammonium sulfate. This apparently contributed inorganic sulfate to the final product. Helbert and Marini (6) found that treatment with IRA-400 removed 11% of titratable acid groups. In tissues extracted with potassium thiocyanate in phosphate buffer by Ottoson and Snellman (19) the final product had a lower sulfate content, which was compensated for by an equivalent amount of phosphate. This suggests that exchange with an inorganic salt can occur in the initial extraction. Anyone who, like me, has engaged personally in the extraction of heparin from various tissues by all the known methods of extraction must be well aware of the possibility that differences in enzymatic processes in the initial extraction stage are responsible for differences in the product, and that bacterial enzymes, including dextran-synthesizing enzymes, contribute to the product at this stage. Even J.E. Scott (20) in the most recent method for extracting heparin from tissues (the elegant potassium acetate extraction that appears to be the method by which all heparin will be prepared in the future), has been unable to replace the autolysis stage discovered by Charles and D.A. Scott (11) to be essential in liberating heparin from minced tissue.

Differences Related to Source

This problem of source was originally reported by Jaques, Waters and Charles (14). The pertinent data are summarized in Table II. The different kinds of heparin were all extracted from lung tissue by the Charles and Scott procedure (11) and were all crystallized as identical barium salts, as shown by the values for sulfur, nitrogen, and optical rotation. When they were examined for relative anticoagulant activity there were two important

TABLE II

Analytic Data on Lung Heparin of Four Species[a]

Source	Sulfur, %	Nitrogen %	$\alpha_D^{20^\circ}$, deg	Relative Potency, u/mg[b]				
				AC	F	AT	USP	MA
Cow	13.8	2.5	+55	154	154	153	122	150
Dog	14.0	2.9	+56	341	284	225	171	158
Pig	14.0	2.3	+53	66	90	75	81	147
Sheep	14.7	2.6	+53	31	54	63	59	166

[a] Data from Jaques and co-workers (6,10,11,13).

Values determined on crystalline barium salt or sodium salt and corrected for content of volatile material and base to give values for equivalent anhydrous free acid.

[b] Compared directly with International Heparin Standard 1.

AC = anticoagulant activity on fresh whole cat blood (Howell).

F = inhibition of clotting of fresh chicken plasma with thromboplastin (Fischer).

AT = inhibition of clotting of blood with thrombin (Jaques).

USP = inhibition of clotting of sheep plasma with calcium and thromboplastin (U.S. Pharmacopoeia XV).

MA = metachromatic test with toluidine blue.

findings. Heparin prepared from tissue of different species was different in relative anticoagulant activity; the dog heparin was most active, and the sheep heparin least active. However, equally important and ignored almost completely until international collaborative study of the assay of heparin, was the finding (12) that the relative anticoagulant activity was markedly affected by the coagulation test system used. (The values found by the metachromatic reaction with toluidine blue were also found by protamine titration). The importance of this observation was that it showed that the relative anticoagulant potency of different heparin preparations depended on the coagulation systems used for the test. Similar differences were observed by Hashimoto *et al.* (4, 5) with whale heparin. Hashimoto and Matsumo (4) and Winterstein (21) showed that, when blood or plasma was used as a test substrate and thrombin as the clotting agent, the values were markedly different, depending on the age of the blood or plasma. In fact, it was Winterstein's finding of a high anticoagulant activity in such systems that led him to adopt the term β-heparin for dermatan sulfate. Differences in anticoagulant activity of heparin from different sources are also seen when the heparin is injected intravenously. Kuo, Millar and I (17) have recently observed, with a particular commercial cow lung heparin and a particular commercial pig intestinal mucosal heparin, no difference in dose response *in vivo* in dogs. This is in contrast with our findings (14) in a similar test in 1940, in which the potencies on intravenous injection in dogs were comparable with the effects observed on fresh cat whole blood. Hence, extraction methods may be as significant as the source in causing differences in potency *in vivo*.

Standardization Problems

In all assays, the activity of a test preparation is compared with that of a reference preparation. Problems of standardization were demonstrated when an international collaborative study of the assay of heparin was organized by the World Health Organization (1). The procedure involved 13 laboratories. Six coded heparin preparations were packaged by the WHO laboratory and shipped to the participating laboratories. The laboratories assayed them against each other. Figure 1 shows some of the data. As has been believed by everyone involved in heparin assays, assaying a given heparin preparation against itself has a high degree of reproducibility as shown by the very small spread in values over a large number of assays. However, when two different heparin preparations were assayed against each other, a tremendous range in values resulted. In other words, the relative anticoagulant activities of two heparin preparations are different with differences in the coagulation test systems. For this reason, Bangham and Woodward (1) recommend that standardization of heparin involve different assay procedures and that the mean or most representative value be selected for a heparin

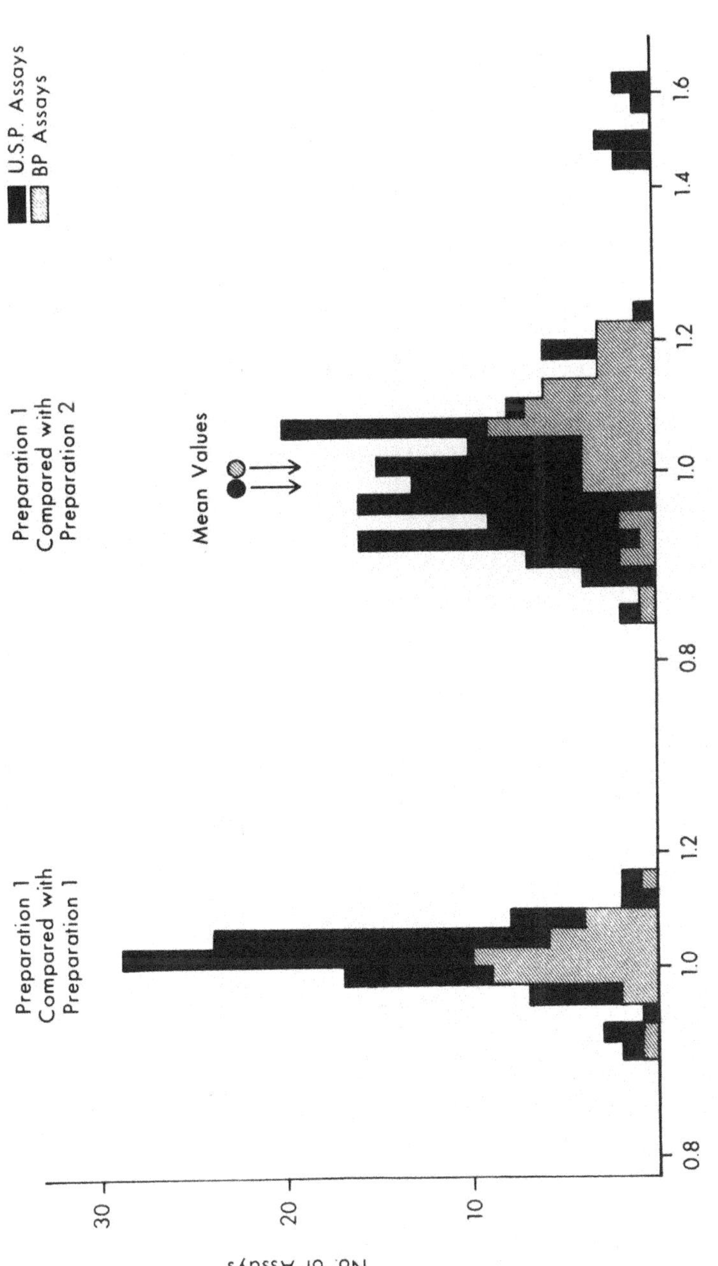

Fig. 1. Frequency histogram of relative anticoagulant potency of two heparin preparations. Left, preparation of porcine intestinal mucosal heparin tested against itself. Right, preparation of porcine intestinal mucosal heparin tested against International Standard Heparin 2. Values based on U.S.P. and B.P. assays. (Data from Bangham and Woodward (1)).

preparation. This means that, if any study is being made of the effects of a particular characteristic or of a fractionation procedure on the biologic activity of heparin preparations, all preparations for comparison must be assayed with the same test system and at the same time. Preferably, a number of different assay procedures would be conducted, and a representative value found.

Because the restrictions on the use of anticoagulant assay methods for heparin demonstrated by the international study as necessary were not observed in standardizing commercial preparations, because the procedure for preparing heparin has changed often in the last 30 years, and because heparin is standardized by biologic assay, it is obvious that the variations that have occurred in chemical composition of commercial heparin are related to the problem of standardizing this drug by anticoagulant assays.

Chemical Characterization of Heparin

Heparin can be crystallized as a barium salt (Fig. 2). It is reasonable to think that "heparin" should refer to this substance. In the early production of heparin at Toronto, we found using the Howell assay, that all the anticoagulant activity in the tissue

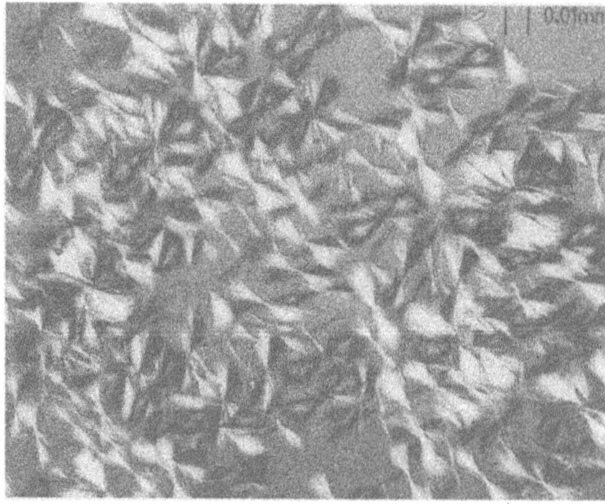

Fig. 2. Photograph of crystals of heparin barium salt. Pig lung heparin extracted by the Charles and Scott (10) procedure in 1969. Photographed under polarizing microscope with gypsum red I compensator.X 350 (Courtesy of Dr. L.W. Kavanagh).

extracts could be recovered in the crystalline salt. Workers in Stockholm used different methods of extraction, purification, and assay and found that they could demonstrate different fractions in this product. The early controversy on the nature of heparin (cf. Jorpes & Bergstrom, 1939; Charles & Todd, 1940) may have been caused by differences in extraction (15, 3) and assay procedures.

Clinicians would prefer what the substance characterized as "heparin" be the component in the commercial drug that has the necessary clinical activity. I reported some years ago that, in testing the effectiveness of heparin in preventing venous thrombosis in rats (Fig. 3), a commercial heparin was about twice as effective as International Standard Heparin 2. This was a relatively old heparin, considered less potent by present standards. We have also observed that most commercial heparin preparations assayed against International Standard Heparin 2 for metachromatic activity have a relatively low activity (about 100 units/mg, instead of the 130 units/mg of International Standard). This suggests that International Standard Heparin 2 has a much higher metachromatic activity than present commercial heparin. Although International Standard Heparin 2 is by definition "heparin", one doubts whether a preparation

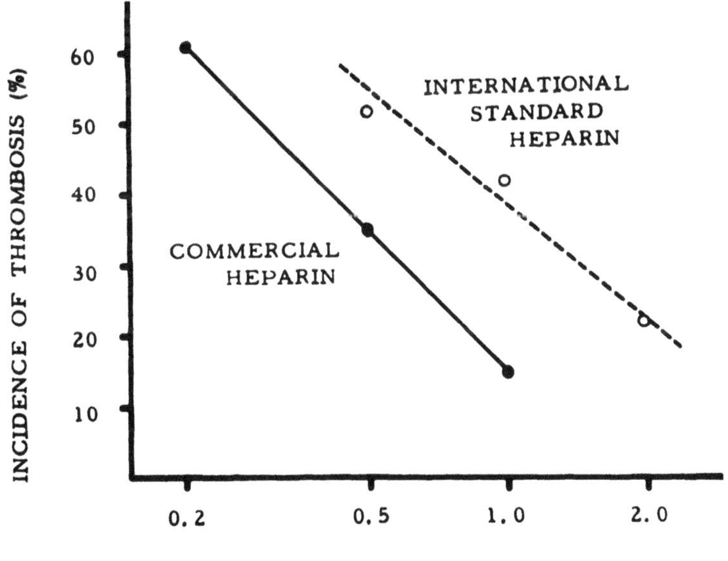

Fig. 3. Relative effectiveness in reducing venous thrombosis in rats. A commercial heparin of 1960 compared with International Standard Heparin 2. Incidence of thrombosis in groups of rats after exposure of jugular vein by procedure of Blake, Ashwin, and Jaques (2). (From Jaques (9)).

with lower antithrombotic activity and higher metachromatic activity necessarily constitutes the most desirable definition.

Another approach to defining heparin as a chemical entity is to assume that the main constituent of the commercial drug is heparin and to remove from it as many of the minor constituents as possible. However, there is no single main component of present commercial heparin. Nader, McDuffie and Dietrich (18) report that electrofocusing demonstrates 21 different fractions or components of commercial heparin, differing only in molecular weight, with no fraction accounting for more than 11% of the total. Such studies are incomplete until there is an extensive examination of biologic properties, particularly in vivo tests in animals and man, and finally a determination of clinical effectiveness.

There is another important way of expressing the chemistry of heparin, as the results of various studies. Heparin is a polyelectrolyte with a great concentration of negative charges spread over the surface of the molecule (Fig. 4). There are significant differences in charge between O-sulfate, N-sulfate, carboxyl, etc. The counterions, whether metals like sodium, low-molecular-weight amino acids and amines, etc. - or high-molecular-weight proteins, will be associated to varying degrees at varying distances from the negative charges on the heparin surface. Co-ions (anions) are similarly associated. The combination of the highest charge density of any biologic substance with relatively low molecular weight gives the special structure that is responsible for heparin's very high

Fig. 4. Sketch of a possible heparin molecule, showing distribution of charges. Charges are shown as a shell around the carbohydrate core of a flexible helical structure. Position of acid groups from a Courtaud molecular model based on the Perlin, Mackie, and Dietrich formula for the biose residue (cf. Jaques (7)). Total length, 100-110 Å; distance across single loop, 17 Å.

solubility in water and its many special biologic activities. Undoubtedly, this structure is also responsible for the variation in chemical composition of the materials as extracted and as modified by various procedures that cause differences in the counterions and co-ions associated with heparin.

It is evident that there are different meanings to this simple term, "heparin." I would urge those using the word to be sure to make clear what they mean by it.

REFERENCES

1. BANGHAM, D.R. and WOODWARD, P.M., Bull. Wld. Hlth. Org., 42 (1970) 129.
2. BLAKE, O., ASHWIN, J. and JAQUES, L.B., J. Clin. Path., 12 (1959) 118.
3. CHARLES, A.F. and TODD, A.R., Biochem. J., 34 (1940) 112.
4. HASHIMOTO, K. and MATSUNO, M., Tohoku J. Exp. Med., 91 (1967) 295.
5. HASHIMOTO, K., MATSUNO, M., YOSIZAWA, Z. and SHIBATA, T., Tohoku J. Exp. Med., 81 (1963) 93.
6. HELBERT, J.R. and MARINI, M.A., Biochemistry 2, 5 (1963) 1101.
7. JAQUES, L.B., Ph.D. Thesis, University of Toronto, (1941).
8. JAQUES, L.B., 22nd Annual Wayne State University Symp. on Blood, (1974) in press.
9. JAQUES, L.B., Thromb. Diath. Haem. Suppl., 46 (1971) 21.
10. JAQUES, L.B., Thromb. Diath. Haem. 9, Suppl. 2 (1963) 27.
11. JAQUES, L.B. and BELL, H.J., Methods of Biochemical Analysis, Interscience Publishers, Inc., New York, 7 (1959) 253.
12. JAQUES, L.B. and CHARLES, A.F., Quart. J. Pharm. and Pharmacol., 14 (1941) 43.
13. JAQUES, L.B., KAVANAGH, L.W. and LAVALLEE, A., Arzneimittel Forsch., 17 (1967) 774.
14. JAQUES, L.B., WATERS, E.T. and CHARLES, A.F., J. Biol. Chem., 144 (1944) 229.
15. JORPES, E. and BERGSTROM, S., Biochem. J., 33 (1939) 47.
16. KAVANAGH, L.W. and JAQUES, L.B., Arzneimittel-Forsch., 23 (1973) 605.
17. KUO, S.H., JAQUES, L.B. and MILLAR, G.J., J. Pharm. Pharmacol., 24 (1972) 858.
18. NADER, H.B., McDUFFIE, N.M. and DIETRICH, C.P., Biochem. Biophys. Res. Comm., (1974) in press.
19. OTTOSON, R. and SNELLMAN, O., Acta Chem. Scand., 13 (1959) 473.
20. SCOTT, J.E., Patent Specification No. 1221784, The Patent Office, London, 1971.
21. WINTERSTEIN, A., Pharmaceutisch Weekblad., 92 (1957) 982.

DETERMINATION OF HEPARIN IN SMALL TISSUE SAMPLES

Louis B. JAQUES

Department of Physiology, University of Saskatchewan, Saskatoon, Saskatchewan (Canada)

The association of heparin and mast cells is on the basis of the metachromatic color change produced in toluidine blue. Among agents from biologic sources, heparin is the most effective in producing this change in a test tube. The mast cells are the most prominent components that show the changes in histologic sections. Many reports have been published on the amine components of mast cells-histamine and 5-hydroxytryptamine (5-HT). Although our studies in 1939 (1) showed that the symptoms of anaphylaxis in dogs were due to the release of both histamine and heparin from liver mast cells, biologic studies of heparin have not been as extensive as those of histamine and 5-HT after the original observation. That is because, although there have been methods for identifying and measuring both histamine and 5-HT in small tissue samples, there have been no corresponding procedures for heparin. A great deal of my research for the last 35 years has been directed at remedying this deficiency, and this I believe we have accomplished by microelectrophoresis (Fig. 1) (2, 4).

The procedure consists of applying 2 µl in the slit of an agarose-coated microscope slide, subjecting this to electrophoresis, and then fixing the slide and staining it with toluidine blue. Heparin and related sulfated mucopolysaccharides appear as purple spots on the slide. It is possible to identify heparin by the distance migrated relative to a reference heparin preparation, as shown in Fig. 1, and by the color and properties of the spot. The heparin can be quantitated by visual comparison or by using a densitometer with a set of reference slides. The method is flexible and sensitive, and is free from interference by the many tissue components that interfere with other tests for heparin, such as thromboplastic activity, lipids, and most proteins. It is free

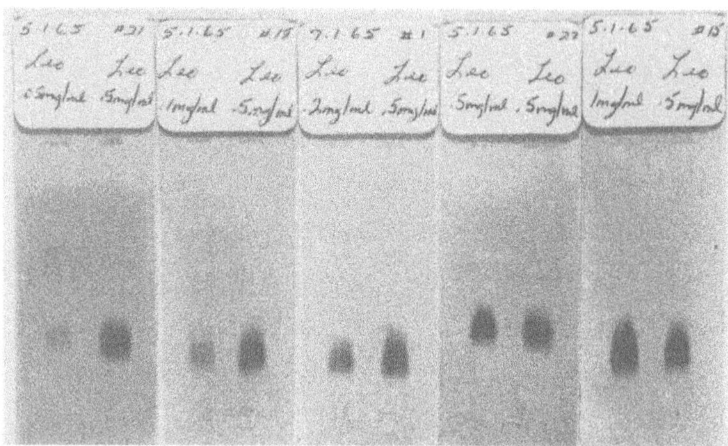

Fig. 1. Microelectrophoresis slides for heparin. Black and white print; original agarose slides show reddish-purple spots on an almost colorless background. Heparin from Leo Pharmaceutische Production N.V. Right-hand spot on each slide, 0.13 units, left-hand spots, from left, 0.006, 0.015, 0.03, 0.06, and 0.13 units applied to point of application (faint line just below label).

from interference, because in the electrophoresis step, the heparin is separated from these other components and therefore stains in the characteristic manner. The sensitivity is shown by the amounts of heparin on the slides in Fig. 1 - 0.1 unit, or 1 μg. The flexibility of the method is indicated by the facts that it can be used for crude tissue extracts containing 1% heparin by weight just as well as for pure heparin and that it can be used to identify other sulfated mucopolysaccharides in a mixture simultaneously and to estimate them quantitatively.

The method can be made more specific for heparin by using J.E. Scott's principle of critical electrolyte concentrations (4). When the procedure was carried out directly, the relations shown in Fig. 2 for optical density versus concentration for heparin, dermatan sulfate, chondroitin sulfates A and C, and heparitin sulfate were found. When the fixation stage with cetylpyridinium chloride was done in molar NaCl, only heparin remained on the slide.

The procedure can be used for many biologic experiments involving commercial heparin (the drug), or natural heparin (as found in tissues). One example will demonstrate the value of the method. Rabbits, as a species, are unique, in that their tissues contain very few mast cells. However, rabbits are the one species with a relatively high basophilic leukocyte count. Michels (3)

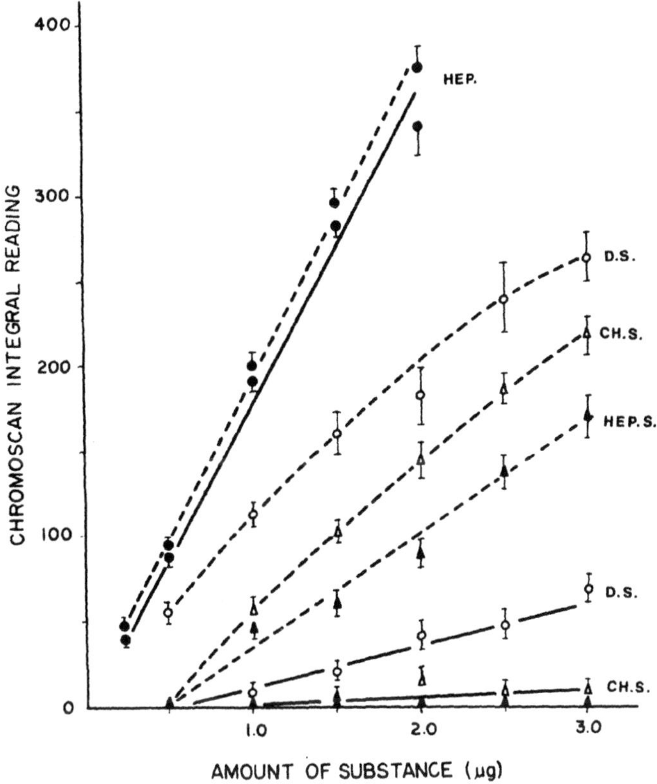

Fig. 2. Relation between optical density and amount of sulfated mucopolysaccharide applied to microelectrophoresis slide. Dashed line slides, fixed with aqueous cetylpyridinium chloride (CPC); solid line slides fixed with CPC in molar NaCl. HEP., heparin; D.S., dermatan sulfate; CH.S., chondroitin sulfates A and C; HEP.S., heparitin sulfate. (From Sue & Jaques (4)).

and others have commented on this. Dr. Tak Sue and I have investigated this problem with microelectrophoresis (5). Table I shows the total sulfated mucopolysaccharide content and the heparin content of five different tissues. For convenience, quantities are expressed in equivalent units of heparin. A considerable amount of the total mucopolysaccharide in the aorta is heparin - over 30%. In contrast is the lack of heparin in the large intestine. A definite amount is found in the small intestine. To see whether the values would respond to experimental modification of the animal, we tested rabbits that received a diet high in cholesterol and olive oil. Both the total sulfated mucopolysaccharide and the heparin fraction in the aorta increased markedly with this diet; there was no change in other tissues. All tissue samples were examined for

TABLE I

Concentrations of Total Sulfated Mucopolysaccharide (S.M.P.S.) and Heparin in Rabbit Tissue by Microelectrophoresis[a]

Tissue	S.M.P.S. Concentration[b]	Heparin Concentration[b]
Skin	116 ± 5	0.0
Large intestine	52 ± 4	0.0
Small intestine	50 ± 8	19 ± 4
Heart	65 ± 6	20 ± 2
Aorta	48 ± 13	15 ± 5
Aorta (8 weeks)[c]	163 ± 21	73 ± 10
Aorta (16 weeks)[c]	181 ± 38	96 ± 32

[a] From Sue and Jaques (4)

[b] Mean values with S.E. for 4 rabbits in equivalent metachromatic units per gram of tissue; 100 units is equivalent to 1.6 mg of mixed sulfated mucopolysaccharide and 1.0 mg of heparin.

[c] Rabbits on diet high in cholesterol and olive oil for 8 and 16 weeks.

mast cells. Mast cells could be found, but only in about one of 20 or 30 fields, showing that the technique was satisfactory for rabbit mast cells, but that there were very few of them.

Because there are few mast cells in the rabbit, Dr. Sue examined the white-cell sulfated mucopolysaccharide(s). Microelectrophoresis (Fig. 3) yielded spots characteristic of chondroitin sulfates A and C. This was confirmed by finding (again with microelectrophoresis) that the sulfated mucopolysaccharide was destroyed by chondroitinase AC. Hence, the metachromasia of the basophilic

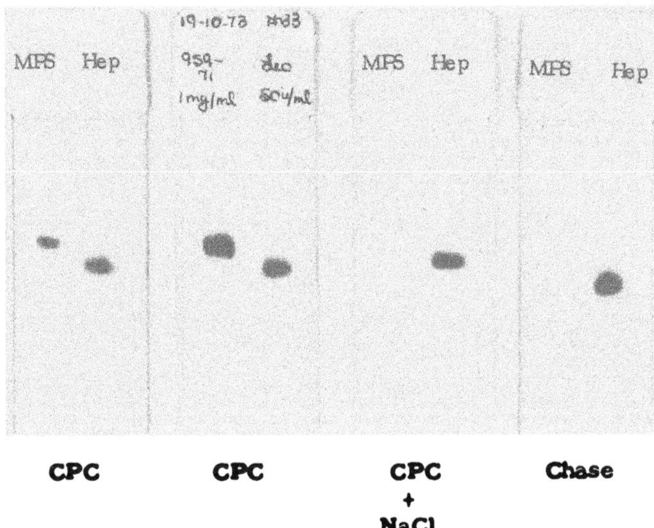

Fig. 3. Microelectrophoresis slides obtained with rabbit leukocyte sulfated mucopolysaccharide fractions on treatment with molar NaCl and chondroitinase. MPS, mucopolysaccharide from rabbit leukocytes; HEP = heparin Leo; CPC, fixation of agarose slide with 0.1% cetylpyridinium chloride; CPC + NaCl, fixation with 0.1% CPC in molar NaCl; Chase, after-treatment of samples with chondroitin A and C (Miles).

leukocytes is due to chondroitin sulfates A and C, not to heparin.

Our procedure permits the examination and identification of heparin and related sulfated mucopolysaccharides in small tissue samples. It thus makes possible the identification and study of these components of the mast cell. Furthermore, results to date indicate differences between these components and histamine and 5-HT, and indicate that the non-mast-cell heparin as the non-mast-cell histamine is of greater physiologic significance - than that derived from mast cells.

REFERENCES

1. JAQUES, L.B. and WATERS, E.T., J. Physiol., 99 (1941) 454.
2. JAQUES, L.B. and WOLLIN, A., Anal. Biochem., 52 (1973) 219.
3. MICHELS, N.A., Downey's Handbook of Hematology, Section IV, (1938) 235.

4. SUE, T.K. and JAQUES, L.B., Can. J. Physiol. and Pharm., 51 (1973) 994.
5. SUE, T.K. and JAQUES, L.B., Can. J. Physiol. and Pharm., 52 (1974) 661.

IDENTIFICATION AND QUANTIFICATION OF TISSUE HEPARIN BY MICROELECTROPHORESIS: A CRITIQUE

Alan A. HORNER

Department of Physiology, University of Toronto,
Toronto, Ontario (Canada)

In 1968, Jaques, Ballieux, Dietrich, and Kavanagh described the use of microelectrophoresis in agarose gel at a pH of 8.6 to quantitate heparin in tissues (2). Using this method, Jaques and Debnath (3) found high concentrations of heparin in rat tissues in which other workers had previously failed to detect any. From their data, which showed considerable heparin in tissues containing very few mast cells, they postulated the existence of "non-mast-cell heparin." Jaques and Roy (4) used the same method to demonstrate high heparin concentrations in blood vessels, especially aorta, in which other workers had previously failed to detect it by accepted biochemical techniques (9).

Recently, Sue and Jaques (8) have modified the microelectrophoresis method to make it more specific for heparin, by precipitating the heparin band in the gel with cetylpyridinium chloride in 1.0 M NaCl and then displacing the cetylpyridinium cation with toluidine blue in aqueous acetone. Dr. Jaques has shown in this symposium how the modified procedure can demonstrate high heparin concentrations in rabbit aorta and dramatic increases in these concentrations that result from feeding cholesterol.

Because the results of Dr. Jaques and his co-workers are contrary to those of others, I feel that the validity of the microelectrophoresis method must be very carefully evaluated.

At a pH of 8.6, the difference between the electrophoretic mobilities of heparin and less highly sulfated glycosaminoglycans, such as dermatan sulfate (DS) and the chondroitin sulfates, is not great. Figure 1 shows the results of electrophoresis in 1% agarose gel at a pH of 8.6 of five purified glycosaminoglycan preparations,

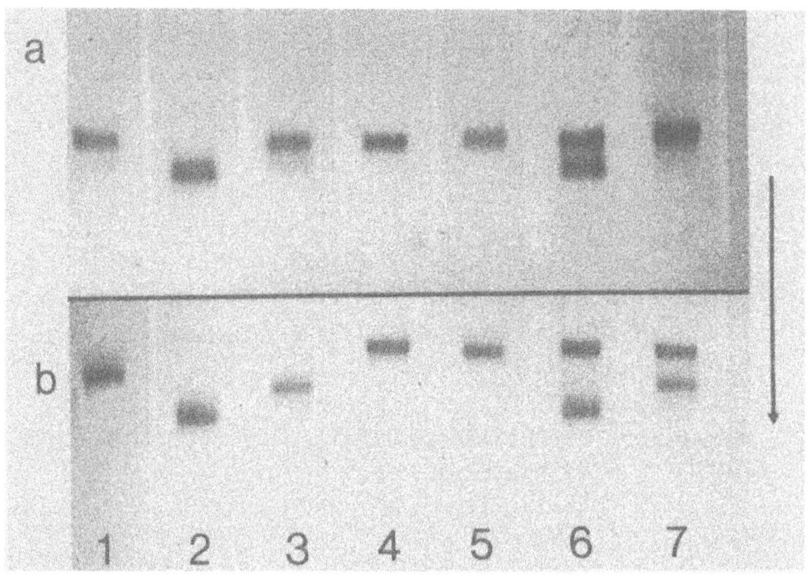

Fig. 1. a. Electrophoresis in agarose gel in 0.05 M barbital buffer (pH 8.6). b. Electrophoresis in agarose gel in 0.87 M acetic acid (adjusted to a pH of 3.0 with lithium hydroxide (1)). In both gels, the samples are: (1) beef mucosal heparin; (2) beef lung heparin; (3) pig mucosal heparin; (4) pig skin dermatan sulfate (DS); (5) chondroitin sulfate from bovine nasal septum; (6) a mixture of pig skin DS and beef lung heparin; and (7) a mixture of pig skin DS and beef mucosal heparin. The concentration of each glycosaminoglycan, including each component of the two mixtures, was 1 mg/ml. The gels were stained directly with toluidine blue (100 mg/100 ml. of water). Commercially prepared 1% agarose gels on plastic film backings (Analytical Chemists Inc., Palo Alto, California) were used. These were washed in distilled water and then equilibrated with buffer. One-microliter samples were applied. Electrophoresis conditions, using a water-cooled Shandon thin-layer electrophoresis tank: at a pH of 8.6, 18 min at 200 V; at a pH of 3.0, 13 min at 300 V. Arrow indicates direction of electrophoretic migration.

including the three most common types of commercial heparin. Only beef lung heparin has a high enough mobility to separate it from pig skin DS at a pH of 8.6. The mixture of beef mucosal heparin and pig skin DS looks like a single component. It is perhaps unfortunate that Jaques _et al_. (1) chose beef lung heparin as a standard in developing their methods, in that its high mobility may well be the exception - it is certainly not the rule.

It must be assumed that failure to obtain good separation of heparin from other sulfated glycosaminoglycans at a pH of 8.6 led Sue and Jaques (8) to introduce treatment with cetylpyridinium chloride in 1.0 M NaCl before toluidine blue staining. Using pig skin DS and beef lung heparin as their standards, they developed a method in which DS gave only 8% of the intensity of staining given by an equal weight of heparin. However, because the specificity for heparin was not absolute, one must ask, when applying the method to such a tissue as aorta, which is known to contain considerable DS (6), whether one is measuring a particular weight of heparin or 12.5 times as much DS.

The problems can be overcome by electrophoresis at an acidic pH at which ionization of uronic acid carboxyl groups is partially suppressed. The use of acidic buffer systems for this purpose was introduced by Mathews (5) in 1961 and has been frequently used since. Figure 1 shows the results of gel electrophoresis in 1% agarose at a pH of 3.0. At a pH of 3.0, even the heparin with the lowest mobility (beef mucosal) is clearly separated from DS. With this degree of separation, manipulations with cetylpyridinium chloride before staining with toluidine blue are unnecessary. It is unfortunate that Jaques and co-workers have consistently used basic buffers. In view of their controversial data and conclusions, I would urge them to attempt to substantiate their results by detecting heparin after electrophoresis at a pH of 3.0.

There are probably very few tissues in which heparin constitutes such a high proportion of the total glycosaminoglycans as to be detected by electrophoresis of a crude tissue extract or digest, even under the most favorable conditions. Some type of fractionation must precede electrophoresis to give a product relatively enriched in heparin. If this is not done, large amounts of less highly sulfated glycosaminoglycans must be applied to the gel in order to apply a detectable amount of heparin. In such cases, resolution is very poor, because of the high viscosity of the sample.

Finally, in applying Scott's principle (7) of critical electrolyte concentration (C.E.C.) to precipitate a glycosaminoglycan with cetylpyridinium chloride, it must be remembered that the C.E.C. depends on both degree of sulfation and molecular size, which can vary with the tissue source. For example, the C.E.C. values for cetylpyridinium complexes in NaCl of four of the glycosaminoglycan preparations used for the present study are: DS (pig skin), 0.9 M; DS (pig intestinal mucosa), 1.2 M; heparin (beef mucosa), 1.3 M; and heparin (beef lung), 1.4 M. Thus, in developing a method with increased specificity for heparin, Sue and Jaques were again perhaps unfortunate in using as standards preparations of DS and heparin with C.E.C. values differing by 0.5 M. At the other extreme, the C.E.C.'s of pig mucosal DS and beef mucosal heparin differ by only 0.1 M. In such a case, which may well be encountered in tissues, the cetylpyridinium chloride method would be totally unreliable.

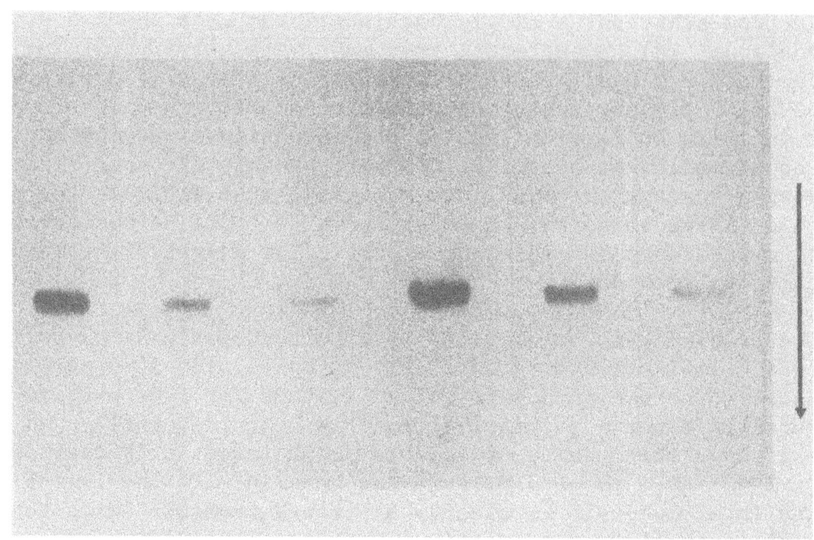

Fig. 2. Electrophoresis in agarose gel in 0.05 M barbital buffer (pH 8.6). The samples are, from left to right, beef mucosal heparin at concentrations of 8, 4, and 2 mg/ml and pig mucosal DS at concentrations of 8, 4, and 2 mg/ml. After electrophoresis, the gel was incubated with cetylpyridinium chloride in 1.0 M NaCl and then stained with toluidine blue, as described by Sue and Jaques (8). Arrow indicates direction of electrophoretic migration.

This is illustrated by Fig. 2, which shows the result of applying the method of Sue and Jaques to samples of DS (pig mucosal) and heparin (beef mucosal), each run at three concentrations at a pH of 8.6. DS and heparin are nearly indistinguishable from each other, in terms of both mobility and intensity of staining. Quantitative densitometry shows that staining intensity could be ascribed either to a unit weight of heparin or to 1.12 times as much DS.

These points are made to illustrate the need to analyze data on tissue heparin concentrations determined solely on the basis of the published microelectrophoresis methods (2, 8) with extreme caution.

REFERENCES

1. HORNER, A.A., Can. J. Biochem., 45 (1967) 1009.
2. JAQUES, L.B., BALLIEUX, R.E., DIETRICH, C.P. and KAVANAGH, L.W., Can. J. Physiol. Pharmacol., 46 (1968) 351.

3. JAQUES, L.B. and DEBNATH, A.K., Amer. J. Physiol., 219 (1970) 1155.
4. JAQUES, L.B. and ROY, P., Fed. Proc., 31 (1972) 248.
5. MATHEWS, M.B., Biochim. Biophys. Acta., 48 (1961) 402.
6. MUIR, H.M., in Amino Sugars, (Eds. Balazs, E.A. and Jeanloz, R.W.) Academic Press, New York, 1965, vol. II A, p. 311.
7. SCOTT, J.E., Methods Biochem. Anal., 8 (1960) 145.
8. SUE, T.K. and JAQUES, L.B., Can J. Physiol. Pharmacol., 51 (1973) 994.
9 THUNELL, S., ANTONOPOULOS, C.A. and GARDELL, S., J. Atheroscler. Res., 7 (1967) 283.

DISCUSSION OF THE PAPERS BY JAQUES AND HORNER

LINDAHL

If I understand your data correctly Dr. Jaques, it appears that a large proportion of the total amount of polysaccharide in rabbit aorta was heparin. Drs. Jansson and Hallén in our laboratory recently fractionated the various polysaccharides of rabbit aorta and found about 80% chondroitin sulfate, 10% dermatan sulfate, 10% heparan sulfate but no heparin.

SILBERT

Buonassisi has shown that cultured endothelial cells from rabbit aorta produce heparitin sulfate. I'd also like to comment at this point that heparin standards obtained from tissues such as mucosa or lung represent material derived from a combination of cell types including connective tissue elements and mast cells. I think one has to be very cautious in calling this material heparin rather than a mixture of heparin and perhaps heparitin sulfate with a high sulfate content.

MARX

In our laboratory we developed a different procedure for microdetermination of heparin up to 0.5 μg levels. We used 8% acrylamide gels run according to the procedure of Loening and merely fixed the heparin in the gels after the development of CPC solution, washed them in water, and then scanned them in the 280 millimicron region. We found the following: for a sample of heparin that was fractionated on a Sephadex G-200 column, we found systematic variation of the electrophoretic mobility with the order of elution from the Sephadex G-200 gel. The heparin which had the highest mobility came out last and the material which had the slowest mobility came out first on the Sephadex G-200, the variation being rather systematic. Secondly, we tested a number of heparins obtained from different sources for their mobility and detectability on these acrylamide gels using this CPC procedure and found the following: Heparins which were obtained from beef lung, beef mucosa, pig mucosa and whale had slightly different mobilities and polydispersities. They also fixed with different amounts of CPC as reflected by the A_{280} peak area. With other mucopolysaccharides such as heparitin sulfate and the chondroitin sulfates, we also found some variability on their ability to fix CPC under the conditions employed. I would suggest that the electrophoretic mobilities are possibly better studied on acrylamide gel procedure which allows a better distinction for molecular weight fractions of heparins. This CPC procedure is sensitive but quantitatively somewhat variable since heparins and other mucopolysaccharides have different chemical compositions and bind variable amounts of CPC.

HORNER

I think the whole reason that the agarose gel electrophoresis system was developed is because polyacrylamide has this molecular sieving effect. If you are looking at all the MPS in a tissue this effect is a great disadvantage because the slower moving heparin molecules, that is the bigger ones, will overlap considerably with the faster moving chrondroitin sulfate or dermatan sulfate molecules. That is why agarose was used - because it will make a gel with much bigger pores and we are trying to get away from the molecular sieving effect, which with this type of work is a disadvantage. But certainly if you want to estimate the molecular weight of a heparin after you are quite sure you got it away from other MPS then it's very useful for that and has been used by several workers. As far as CPC binding goes, your findings support the work of Scott who has shown the critical electrolyte concentration to be proportional both to molecular size and to sulfate content which are both known to be variables in heparin.

JAQUES

First in regard to Dr. Horner's comments, I'd say I couldn't agree more. We have followed each other's work for many years, but given the couple of thousand miles between Toronto and Saskatoon, we don't always know what each has been doing recently. Secondly, while we used a beef lung heparin as our reference through all this work, I would point out that this was simply because it was much more convenient to have a single component heparin on the agarose slide to which we could refer distance of migration. We could have used as the reference, some other material. For example, we have done work with S-35 inorganic sulfate, which migrates twice the distance of this reference heparin, but of course for the routine work we do, a radioisotope reference would be inconvenient. Thirdly, all this work of checking of the procedures has been done with a number of different commercial heparins including pig intestinal and mucosa heparin, dog liver heparin prepared by ourselves, and other heparins. A fourth point is that in our own thinking, Dr. Horner and ourselves appear to have started from opposite ends and sort of passed each other. Dr. Horner, I think, has been zeroing in on making this method more specific for heparin. We started at that end (as a method initially for purified heparin) and we have gone more or less as a rapid, quick, flexible, and sensitive method for getting information quickly on many biological samples. The information must include some information on the total sulfated mucopolysaccharides and various sulfated mucopolysaccharides and we hope in the process to obtain some information on heparin. As I said before, my thinking in all this has been, because of what happened on finding that the release of histamine and heparin was responsible for symptoms of anaphylaxsis and other phenomena. This meant that the biological significance of histamine and heparin was discovered simultaneously. As we know there has been great development with histamine and later 5-HT, even catecholamines, the other biologically active components in the mast cell. Now there are thousands of papers on this aspect, yet for the component of the mast cell which is used to define the mast cell, that is for the component of the mast cell that gives it its metachromatic staining in histology, we still have no information on biological changes in that component other than what we see on the histological slide. Therefore, we need to have some routine method for studying small tissue samples. As I pointed out and of course known for a long time, probably heparin is not the only sulfated mucopolysaccharide in the mast cell. We require some means of carrying out investigations at this level. Again, I agree entirely with Dr. Silbert. It is a great mistake to think of the lung as if we extract heparin from the whole lung. This again is one of the reasons I spent so many years developing this method. Up until quite recently, if you wanted to study the heparin in the tissue you had to start with 50 pounds of the tissue and even with methods, the results of which were presented this morning and yesterday afternoon, one cannot take the quite small tissue samples needed for the examination of the sulfated mucopolysaccharides at the anatomical level indicated by Dr. Silbert. The flexibility of the method means that one can carry out enzyme studies, radioisotope studies, etc., with microgram quantities of material. I emphasize that we have been using this method in different types of biological studies. The method can be used for experiments pharmacological in nature, where the drug heparin is injected. We have a paper in press, for example, showing the uptake of the injected heparin on the endothelium. So this is where we are using the method to study and follow the drug, heparin. The other type of study is that of endogenous heparin and the related materials in tissues. Up to now, because the methods require larger tissue samples, studies on endogenous heparin and sulfated mucopolysaccharides consisted of taking one or two large samples of tissue, extracting the sulfated mucopolysaccharides and then running a profile on this. What I am suggesting is that it is now possible to carry out studies where one can obtain figures from a number of animals and organs to see what the distribution is not only between different tissues but since the values are means ± standard errors, the variation there is between individuals. To answer Dr. Lindahl's question regarding our values for aorta. When we prepared the table, we had two alternatives. We could have converted the readings using the standard curves I showed on the second slide, and could have converted the values to micrograms of components. Since heparin gives a spot on the slide more optically dense than dermatan sulfate, then on such a conversion converting to micrograms, heparin would have been significantly smaller percentage of the total sulfated mucopolysaccharide. For convenience and largely because this is an audience that I think, is more used to dealing with units than with micrograms of these components, we expressed the data in units. We are aware of the papers showing no heparin in the aorta. On the other hand, two papers in the recent literature have reported values for sulfated mucopolysaccharides in the aorta with significant anticoagulant activity and these authors suggested that this represents the heparin component. At present, our definition of the heparin fraction is the amount of sulfated mucopolysaccharides seen on the agarose gel slides of which the CPC complex is not eluted by molar NaCl, and which does appear in position in the slide similar to heparin. Regarding Dr. Horner's last point, since this is a flexible, sensitive method, what one can do is quickly obtain a considerable amount of biological data and then carry out a series of experiments with critical electrolyte concentrations, pH changes of the gel, enzymes, etc., to complete the identification of the various sulfated mucopolysaccharides making up the sample obtained from the tissue.

STANDARDS FOR HEPARIN

Milica BROZOVIC and D.R. BANGHAM

National Institute for Biological Standards and Control, Holly Hill, Hampstead, London (United Kingdom)

The benefits of reference materials are generally accepted, but their correct use and particularly the limitations on their use are not so well understood. The need for a reference preparation of heparin was recognized before World War II, and the first International Standard for heparin was established in 1942 (20). Over 30 years later, we are still debating, with perhaps greater uncertainty than ever before, which heparin, what assay, which substrate, and which reference preparation should be used.

There is a pattern in the development of the measurement of complex biologic substances (3). First, a particular specific biologic effect is observed. The anticoagulant activity of heparin was noted in 1916 (23).

Second, we try to measure the biologic activity. The attempts are often in terms of a defined amount of response in a biologic system - this is the stage of "mouse units" for insulin, "frog units" for digitalis, "clotting units" for thrombin, and percentages of the elusive "normal." Because of the almost infinite variability inherent in biologic systems, such attempts are soon found to be unsatisfactory. During this stage, the material may become available in quantity, especially if it is of possible clinical use and thus accessible to many research workers.

At the third stage, an amount of active but generally impure material is set aside in a stable form as a standard to which is assigned a potency (or "strength") in terms of units of biologic activity per milligram. The activity of other samples can then be estimated by comparative bioassays against known amounts of this stable yardstick. This third stage was reached by heparin in 1942

(20). At this stage, the standard preparation is often impure, and the validity and precision of methods not yet proved or tested. However, once a common standard is available, different methods for assay can be studied, and such differences as those due to the species and tissues of origin may be recognized. The search for purer material is made easier.

A fourth stage may occur with the replacement of the standard by another, of purer material with much higher specific activity. Use of this purer standard may lead to better assay techniques of higher precision and greater specificity, which themselves help in the isolation of the active principle(s) - a fifth stage.

Some substances may progress to two further stages: they may be synthesized, and chemical and physical methods may be developed for measuring the isolated or synthesized substance. An eighth and ultimate stage is reached when such methods can be used to measure the substance in crude material without prior purification. It may be only when a substance has reached this last stage that it is safe to discontinue relying on bioassay against a standard (3). These steps in establishing a biological standard are summarized in Table I.

Some substances pass relatively quickly from one stage to another. Why, then, did it take a quarter of a century for heparin to reach the third stage (establishment of a standard), and why has it remained there for another quarter of a century? To answer these questions, we have to discuss two aspects: the validity of the bioassay methods we use, and the problems of suitability of the reference material.

Validity of Bioassay

Let us turn first to the bioassay of heparin. Many methods have been described, and some are listed in Table II.

All but the last two are based on the heparin-caused delay in coagulation of blood or plasma. To assess the validity of these methods, we must consider the general requirements for valid bioassay:

Existence of a standard
Assays of like against like
Measurement of a specific biologic effect
Potency ratio that is independent of biologic system
Similarity of dose, parallelism, and statistical validity

TABLE I

Steps in Establishing a Biological Standard

Steps	Examples
1. Biological effect observed.	Platelet aggregation.
2. Bioassay. Quantitation.	Old Mouse Unit of Insulin. Old Frog Unit of Digitalis. Thrombin Unit "that which clots..."
3. Standard of impure material.	Heparin, Thromboplastins, Streptokinase, Erythropoietin.
4. Standard of purified material.	Porcine corticotrophin, Growth hormone, Factor VIII.
5. Isolation of active principle.	Insulins, porcine calcitonin.
6. Synthesis.	Oxytocin, Vasopressin, Insulin, Porcine corticotrophin.
7. Chemical/physical measurement of pure material	Vitamin B_{12}, Steroids.
8. Chemical/physical measurement of the active principle in crude mixture.	Steroids, Vitamins.

Assay of like against like. - Whatever the form in which heparin may exist in its native intracellular state or when released naturally, heparin in preparations for clinical use is notoriously heterogeneous; preparations vary according to the tissue source (lung, gut mucosa, skin), the species (ox, pig, sheep, whale), and the methods of isolation and purification, which may affect molecular structure and biologic activity. Even preparations from a single tissue from a single species are heterogeneous (21, 31). Thus, experimental findings are related not to heparin in general, but only to the specific heparin preparation used.

Are heparin preparations obtained from different sources, from different species, and with different methods comparable? Using a

TABLE II

Some Methods for Assaying Heparin

British Pharmacopoeia method
Whole-blood clotting time
Whole-blood activated partial thromboplastin time
Howell method
United States Pharmacopoeia method
Thrombin titration
Protamine, polybrene neutralization
Activated partial thromboplastin time
Plasma recalcification time
Anti-Xa assay
In vivo assay in dogs
Lipoprotein lipase activity
Metachromatic methods

very sensitive anti-Xa assay, Yin, Wessler, and Butler (35) reported that the 2nd International Standard and the 1st British Standard - both bovine and lung and mucosal, respectively - yielded identical clotting times for the same number of units of heparin tested, whereas two preparations of porcine origin had a steeper slope and, on the basis of this difference, according to the authors, approximately 25% more activity than the former pair (Fig. 1).

Dissimilarity, however, is due not only to the species of origin. Further evidence that there are differences, both between porcine and bovine and between mucosa and lung as sources, comes from a number of reports. For example, in an assay system using the binding of a preparation of purified platelet factor 4 performed by Pepper in Edinburgh (Fig. 2), the same concentration (expressed in nominal units) of different heparin preparations neutralized different amounts of platelet factor 4 or platelet antiheparin activity. In this system, the 2nd International Standard (bovine lung preparation) appears the most potent, the 3rd International Standard (porcine mucosa) next most potent, and the British Standard (bovine mucosa) the least potent. Even such relatively simple assay procedures as those recommended by the British Pharmacopoeia (B.P.) and the U.S. Pharmacopoeia (U.S.P.) show differences between lung

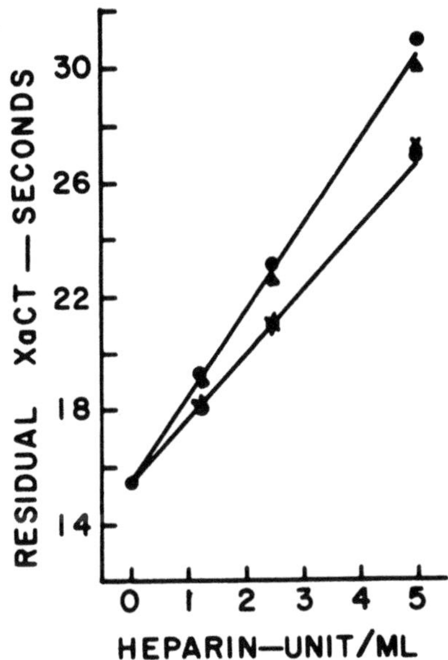

Fig. 1. Response of the heparin assay to various standardized heparin preparations: O, Liquaemin, sodium "10"; ●, first British Standard; X, second International Standard; Δ, third International Standard. (From Yin et al. (35)).

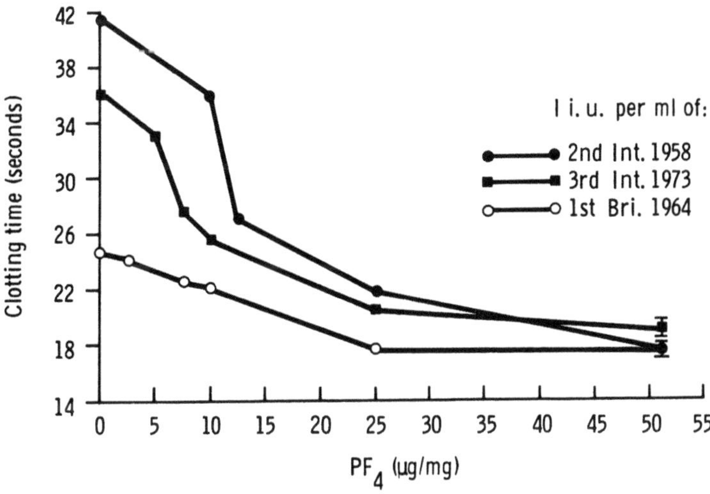

Fig. 2. Behavior of heparin standards in the heparin-thrombin-platelet factor 4 assay. Assays were performed by Dr. Duncan Pepper, Edinburgh.

and mucosal heparin, as shown in Table III. It is evident from the table that the U.S.P. assay generally gives lower relative values than the B.P. assay when mucosal heparin is assayed in terms of lung heparin. The opposite occurs when mucosal heparin is assayed by these methods in terms of a standard of mucosal material. Furthermore, the relative potency of preparations of whale heparin compared with bovine lung preparations is reported to be higher with B.P. assays than with U.S.P. assays (16).

On the basis of these few examples, we may conclude that like was not assayed against like. In addition, as more specific and sensitive assays become available, or when more than one method is used, species differences and the difference between lung and mucosal heparin preparations may become even more obvious.

Measurement of specific biologic effect. - A third requirement for a valid bioassay is that the characteristic measured should be an effect relevant to and preferably biologically specific for the activity for which the substance is used. What is the "relevant" biologic activity of heparin? Are we using it for its anticoagulant or its antithrombotic activity? Heparin interferes with blood coagulation sequence at a number of different stages: it neutralizes

TABLE III

Results of a Collaborative Study of Heparins from Different Sources[a]

		Source of Heparin			
	Method	Bovine Mucosa	Porcine Mucosa	Sheep Mucosa	Mixed Mucosa
Potency in terms of lung heparin, iu/ampule	B.P.	1,264	1,322	512	1,385
	U.S.P.	1,142	1,283	504	1,271
Potency in terms of porcine mucosal heparin, iu/ampule	B.P.	1,264	--	--	1,137
	U.S.P.	1,332	--	--	1,283

[a]Data from Bangham and Woodward (5).

thrombin by potentiating the effect of the natural inhibitor, antithrombin III (1, 2, 6, 7, 22, 27, 28), and possibly interfering directly with the thrombin-fibrinogen reaction (18); it increases the neutralization of factor Xa by antifactor Xa and its rate (6, 12, 33, 34, 36); it interreacts with factor IX and factor XI (11, 26); and it has antithromboplastin activity (9).

Rosenberg and Damus (27) have recently shown how the complex of heparin and antithrombin III inhibits thrombin in a pure system; they postulate that heparin binds specifically to antithrombin III (or heparin cofactor) to form a highly stable complex. The isolated complex appears to thave a wide range of specificities against plasma proteinases. However, more than one heparin cofactor may be present in impure systems, such as plasma or blood (7), and heparin may interreact directly with the components of the clotting sequence.

At the same time, one must remember that the amount of heparin available to exert an anticoagulant effect in vivo and in impure systems in vitro depends on the relative affinities of different plasma and blood constituents for heparin. In 1971, Yen et al. (32) suggested the following decreasing order of affinities for heparin: platelet factor 4, factor IX, "thrombin" system, and β-lipoproteins; the experimental data to substantiate this claim are not yet available. If this is the true order, the anticoagulant effect of heparin will be seriously hindered or even reversed by the presence of platelet factor 4. Both clinical and experimental data (10, 13, 24, 25, 30) suggest that this is the case. In addition, it is also possible that different preparations of platelet factor 4 or antifactor Xa have different affinities for various preparations of heparin - in analogy with the well-known relation of heparin to protamine (31), shown in Table IV.

One answer to these difficulties may be to test the potency of heparin in vivo, in the whole intact animal. The best documented in vivo assay of heparin (19) uses dogs to compare a number of clotting characteristics after intravenous injections of standard and test heparin preparations. The method yields reproducible and statistically valid estimates. Unfortunately, the hemostatic system in each species is different, and these differences become more marked in the species further removed in the phylogenetic order. Hall (15) has summarized some of the differences between dog and man:

> Thrombin fibrinogen reaction with canine substrate 4.4 times faster than with human substrate; three fibrinopeptides released from dog fibrinogen, two from human
>
> Factor V content in human 0.5 - 1.5 u/ml; in dog, 1 - 32 u/ml

TABLE IV

The Neutralization of Heparin by Protamine Sulfate[a]

Source of Heparin	Specific Activity, iu/mg	Units Neutralized by 1 mg of Protamine	
		Int. Ref. Protamine	Salmine
Mixed lung	130	76	81
Lung	113	77	85
Sheep mucosa	154	120	134
Bovine mucosa	167	112	127
Porcine mucosa	211	148	162
Porcine mucosa	266	133	151

[a]Modified from Walton *et al*. (31).

Factor VII in human 0.5 - 2.0 u/ml; in dog, 1 - 8 u/ml

Human platelets aggregate readily with collagen; dog platelets do not.

In view of such differences, we can expect that, sooner or later, heparin preparations will be found that display different *in vivo* activity in dogs and men.

It is at least as important that each person will have his own hemostatic equilibrium; and the effects of heparin in any one person at any one time will depend on the idiosyncrasies of that equilibrium. The differences between normal, healthy, young people may be minor, but the response to heparin in the elderly or sick may be significantly different.

This is illustrated in Fig. 3, which shows what happened when heparin preparation was used in fresh platelet-poor plasma of a young man and of an elderly man in two different assay systems. In both systems, the slopes of the dose-response curve were much steeper

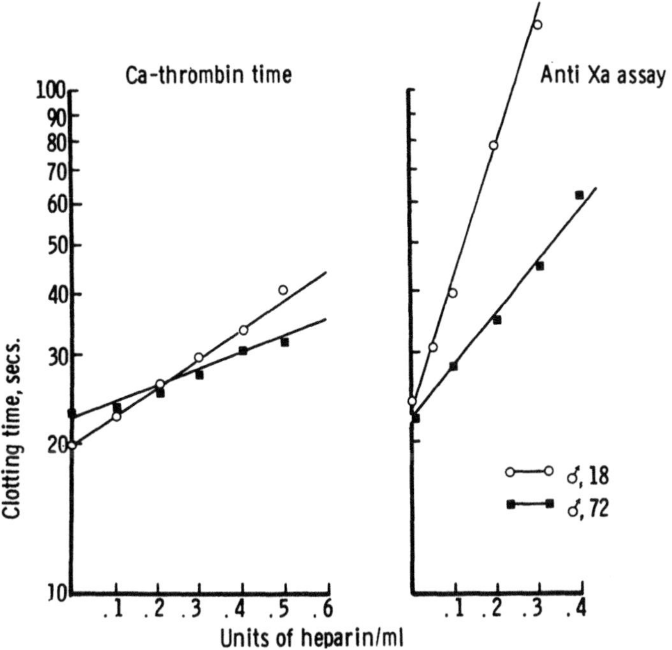

Fig. 3. Comparison of the same heparin preparation in two men.

with the younger man's plasma. Similar results were reported by Estes (14) who used the whole-blood partial thromboplastin time to measure heparin levels and by Brodows and Campbell (8), who noted that postheparin lipase content appeared to be relatively unresponsive to small doses of intravenous heparin in a group of 20 elderly subject, compared with 28 young controls. For this reason, laboratory control of heparin treatment is desirable, with a variety of different tests. Fig. 4 shows the relative sensitivites of the tests commonly used.

We can conclude that we are indeed measuring some effects of "heparin," but that there is probably no single specific biologic effect that will account for either its overall anticoagulant or its overall antithrombotic effect _in vivo_. Moreover, the search for a single specific effect and a too-specific assay may prove unhelpful.

Potency ratio independent of biologic system. - Another requirement for a valid bioassay is that the potency ratio be independent of the biologic system. We already know that each heparin preparation behaves differently in each assay system used. As the precision of these new assays improves, we may discover that some preparations are particularly good in combining with platelet factor 4 but less good in their anti-Xa effect, or that they are excellent in blocking

(1973)

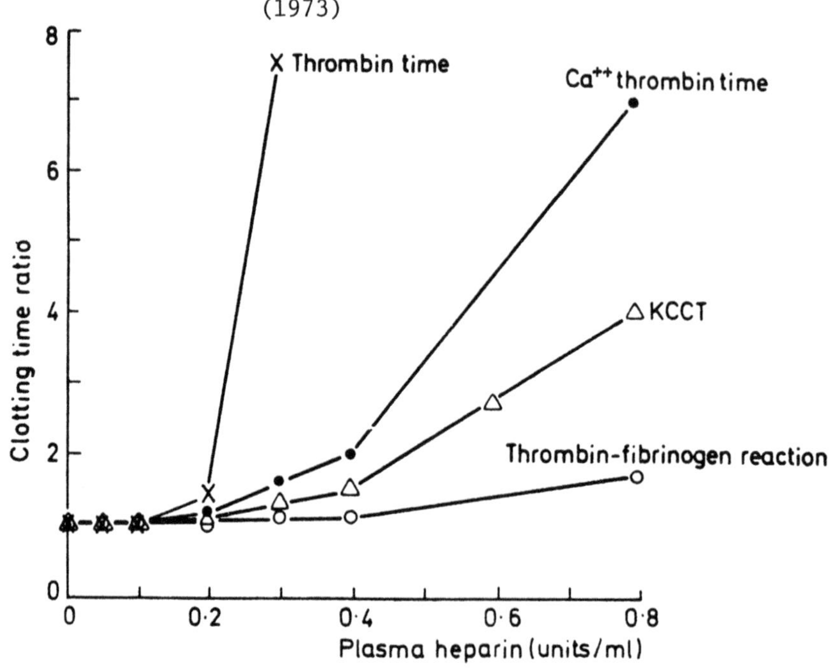

Fig. 4. The sensitivity of three tests for the control of heparin, based on the clotting-time ratio of the heparinized sample to normal plasma. Whole-blood clotting time has about the same sensitivity as the calcium thrombin time. Effect of heparin on the thrombin-fibrinogen reaction is shown for comparison. (From Denson and Bonnar (12).)

factor IX but less potent in the reactions that require the presence of plasma cofactor. It may well be that a therapeutically effective heparin must be capable of interreacting at a number of points in the hemostatic system and is thus by definition relatively nonspecific.

Similarity of dosage. - The shape of the dose-response curve for each of the preparations in the assay system should be known; roughly similar dosages should be used for evaluation of linearity and parallelism and for the calculation of potency. As shown in Fig. 5, taken from a collaborative study of heparin in 19 laboratories, only a small proportion of assays (denoted by a dot within the square) did not satisfy the statistical criteria of validity (5). In this particular study, it became apparent that there was little or no variation in response when any one preparation was assayed, in any one laboratory, using the same batch of reagents. Such data are not normally distributed and cannot be analyzed with the conventional analysis of variance. The apparent high reproducibility of assays with one batch of reagents is misleading, and the apparent high

Erratum

The text in the paragraph ending at the top of page 168 and the data in Table III on that page contain a number of inaccuracies. A corrected version of page 168 appears below.

and mucosal heparin, as shown in Table III. It is evident from this table that the USP assay generally gives lower relative values than the BP assay when mucosal heparin is assayed in terms of lung heparin. The same is reported for whale heparin compared to bovine lung preparation (8).

On the basis of these few examples, we may conclude that like was not assayed against like. In addition, as more specific and sensitive assays become available, or when more than one method is used, species differences and the difference between lung and mucosal heparin preparations may become even more obvious.

Measurement of specific biologic effect. - A third requirement for a valid bioassay is that the characteristic measured should be an effect relevant to and preferably biologically specific for the activity for which the substance is used. What is the "relevant" biologic activity of heparin? Are we using it for its anticoagulant or its antithrombotic activity? Heparin interferes with blood coagulation sequence at a number of different stages: it neutralizes

TABLE III

Results of a Collaborative Study of Heparins from Different Sources[a]

Heparin preparation	Method	Potency (iu/ampoule) when assayed against		
		Lung heparin	Bovine mucosal heparin	Mixed mucosal heparin
Porcine mucosal heparin	B.P.	1,322	1,264	1,137
	U.S.P.	1,283	1,339	1,228
Bovine mucosal heparin	B.P.	1,355	-	-
	U.S.P.	1,142	-	-
Mixed mucosal heparin	B.P.	3.28[b]	-	-
	U.S.P.	2.75[b]	-	-

[a] data from Bangham & Woodward
[b] iu/mg, tablets containing excipient

Advances in Experimental Medicine and Biology, Volume 52, *HEPARIN*
edited by Ralph A. Bradshaw and Stanford Wessler

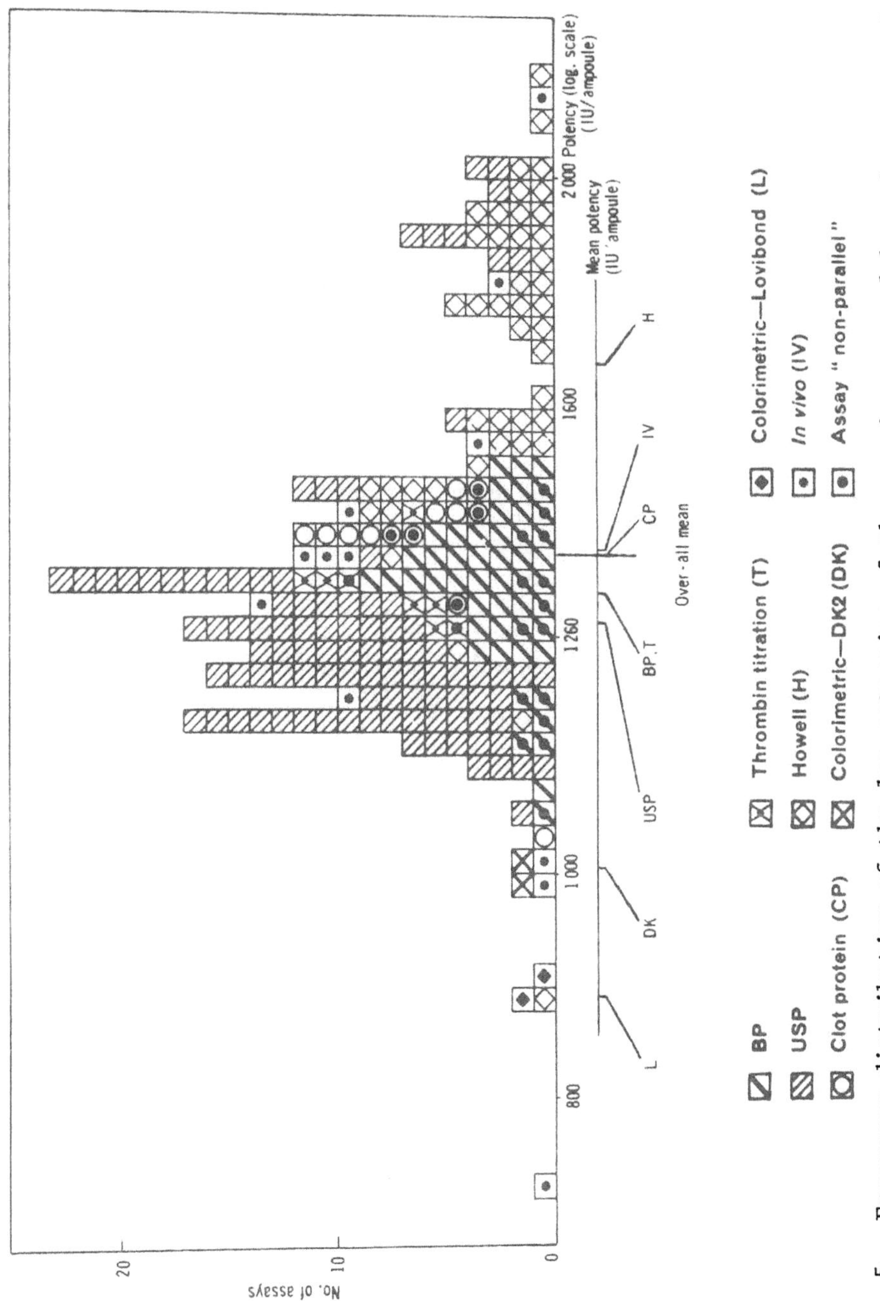

Fig. 5. Frequency distribution of the log potencies of the porcine mucosal heparin preparation in terms of the International Standard Heparin established in 1958. (From Bangham and Woodward (5).)

TABLE V

Requirements for a Biological Standard

1. Possesses specific biological activity; demonstrable dose/response relation.
2. Similar type of preparation; Species specific.
3. Known stability of biological activity.
4. Unit

precision of the results gives an incorrect impression of accuracy. For this reason, a reliable estimate of heparin potency requires the combination of several completely independent assays, that is, assays using different batches of reagents, as well as fresh dilutions of test and standard heparin.

Suitability of standards. - The most important necessity for a valid bioassay is a standard. Three successive international standards have been used to define the international unit of heparin activity. The first two were prepared from bovine lung, and crystallization as a barium salt was used in their purification. The current one, the 3rd International Standard, was made from porcine mucosa.* A biological standard should fulfil certain requirements, as those listed in Table V.

Appropriate biologic activity. - Clearly, a standard must possess the particular biologic activity for which it is used, to define the unit. All three International Heparin Standards have been extensively used over many years; they all have anticoagulant activity of about 130 iu/mg (5, 20, 30). The dose-response relation was documented in each successive collaborative assay (4, 5, 20). For example, the calibration of the 3rd International Standard in terms of the 2nd International Standard, on the basis of 224 assays by 13 laboratories from eight different countries (5), yielded a potency of 1,371 iu/ampule (range, 887-2,195) with confidence limits (p = 0.05) of 1,342-1,401 iu/ampule.

*These international standards are kept at the WHO Centre, now the National Institute for Biolgoical Standards and Control, London.

Biologic identity with standard. - The second requirement is that the biologically active constituent in the standard be identical with the preparation(s) of heparin assayed against it; this seems unlikely to be realized. The evidence already discussed suggests that there are differences between heparin preparations derived from different tissues and species. The widespread use of more sensitive assay systems may reveal consistent differences between heparin preparations of different molecular-weight distribution or subjected to different bleaching procedures. Until we can identify these differences clearly, it may be safer to assay lung heparin against lung heparin, mucosal against mucosal, and, possibly, bovine against bovine and porcine against procine.

Nevertheless, it is evident from many years of extensive study and testing of heparin preparations that statistically valid and consistent results can be obtained when dissimilar preparations are assayed with only one method. This is best illustrated in Fig. 5, which shows the results of estimating a porcine mucosal heparin preparation in terms of a bovine lung preparation; all laboratories using the same method - be it B.P., U.S.P., or Howell assay - obtained very similar results.

Stability. - One of the essential attributes of a standard is that it is stable, and the stability should always be tested. It may be estimated by comparing the rate of loss of activity in unopened ampules stored at high temperatures (17). Some recent experiments on the stability of freeze-dried heparin are summarized in Table VI. To test whether the assay used (anti-Xa potentiation) was sensitive to the loss of anticoagulant action, heparin preparations that had been subjected to autolysis in the free-acid form were compared at the same time.* It is evident from Table VI that freeze-dried material lost less than half its anti-Xa potentiating effect after 12 years at 37°, whereas, after 25 days of free heparinic acid autolysis, four-fifths of this activity was lost. Whether all its anticoagulant and antithrombotic actions are as stable in freeze-dried preparations remains to be determined.

International unit. - With the establishment of the 1st International Standard of heparin, the international unit of heparin was

---

*An aqueous heparin solution (e.g., 750 mg in 5 ml of water) was passed through a thoroughly washed 1.3 x 25 cm column of Amberlite IR 120 resin in the H^+ form, and the heparin-containing effluent made up to 15 ml. The aqueous heparinic acid solution (29) was allowed to autolyze at room temperature (20°-23°) and aliquots of 2 ml were withdrawn at intervals of days or weeks. These were adjusted to a pH of about 11 with normal NaOH, dialyzed, concentrated, and precipitated with ethanol. (E.A. Johnson, personal communication).

TABLE VI

Stability of Heparin Using (Anti-Xa Assay)

Freeze-Dried Heparin (53/11) Stored for 12 Years			
Storage Temperature, C	-20	+20	+37
Potency, iu/ampule	2,600	2,160	1,540
Heparin in Solution			
Days of storage at room temperature	0	11	25
Potency, iu/ampule	1,980	580	260

defined as the activity contained in 1/130 mg of that standard. It is sobering to recognize that the conclusions about heparin measurement reached in 1943 are still completely valid in 1974:

> It is essential that the assay of preparations of heparin should be carried out in direct comparison with the Standard Preparation. In conformity with usual practice, no particular method or methods for the assay of heparin are prescribed in this Memorandum. It is expected that individual workers will employ the methods of which they have experience and in which they have confidence. It is suggested, however, that the method employed should measure the specific anticoagulant activity of heparin and not some chemical or physical property which may be associated with it. Provided the method employed is capable of distinguishing potency differences of something like 10 per cent, the questions as to whether fresh or oxalated whole blood or plasma should be used, whether extracts containing thrombin or thrombokinase (thromboplastin) should be added, or whether the actual clotting time should be measured or merely the presence or absence of a clot should be observed, may be left to the decision of the individual worker.

One source of uncertainty in the standardization of heparin must, however, be pointed out, though it appears unlikely that it will often lead to serious practical difficulty. This lies in the fact that the potency of a given preparation of heparin, stated in terms of the potency of another preparation, is often not a fixed quantity, but depends to some extent on the particular biological test used for the comparison. Discrepancies thus arise, when the results of one kind of biological assay are compared with the results of another, although either method by itself may give results which are sufficiently precise and reproducible.

Similar discrepancies have been met with in the standardization of other drugs; they are due to the presence, in the standard or the preparation to be tested, or in both, of more than one substance which can affect the biological test. In the case of heparin, as elsewhere, it is unfortunately not yet clear which of the methods available measures most accurately the therapeutic efficacy of the material under test. These uncertainties are of little practical importance so long as the preparations being tested are, like the Standard, of good quality (20).

Conclusion

What, then, is the future of heparin reference materials? Do we now require a new standard?

One would wish to relate chemical structure or physicochemical properties of a heparin preparation to its effects on blood coagulation and platelets. Until this becomes possible, materials that have been extensively studied and are known to be stable should be retained and used in many different assay systems *in vitro* and *in vivo*. In this way, we will accumulate a store of information on the present standards that may help to define optimal criteria of purity and specificity for the reference materials of the future.

REFERENCES

1. ABILGAARD, U., Scand. J. Clin. Lab. Invest., 21 (1968) 89.
2. ASTRUP, T. and DARLING, S., Acta Physiol. Scand., 5 (1943) 13.
3. BANGHAM, D.R., in Erythropoiesis, (Ed. Jacobson, L.O. and Doyle, M.), Grune and Stratton, (1962) p. 23.
4. BANGHAM, D.R. and MUSSET, M.V., Bull. Wld. Hlth. Org., 20 (1959) 1201.

5. BANGHAM, D.R. and WOODWARD, P.M., Bull. Wld. Hlth. Org., 42 (1970) 129.
6. BIGGS, R., DENSON, K.W.E., AKMAN, N., BORRETT, R. and HADDEN, M., Brit. J. Haemat., 19 (1970) 283.
7. BRIGINSHAW, G.F. and SHANBERGE, J.N., Thromb. Research, 4 (1974) 463.
8. BRODOWS, R.G. and CAMPBELL, R.G., New Eng. J. Med., 287 (1972) 969.
9. CHAGRAFF, E., ZIFF, M. and COHEN, S.S., J. Biochem., 136 (1940) 257.
10. CONLEY, C.L., HARTMANN, R.C. and LALLEY, J.S., Proc. Soc. Exp. Bio. Med., 69 (1948) 284.
11. DANISHEFSKY, I. and TZENG, F., Thromb. Res., 4 (1974) 237.
12. DENSON, K.W.E. and BONNAR, J., Throm. Diath. Haem., (1973).
13. DEUTSCH, E. and KAIN, W., in Blood Platelets, (Ed. Johnson, S.A. et al.), Little Brown Co., Boston, (1960), p. 337.
14. ESTES, J.W., Ann. N.Y. Acad. Sci., 179 (1971) 187.
15. HALL, D.E., in Blood Coagulation and its Disorders in the Dog, Bailliere and Tindall, London, (1972) p. 1.
16. HASHIMOTO, K. and MATSUNO, M., Tohoku J. Exp. Med., 9 (1967) 295.
17. JERNE, N.K. and PERRY, W.L.M., Bull. Wld. Hlth. Org., 14 (1956) 167.
18. KLEIN, P.D. and SEEGERS, W.H., Blood, 5 (1950) 742.
19. KUO, S.H., JAQUES, L.B. and MILLAR, C.J., J. Pharm. Pharmac., 24 (1972) 858.
20. League of Nations, Bull. Hlth Org., L.O.N., 10 (1943/1944) 151.
21. LINDAHL, U., BACKSTROM, G. and MALMSTROM, A., Biochem. Biophys. Res. Com., 46 (1972) 985.
22. LYTTLETON, J.W., Biochem. J., 58 (1954) 8.
23. McLEAN, J., Amer. J. Physiol., 41 (1916) 250.
24. NATH, N., NEIWIAROWSKI, S. and JOIST, H.J., J. Lab. Clin. Med., 82 (1973) 754.
25. PEPPER, D., MOORE, S. and CASH, J.D., in Erythrocytes, Thrombocytes and Leukocytes, Second International Symposium, Vienna, (1972) p. 213.
26. PITLICK, F.A., PORTER, M.C., LUNDBLAD, R.L. and DAVIE, E.W., J. Biomed. Mater. Res., 3 (1967) 95.
27. ROSENBERG, R.D. and DAMUS, P.S., J. Biol. Chem., 248 (1973) 6490.
28. SNELLMAN, O., SYLVEN, B. and JULES, C., Biochim. Biophys. Acta, 7 (1951) 98.
29. VELLUZ, L., Bull. Soc. Chim. Biol., 41 (1959) 415.
30. WALSH, P.N., BIGGS, R. and GAGNATELLI, G., Brit. J. Haemat., 26 (1974) 405.
31. WALTON, P.L., RICKETTS, C.R. and BANGHAM, D.R., Brit. J. Haemat., 12 (1966) 310.
32. YEN, T.F., DAVAR, M. and REMBAUM, A., in Biomedical Polymers, (Ed. Rembaum, A. and Schen, M.), Marcel Dekker, New York, (1971).

33. YIN, E.T. and WESSLER, S., Biochim. Biophys. Acta, 201 (1970) 381.
34. YIN, E.T. and WESSLER, S., J. Biol. Chem., 246 (1971) 3703.
35. YIN, E.T., WESSLER, S. and BUTLER, J.V., J. Lab. Clin. Med., 81 (1973) 298.
36. YIN, E.T., WESSLER, S. and STOLL, P.J., J. Biol. Chem., 246 (1971) 3712.

APPLICATION OF THE KINETICS OF HEPARIN TO THE FORMULATION OF DOSAGE SCHEDULES

J. Worth ESTES

Boston University School of Medicine
Boston, Massachusetts (USA)

Because of the difficulties commonly encountered in evaluating the anticoagulant effect of heparin at the bedside and in determining appropriate dosage schedules for its administration, we have been studying the rate of heparin's disappearance from blood. However it is measured, the half-life of heparin is about 1.5 hr (4). This suggests that giving the drug by intermittent intravenous injection every 4-6 hr, as is usual, does not ensure a sustained degree of anticoagulation, even when the dose is adjusted by, say, the Lee-White clotting time. Furthermore, in most reports of excessive bleeding after administration of heparin, the patients' body weights have not been taken into account in reporting the dose of drug administered (9,15), so that the possibility of having administered an anticoagulating overdose of heparin can have been overlooked (3, 6).

Because an adequate method for the direct assay of heparin in blood is not yet available, we have studied its pharmacokinetics with methods that yield data pertaining to the kinetics of its anticoagulant effect. Early experiments, summarized in Table I, led us to conclude that the whole blood activated partial thromboplastin time (WBPTT) (1) is the method for evaluating heparin therapy that combines the greatest precision of bioassay and information yield with the greatest economy. The WBPTT's value is augmented by its ability to yield results on the basis of which clinical decisions can be made at the bedside with minimal delay (4). In these studies, we were concerned, not with bioassaying different heparin preparations (we used the same lot of a mucosal preparation throughout), but with comparing the relative potencies of the assays themselves. Because of the decisive advantages offered by the WBPTT, we have continued to use it to assess heparin activity.

TABLE I

Relative Precision, Information Yield, and Economy of Four Assays of Heparin Effect

Methods of Estimating Heparin Effect	Relative Precision,[b] Calculated from Relation between Dose and:		Relative Information Yield,[c] Calculated from Relation between Dose and:		Relative Economy,[d] Calculated from Relation between Dose and:	
	Maximal Clotting Time	Duration of Action	Maximal Clotting Time	Duration of Action	Maximal Clotting Time	Duration of Action
WBPTT (6)	1	1	1	1	1	1
Activated plasma PTT (7)	1.24	1.56	1.20	1.56	0.33	0.42
Plasma PTT (8)	0.33	0.58	0.35	0.57	0.09	0.15
Whole-blood clotting time (9)	0.43	0.61	0.40	0.61	0.11	0.16

[a] Calculated from raw data.
[b] In raw data, calculated as ratio of mean square deviation from regression to square of the regression coefficient.
[c] In raw data, calculated as reciprocal of square of precision.
[d] In raw data, calculated as ratio of information yield to cost per replication.

Heparin was bioassayed with the WBPTT in 19 normal subjects. Aliquots of each subject's blood were added to five concentrations of heparin *in vitro*, and the WBPTT was determined for each concentration. This provided calibration curves for each subject, with which his WBPTT at intervals after heparin injection could also be converted to heparin concentration. As the representative calibration curves in Fig. 1 show, the slopes of the curves decreased significantly with increasing intercepts; i.e., as the baseline clotting time, with no heparin added, increased through the normal range of 85 - 140 sec, the rate of change of clotting time decreased with heparin concentration. It is clear, then, that the same WBPTT in two heparinized subjects does not necessarily represent the same amount of heparin in the blood of both subjects. This lack of predictive value among subjects demonstrates only one problem in evaluating a single clotting time in a heparinized patient, even when his preheparin (baseline) clotting time is known.

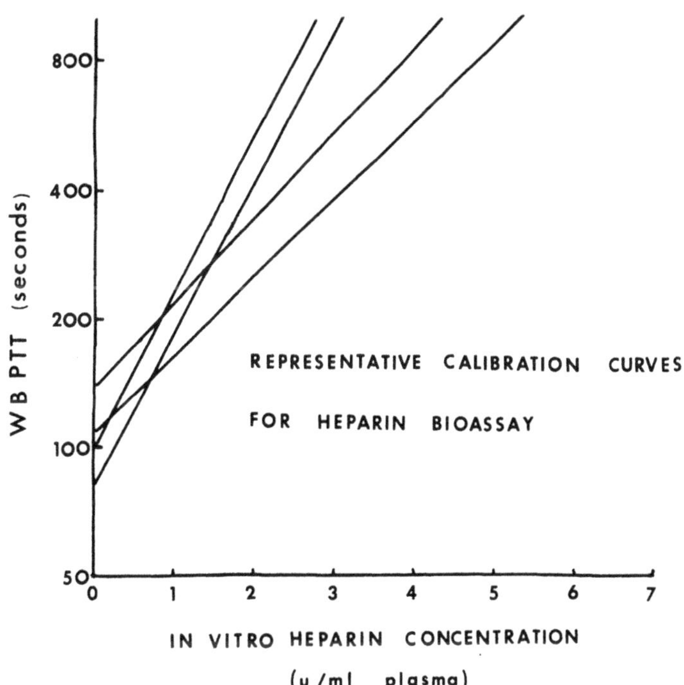

Fig. 1. Representative regression lines showing the relation between whole-blood activated partial thromboplastin time (WBPTT) and heparin concentration *in vitro*, used in the bioassay of plasma concentrations of heparin *in vivo*.

Figure 2 shows (in the upper graph) that the half-life of heparin's anticoagulant effect on the WBPTT, when drug concentration is not taken into account, does not vary with dose. However, when the half-life of heparin was determined by bioassay - i.e., by interpolating each subject's WBPTT at intervals after drug injection on his own calibration curve - we found that the half-life of heparin does vary with dose (Fig. 2, lower graph). This relationship is unusal among pharmacologic agents.

Furthermore, the half-lives of heparin that were calculated from the two ways of utilizing the same raw data for each subject were not correlated with each other. Nor did the properties of his own dose-response curve appear to influence the kinetics of heparin in each subject. In addition, the mean half-life of bioassayed heparin was about 80% of the mean half-life of heparin's effect - the same ratio we found when we compared previously published bioassay data with our own prospective data on the disappearance of effect (5). However, in neither case was the 20% difference statistically significant. It would be anticipated that the drug's effect would disappear more slowly than the drug itself, if it is bound to its site of action in plasma. The reactions of heparin with proteins have not yet been elucidated in terms of their dissociation constants, because of the diverse range of sizes and constituents of the molecular moieties that constitute heparin. Heparin is known to be bound extensively to globulins and fibrinogen (8). Its binding spectrum would also account for its distribution in the plasma volume at all but the largest doses (7).

Sensitivity to a drug is inversely proportional to its minimal effective threshold dose to produce a desired effect. Sensitivity should not vary with reactivity to the drug, which is measured by the slope of its dose response curve. The inverse relationship between the baseline clotting time and the subjects' reactivity to heparin, exemplified in the calibration curves, led us to explore how best to measure heparin-induced hypocoagulability, without having to take into account either sensitivity or reactivity to the drug.

The first column in Table II shows several criteria that might be chosen to define the presence of appropriate clinical effect of heparin. They are analogous to the common practice of adjusting doses of heparin to produce some multiple of the baseline Lee-White clotting time.

The threshold doses required to achieve these possible effects, shown in the second column, were determined from the *in vitro* calibration curves. The smallest threshold doses indicate the greatest sensitivities. The third column shows that, as the threshold dose required to produce a given degree of anticoagulation increased, the rate of response to heparin decreased. This confirms that sensitivity

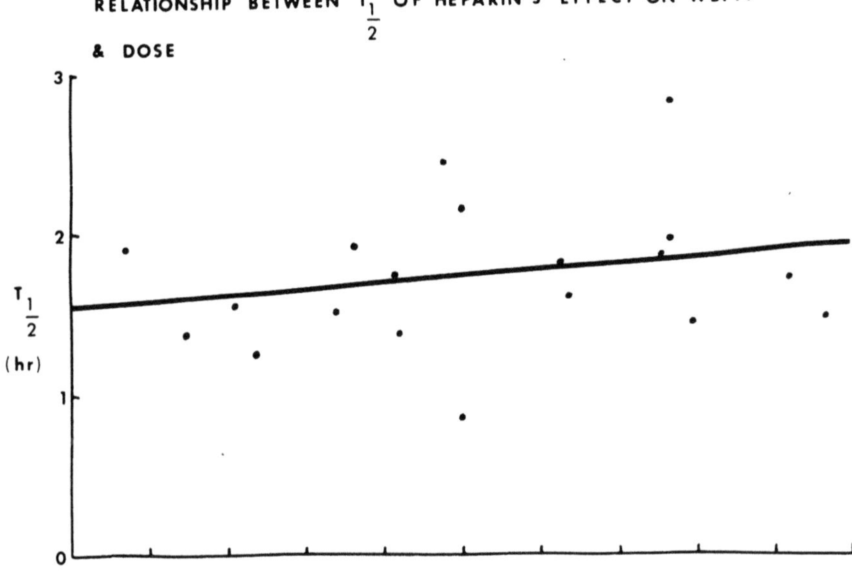

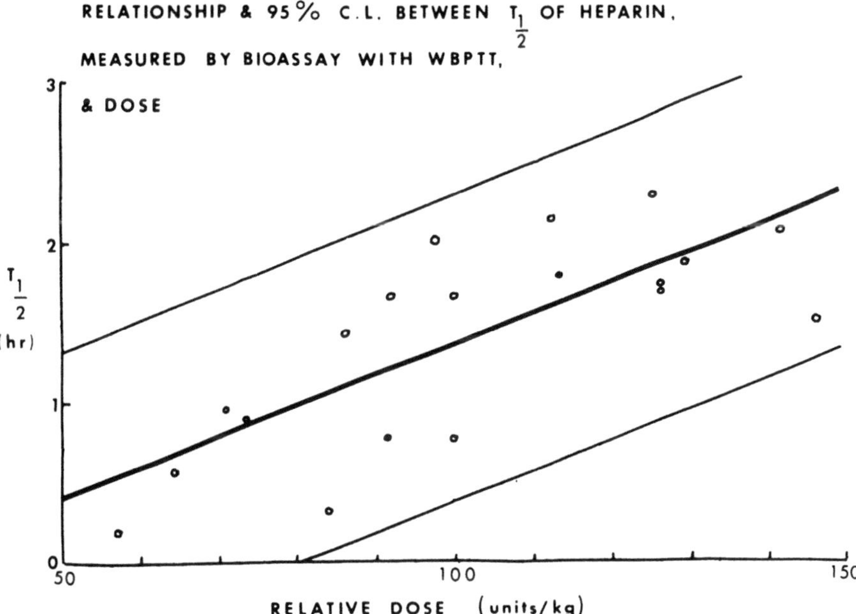

Fig. 2. Regression lines showing the relation between the dose of heparin and its half-life measured for its anticoagulant effect (upper graph) and for its concentration bioassayed in plasma (lower graph).

TABLE II

Relation among Sensitivity and Reactivity to Heparin *in Vitro* and Baseline Clotting Status[a]

Criteria of Anticoagulant Effect *in Vitro*	Mean Threshold Dose ± S.D., units/ml (Sensitivity)	CORRELATION COEFFICIENTS AMONG: Threshold Dose (Sensitivity) and Slope of Dose-Response Curves (Reactivity)	Threshold Dose (Sensitivity and Baseline WBPTT)	Threshold Dose (Sensitivity) and Baseline WBPTT, with Slope of Dose-Response Curves (Reactivity) held constant
Increase WBPTT *to* 150 sec	0.592 ± 0.166	+ 0.053	- 0.744[b]	- 0.953[b]
Increase WBPTT *to* 200 sec	1.134 ± 0.278	- 0.182	- 0.393	- 0.702[b]
Increase WBPTT *to* 300 sec	1.772 ± 0.314	- 0.831[b]	+ 0.118	- 1.000[b]
Increase WBPTT *to* 500 sec	2.642 ± 0.498	- 0.923[b]	+ 0.311	- 1.000[b]
Double WBPTT	1.089 ± 0.287	- 0.882[b]	+ 0.736[b]	+ 0.421
Triple WBPTT	1.778 ± 0.438	- 0.938[b]	+ 0.695[b]	+ 0.269
Increase WBPTT by 100 sec	1.088 ± 0.207	- 0.867[b]	+ 0.397	- 0.487[b]
Increase WBPTT *by* 10 S.D. (=150 sec[c])	1.466 ± 0.268	- 0.941[b]	+ 0.390	- 0.943[b]
Increase WBPTT *by* 20 S.D. (=300 sec[c])	2.265 ± 0.437	- 0.945[b]	+ 0.395	- 0.964[b]
Increase WBPTT *by* 30 S.D. (=450 sec[c])	2.806 ± 0.551	- 0.955[b]	+ 0.415	- 1.000[b]

[a]For every statistic in the table, N = 19.
[b]$p < 0.05$.
[c]Calculated from standard deviation of WBPTT's determined for 55 unheparinized healthy persons, 14.9 sec.

to heparin varies with reactivity to the anticoagulant. The next column shows that sensitivity to heparin does not vary with baseline coagulability, except when baseline clotting status was used to define the criteria of effect. However, as the last column shows, when reactivity was held constant statistically with partial correlation, sensitivity to heparin did, in fact, vary with the baseline clotting time. This means that sensitivity was significantly related to baseline clotting status, as well as to reactivity.

For most drugs, and in all ideal pharmacological situations, sensitivity to a drug is not, and should not be, related to reactivity to the drug. Because we found that sensitivity and reactivity to heparin were related, we studied other metameters of dose and effect. Listed in the first column of Table III are six possible ways of measuring heparin's effect. The top three take into account the subject's WBPTT 2 hr after heparin injection, with or without the baseline clotting time. The lower three take into account the theoretical maximal clotting time, extrapolated to time zero after drug

TABLE III

Statistical Index of Precision and Information Yield for Dose and Response Metameters of Heparin, Determined from Curves Expressing Rate of Disappearance of Anticoagulant Effect

	TOTAL DOSE units		RELATIVE DOSE units/kg	
	Precision	Information	Precision	Information
Metameter of Effect	$\times 10^{2}$	$\times 10^{-15}$	$\times 10^{2}$	$\times 10^{-8}$
WBPTT at t = 2 hr	16,088	386.4	2.6	1482.6
WBPTT at t = 2 hr MINUS Baseline WBPTT	10,770	862.1	3.2	991.6
$\frac{\text{WBPTT at t = 2 hr}}{\text{Baseline WBPTT}}$	10,443	918.7	5.5	330.1
WBPTT at t = 0	76,931	16.9	989.0	0.001
WBPTT at t = 0 MINUS Baseline WBPTT	70,062	20.4	9.8	103.4
$\frac{\text{WBPTT at t = 0}}{\text{Baseline WBPTT}}$	38,275	68.3	9.8	105.0

administration on the time-concentration curves, as well as the preheparin clotting times. The upper three require the determination of only one clotting time after drug administration; the lower three require at least two.

We compared the six metameters of heparin effect and two dose metameters -- total and relative dose -- by computing regression coefficients for the *in vivo* dose-response curves for the 19 normal subjects. All 12 curves had positive slopes that were statistically different from zero and values of F demonstrating that the curves were linear.

Table III shows the statistical indices of precision and information yield for each metameter of dose and effect, calculated from the regression lines relating dose and effect. The greatest precision -- that is, the greatest ability to discriminate among doses by measuring their effects -- is indicated by the lowest values, and occurred when the 2-hr WBPTT was the measure of effect of relative dose, in the first row. This dose-and-effect combination also provided the greatest relative information yield, indicated by the largest values. For all six metameters of effect, greater precision and information may be expected when dose is calculated in terms of body weight, rather than when expressed only as total dose. (The relative precisions, information yields, and economies given in Table I are calculated to the base 1 for the WBPTT as the standard, and are not given in absolute values of these measures, as they are in Table III.)

Information and precision were greater for all three metameters of effect that take into account the 2-hr WBPTT. Greater economy would also be expected for these metameters of effect, inasmuch as only the 2-hr clotting time, with or without the baseline clotting time, is necessary, whereas for the other three metameters of effect, at least two clotting times after heparin injection are required, regardless of the additional use of the baseline WBPTT.

The data show, then, that the 2-hr postheparin WBPTT is the most precise and economical measure of the effect produced by a single dose of heparin. Its predictive value is not improved by taking the baseline clotting time into account. And it is clear that greater predictability and smaller variability of effect occur when dose is calculated in terms of body weight, rather than as absolute dose.

For drugs administered intravenously, it has been shown that, if the interval between doses of any drug is chosen to approximate the half-life, then the initial dose must be approximately twice the maintenance dose to maintain a minimal amount of drug in the body for effective therapy. In fact, for most drugs requiring repeated doses, empirical adjustment of dose has led to therapeutic success at intervals approximating the half-lives of such drugs (14).

Figure 3 shows the effect of the dose ratio and the dose interval on the maintenance of a therapeutically effective concentration of heparin in the body. The minimal effective dose range was computed as the concentration of drug in the body necessary to increase the WBPTT by 100%; this is an arbitrary approximation but one that is consistent with the smallest heparin effect sought in usual clinical practice, one that will demonstrate the factors that must be considered in choosing a dosage schedule. Curve A represents the usual clinical dose schedule. It demonstrates that the administration of a fixed dose at 4-hr, or 6-hr, intervals cannot be shown to maintain anticoagulation to a degree meeting commonly accepted criteria for more than about half the interval between doses. Curve B represents the situation that occurs when equal doses are administered at equal intervals approximating the half-life. As predicted for

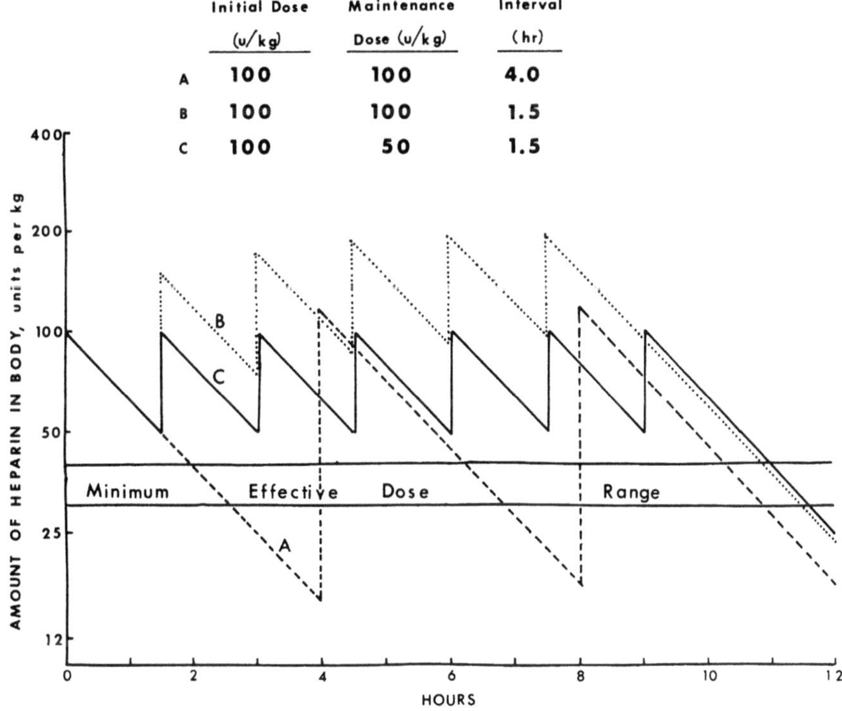

Fig. 3. Theoretical curves showing the effect of the dose ratio (initial dose: maintenance dose) and dose interval on the maintenance of heparin effect when the half-life is 1.5 hr.

most drugs when dose ratios of less than the experimentally determined ones are used (11), this schedule would provide adequate continuous anticoagulation, but to a perhaps hazardous extent, in that drug accumulation would be expected. Curve C represents the situation in which the experimentally determined dose ratio is used - namely, 2:1. This schedule not only provides continuous adequate anticoagulation; it also permits a more rapid return to normal clotting status when the drug is withdrawn than does the schedule represented by curve B. The relations shown in the figure strongly suggest that heparin should be given by continuous intravenous infusion in order to provide clinically useful antithrombotic anticoagulation, while permitting a reasonable and practical form of drug administration on the hospital ward.

In summary, the data in the aggregate permit us to take into account, in assessing a patient's response to heparin, the drug's rate of disappearance from his blood, inasmuch as sensitivity to heparin is not related to its half-life. And we can now ignore the patient's preheparin clotting status, which is related to both sensitivity and reactivity to the drug. However, it is still painfully clear that we cannot yet make any firm rules for determining the "correct" dose of heparin for any given patient.

ACKNOWLEDGEMENTS

This work was supported by a General Research Support grant from the National Institutes of Health, by a grant-in-aid from the Northeast Chapter of the Massachusetts Heart Association, and by a grant from the National Heart and Lung Institute.

REFERENCES

1. BLAKELEY, J.A., Canad. Med. Assoc. J., 99 (1968) 1072.
2. ESTES, J.W., Ann. N. Y. Acad. Sci., 179 (1971) 187.
3. ESTES, J.W., JAMA, 214 (1970) 1122.
4. ESTES, J.W., JAMA, 212 (1970) 1492.
5. ESTES, J.W., J. Clin. Path., 25 (1972) 45.
6. ESTES, J.W., New Engl. J. Med., 279 (1968) 887.
7. ESTES, J.W., PELIKAN, E.W. and KRUGER-THIEMER, E., Clin. Pharmacol. & Therap., 10 (1969) 329.
8. HOCH, H. and CHANUTIN, A., J. Biol. Chem., 197 (1952) 503.
9. JICK, H., SLONE, D., BORDA, I.T. and SHAPIRO, S., New Engl. J. Med., 279 (1968) 284.
10. KRUGER-THIEMER, E., J. Theoret. Biol., 13 (1966) 212.

11. KRUGER-THIEMER, E., in Physico-Chemical Aspects of Drug Actions, Pergamon Press, New York, 1968, p. 63.
12. McGOVERN, J.J., BUNKER, J.P., GOLDSTEIN, R. and ESTES, J.W., JAMA, 175 (1961) 1011.
13. RODMAN, N.F., BARROW, E.M. and GRAHAM, J.B., Amer. J. Clin. Path., 29 (1958) 525.
14. SPECTOR, I. and CORN, M., JAMA, 201 (1967) 157.
15. VIEWEG, W.V.R., PISCATELLI, R.L., HOUSER, J.J. and PROULX, R.A., JAMA, 213 (1970) 1303.

DISCUSSION OF ESTES PAPER

RATNOFF

This is all very pretty and I like it, and of course fits in with long-time knowledge provided by Conley and others that measures involving whole blood are more accurate representations of heparin effects than of plasma because, I suppose, of what now would be called platelet factor 4 activity. The problem I see at the end of your talk is that term "minimal effective therapeutic dose". Could you tell me how you know what the minimal effective therapeutic dose is? How do you know that you get any better results clinically, since that is what therapy means, if you have your patient anticoagulated at a steady pace or at a zigzag pace, reaching at the nadir a level above an arbitrary point, or going up above and below an arbitrary point, repeatedly. Are there any data with which you could make that statement?

ESTES

The lines describing the "minimum effective dose range" in Figure 3 were estimated from the data as the dose required to increase the pre-heparin clotting time by a factor of 2.5. It's admittedly somewhat artificial, but it helps to demonstrate some of the problems we must face in ascertaining rules and principles for determining appropriate dose schedules for this potent anticoagulant. A major problem in interpreting both the experimental data and these theoretical curves is that we still have no information about heparin's binding to plasma proteins in terms of dissociation constants, so that we cannot dissect the relationships among clotting times and heparin's distribution in the body. Until we can do so, we are hard pressed to evaluate even the most accurately measured concentrations of the drug in the blood, much less to use those evaluations in adjusting doses.

WRIGHT

I was interested in Dr. Ratnoff's problem and I'd like to comment in terms of the last sentence of the speaker. He stated that it would be very much better to use heparin by continuous intervenous drip. I want to point out that this is the way heparin was used when it was first administered in 1938 and thereafter in the United States. We tried to enforce this kind of a program. But a problem almost always arose as follows. The house staff could manage this or the nurses could manage it for a day or two for a single patient but if we had a number of patients requiring I.V. heparin therapy, the system frequently broke down because of staff fatigue. They didn't follow the clotting time closely enough. Someone would forget to do the test, or go off-duty, and his successor wouldn't follow through. The result was that suddenly it would be realized that the patient had a clotting time of 140 minutes and then it was usually a matter of who had the hemorrhage first, the house staff or the patient. This led to the more widespread adoption of the intermittent subcutaneous technique which was then in use in Sweden. They found as we did that the therapeutic results were essentially the same. This method is safer for use especially in hospitals where there is no special anticoagulant team. As far example, community hospitals generally - it may be less precise theoretically but it is practical clinically.

SECTION 2

FUNCTIONAL ASPECTS

ANTITHROMBIN III: A BACKWARD GLANCE O'ER TRAVEL'D ROADS

Walter H. SEEGERS

Department of Physiology, Thrombosis Specialized Center of Research, Wayne State University, School of Medicine, Detroit, Michigan (USA)

Having been given the rare privilege of reviewing my own work on antithrombin, I will try here to be informative both about antithrombin and about how I found my scientific endeavor to function. Except for "antithrombin III," I borrowed the title from Walt Whitman, who used it for reflections on his great work, Leaves of Grass. One of his critics thought that poetry would cease to be read in 50 years, owing to the tendency toward science and its all-devouring force. Whitman rightfully anticipated the opposite. There was another problem, which he expressed in the following phrase: "Public criticism on the book and myself as author of it yet shows mark'd anger and contempt more than anything else." Today, any literary scholar thinks of Leaves of Grass when Whitman is mentioned, and he would be embarrassed if he made the mistake of saying that Ralph Waldo Emerson was the author. Similar associations of author and his work are not common in science.

I ask myself why an author of scientific papers becomes dissociated from his contributions as they are incorporated into world culture. Technical work appears fragmented in short articles and is read for information, rather than pleasure or entertainment. I have read each one of my papers on antithrombin several times, and with substantial satisfaction. I doubt that anyone else has even done it once. At the outset, it would be difficult because of the distribution among many different journals. But there are other reasons for separation of an author from his contributions. In terms of the whole field, the volume of papers is tremendous. Investigators must contend with a competitive spirit that tends to promote repression. To a great extent, scientific literature consists of papers that scientists write, but do not read, in the classical sense of "reading". Apparently, even the writer of a review cribs references from another review without reading the original and passes on errors, as well as accurate information. The language is sufficiently specialized to be regarded as foreign for all those outside the specialty. By being forced to comply with the format of a journal, an author is restricted and can hardly develop his own individual style. There is resistance to first-person or autobiographical reporting, and each reader of this paper can judge his own propensities to mobilize that feeling. To understand oneself in that regard is another matter. In my country and some others, political orientation honors, respects, and defends individual primacy. Each person is unique. World science orientation is toward depersonalization, collective consensus, and disclaimer of personal authority. In a free country, the accused can face his opposition. In the scientific review system, the reviewer is a disguised authoritarian. This orientation is not biologically adequate and contributes to repression.

Much of the information in my original papers on antithrombin has been utilized, although only a few know the source of the information. Now I understand the remark that Leandro Tocantins made when we were on the river bank in Wiesbaden: "You are planting fruit trees, and others will harvest the fruit without thinking about the plantings." With success, a scientist gets lost in the phrase, "it is well known." Another significant displacement occurs when an observation is referred to many years later by the citation of some other author who eventually said the same or made a modification. This brand of careless scholarship is also associated with the most prominent authors. They are sometimes so consistently inaccurate that I suppose they use their old reference cards without any further reflection. With respect to my own work, I could document innumberable instances of such occurrences.

Thus, the author and his work become separated, and there is repression and discomfort associated with the process. To see all of my living as the wonderful personal living that it is solves my problem of humanizing my scientific endeavor.

Antithrombin Experiments

A prerequisite for the study of antithrombin was to have thrombin and prothrombin in concentrated and stable form. I began to devise methods for that purpose in 1937 at the State University of Iowa (36, 22). In that laboratory, there were several advantages contributed by H. P. Smith, Emory D. Warner, and Kenneth M. Brinkhous. They were familiar with the literature and knew many of the personalities who created it. I could learn from them without spending time in a library. They had just introduced methods for the quantitative determination of prothrombin and thrombin, and quantitative measurements were essential for the research work. I helped to perfect these methods (34, 42).

In the first paper, we reported on the inhibition of prothrombin activation by heparin. The conversion of our purified prothrombin complex preparation to thrombin was not blocked by heparin unless serum was also added (1). It seemed possible that the essential substance in serum might be antithrombin, but the new factor remained unidentified. (An attempt to give credit for this contribution to another laboratory is made in a recently published treatise on blood coagulation.) We next studied the inactivation of purified thrombin preparations by plasma. The latter was limited in its capacity to neutralize thrombin, and this capacity was not altered by heparin (39). Only the rate of inactivation was found to be accelerated. On this latter point, I soon got contrary results (8); as a consequence, I sometimes unfortunately was not explicit at one place or another in subsequent years. I repeated the work especially for this paper and encountered problems. In the majority of the conditions, heparin did not accelerate the rate of thrombin inactivation by antithrombin. Apparently, a better understanding of the underlying mechanisms is needed through a new experimental approach.

Up to this point, two effects were certain: heparin and a cofactor blocked thrombin formation, and the capacity of plasma to destroy thrombin was limited. Although it was implicit in the data of the first papers, a few years passed before I was able to observe that thrombin and antithrombin neutralize each other (38). The term "mutual depletion system" was used for the first time only much later (3).

I wanted to produce good quantitative data for the purpose of relating the logarithm of thrombin units destroyed to the logarithm of thrombin units remaining (35). From the linear relation, it was clear that the amount of thrombin destroyed increased when the initial thrombin concentration was increased. This unusual relation (Fig. 1) presented difficulties for the quantitative work, but an equation was given to fit the data (35).

While still feeling satisfaction about the quantitative relation described, I found that the regression-line constants had been modified in the experiments by adsorption of thrombin on fibrin. I was quite unhappy about that observation, because I thought all loss of thrombin activity in the experiments was due to antithrombin, and here I was mistaken at the beginning of my career. With a feeling of wanting to set the record (Fig. 2 and 3) in order, I proceeded to add facts - namely, that the thrombin adsorbed on fibrin could be recovered by lysis of the clot with fibrinolysin (14, 31); that albumin interfered with the adsorption, whereas heparin favored it (8); and that heating to 56° for 3 min removed fibrinogen from plasma without altering antithrombin activity (30).

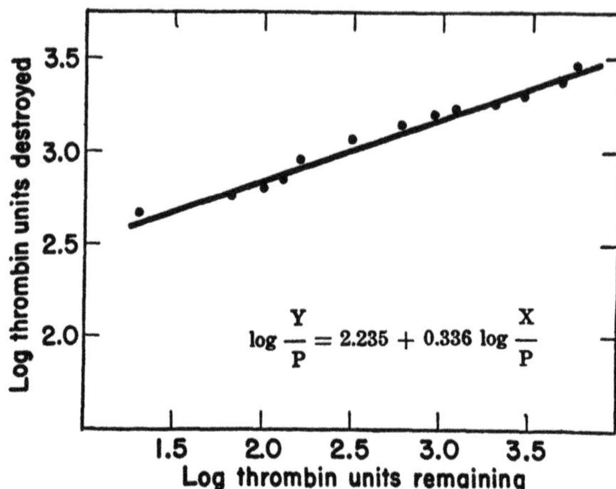

Fig. 1. When 0.8 ml of oxalated bovine plasma was mixed with various amounts of thrombin, a portion of the thrombin activity disappeared, owing to the antithrombin of the plasma. The logarithm of thrombin units destroyed by antithrombin is represented by the Y axis. The logarithm of the thrombin units remaining is represented by the X axis. A given amount of plasma destroys more thrombin in the concentrated solutions than in the dilute solutions. In the equation, P represents milliliters of bovine plasma in the reaction mixture.

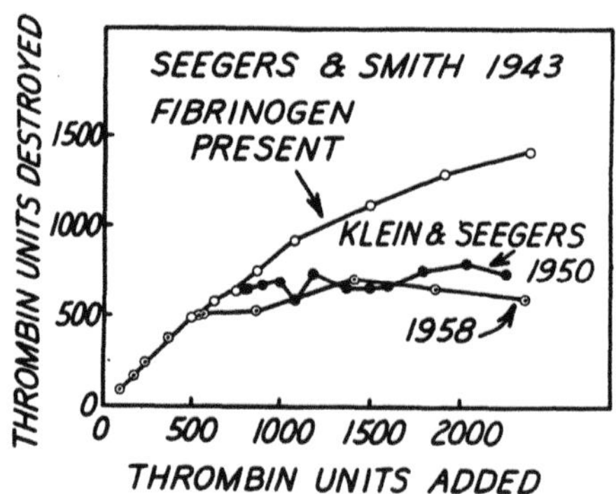

Fig. 2. One milliliter of thrombin solution (concentration increased from one tube to next) was mixed with 1 ml of plasma. At the end of 2 hr, total units of thrombin remaining were determined. Without fibrinogen removed, amount of thrombin destroyed was proportional to thrombin at beginning, but with fibrinogen removed, this relation held only up to a point, which is antithrombin capacity of plasma. Beyond antithrombin capacity, original thrombin retained its activity.

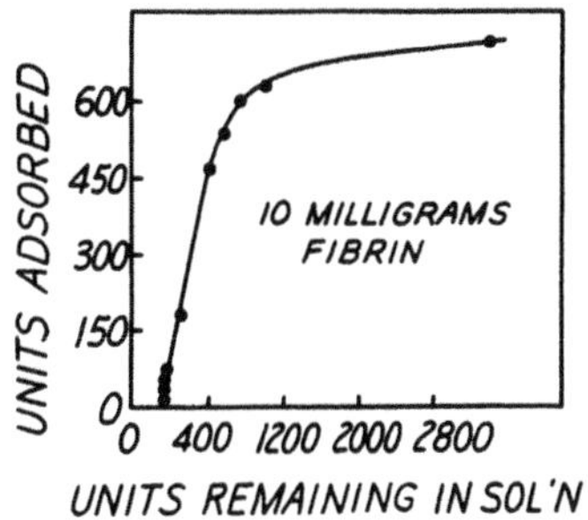

Fig. 3. Adsorption of thrombin on 10 mg of fibrin. Solution volume, 2 ml; temperature, 28°. Clot was removed mechanically. Amount adsorbed depended on initial concentration of thrombin solution.

It was important to know how much thrombin is required to clot heparin-plasma mixtures. For this purpose, I obtained sheep heparin from Louis B. Jaques in Toronto. The clotting time was prolonged in proportion to the heparin concentration (Fig. 4); the relation was linear (13). This prolonged clotting time was evidently due to interference with the thrombin-fibrinogen reaction, and such interference was clearly demonstrated (8). I emphasized that the facts could be the basis for quantitative measurement of heparin activity.

Nature of Antithrombin III

Antithrombin III proved to be relatively stable in plasma on storage in a refrigerator for up to 4 weeks (30). This was true for human blood, plasma, and serum, as well as for bovine blood, plasma, and serum. A quantitative method for analysis was devised. In the procedure, 1,100 units of thrombin were mixed with 1 ml of defibrinated plasma or serum. The percent of thrombin destroyed in 2 hr was taken as a quantitative measure of antithrombin III activity. Under the specified conditions, about 700 units of thrombin were neutralized, or about twice the thrombin potential of 1 ml of plasma. There was less in serum than in plasma, because all thrombin produced during the clotting process combined with a fraction of antithrombin III. In serum from patients taking Dicumarol, the activity seemed to increase, because very little intrinsic thrombin formed to combine with antithrombin III during clotting. Dicumarol, as used in the clinics, did not change antithrombin III concentration in plasma (7). No antithrombin III was adsorbed from plasma by kaolin, tricalcium phosphate, Fuller's earth, magnesium oxide, activated charcoal, asbestos, Lloyd's reagent, or barium sulfate. In accordance with what Wöhlisch had said, ether extraction of plasma or serum or the

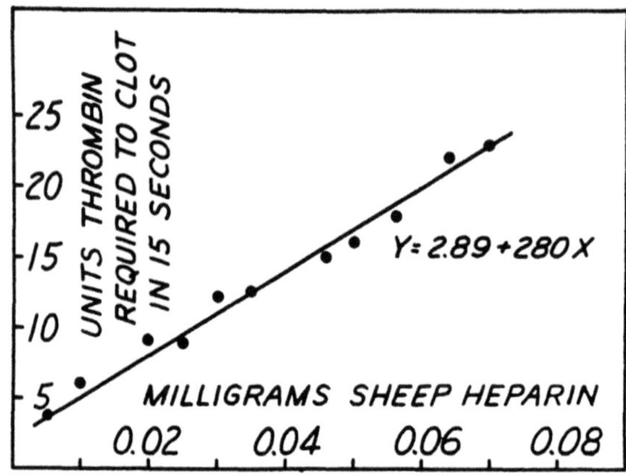

Fig. 4. To a tube containing 0.8 ml of oxalated plasma were added 0.1 ml of saline (0.9% salt solution) and 0.1 ml of thrombin solution, dissolved in saline, and containing a total of 2.9 units of thrombin. Clotting occurred in 15 sec. To other tubes, 0.8 ml of oxalated plasma was added, with 0.1 ml of heparin solution of progressively increasing concentration. At each heparin concentration, 0.1 ml of thrombin solution of variable concentration was added. By repeated trial, it was possible to determine the number of units of thrombin needed to obtain clotting in 15 sec. This was done at a room temperature of 26-28°. Amounts of thrombin in the final plotting mixture are plotted against corresponding amounts of heparin.

extraction of dried plasma or serum destroyed antithrombin III activity.

The new quantitative method was applied in a survey of 306 hospital patients taken at random (20, 30). There were a few with very low antithrombin III concentration (Fig. 5). I realized that it would be wise to study these cases, but I could not do the work, owing to my heavy academic assignments. Most likely, one or more of these persons had thrombotic difficulties. It seems certain that an opportunity to describe pathology due to antithrombin III deficiency was missed.

Recognition of Associated Phenomena

During the coagulation of blood, inactivation of thrombin is accelerated (15-17). This was demonstrated as follows: Plasma was ether-extracted to take away antithrombin III activity. Such plasma did not neutralize purified thrombin. However, if thromboplastin

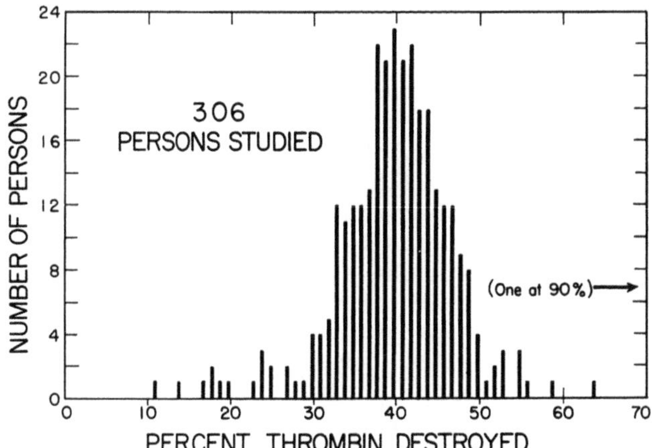

Fig. 5. Distribution diagram shows percentage of thrombin neutralized in 2 hr by 1 ml of the patient's defibrinated plasma when mixed with 1,100 U of thrombin as substrate.

and calcium ions were added, the new thrombin that formed was inactivated rapidly. This was called the "antithrombin IV effect" (25). Serum did not have this property unless it was first adsorbed with barium carbonate. To differentiate other antithrombin effects, Roman numerals were used (5). "Antithrombin I" referred to adsorption of thrombin on fibrin. "Antithrombin II" referred to heparin cofactor required for interference with the thrombin-fibrinogen reaction and blocking thrombin formation. "Antithrombin III" applied to the capacity to inactivate thrombin. "Antithrombin IV" referred to accelerated thrombin inactivation while thrombin was being formed. Although only one anticoagulant protein is involved, three distinct effects were noted: blocking the activation of prothrombin, interference with the thrombin-fibrinogen reaction, and inactivation of thrombin. The last two effects depend on heparin, and the first can be accelerated by heparin.

The conditions for observing the antithrombin IV effect are complicated, and some day it should be possible to clarify details with the use of purified components. For me, it is an important effect requiring much more study. I do not agree with one observer, who remarked: "The work has not been substantiated." On the contrary, I feel that it would be rewarding to investigate further along these lines. The critic probably fabricated the negation without doing even a few simple experiments.

Toward Finding Mechanisms of Antithrombin III Function

In efforts to purify antithrombin III, plasma was defibrinated by heating to 56° for 3 min, and adsorbed with barium carbonate to remove the prothrombin complex. The antithrombin was then adsorbed on aluminum hydroxide, from which it was eluted with phosphate buffer (12). This work was begun when Frank C. Monkhouse served as visiting professor at Wayne State University. On his return to the University of Toronto, he continued with the purification work. Several others have included these steps in their methods for purification and have added the use of modern resins and techniques for obtaining what is probably pure antithrombin III. Advances in technology have also made it possible to measure the purity. In our purification work, antithrombin III and heparin cofactor (antithrombin II) appeared in the same fractions and seemed to have the same properties. Therefore, I correctly assumed that each effect was due to one and the same substance, but I made appropriate allowance for the remote possibility of evidence to the contrary. Seitz filtration of plasma removed neither heparin cofactor nor antithrombin III (6).

The desire to have an understanding of the underlying mechanisms was constantly felt. At the first Macy Conference, I mentioned that thrombin and antithrombin III neutralized each other (38). I was puzzled by the fact that no one seemed impressed by this. In retrospect, there was no reason why anyone should, inasmuch as one piques research persons primarily when they see themselves as a part of the action. I was way ahead. I kept thinking that thrombin, being an enzyme, had to function as an enzyme in the partnership. Ricardo Landabru then performed a significant experiment: An antithrombin III concentrate was combined with purified thrombin, so that a high proportion of thrombin was in the mixture. We were able to demonstrate regeneration of thrombin activity (40). Antithrombin III activity did not regenerate (Fig. 6). It took a long time for the thrombin activity to regenerate. It thus seemed evident that antithrombin III functioned as a special substrate for thrombin. We presented the following summary (40):

> In the interaction of thrombin (Th) and antithrombin (A), the activity of both is temporarily neutralized. Later, thrombin activity regenerates and antithrombin is altered (P). Thrombin activates antithrombin. The experiments are interpreted on the basis of the Michaelis-Menten concept:

$$\mathrm{Th} + \mathrm{A} \underset{k_2}{\overset{k_1}{\rightleftarrows}} \mathrm{ThA} \overset{k_3}{\rightleftarrows} \mathrm{P} + \mathrm{Th_o}$$

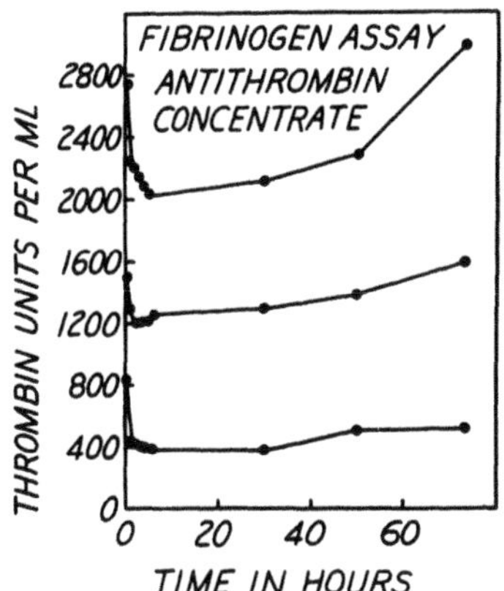

Fig. 6. Three different solutions of purified thrombin were mixed with an equal volume of antithrombin concentrate. Final concentration in units per milliliter is indicated at zero time. At room temperature, clotting power diminished and later was recovered completely, except for one combination of reagents, with which recovery was partial. Ratio of thrombin to antithrombin must be within a specific range to get recovery of thrombin.

> The ThA complex is quite stable and tends to form rapidly when the concentration of A is relatively high. Dissociation and formation of inactive antithrombin (P) is favored with relatively high concentrations of thrombin.

It remains to be demonstrated that antithrombin III fragments are produced. I imagine that, when the amino acid sequence of antithrombin III is known, there will be special segments in which a peptide bond(s) is split by thrombin. On the basis of what is known about the specificity of thrombin, it should be an arginyl bond followed to the right at position 6, 7, or 8 by glutamic acid (Fig. 7), aspartic acid, or the respective amides (41). I suppose that acetylated thrombin would not regenerate its own activity, as described for thrombin, because the regeneration presupposes a proteolytic function for thrombin.

The next advance depended on a fundamental observation about thrombin. We found that the esterase function could be separated from the proteolytic function. This was done by autolysis (9) and by acetylation (10). It is interesting that a comparable separation was made for trypsin 12 years later.

The question was whether antithrombin III would inactivate acetylated thrombin. When the experiments were done, esterase

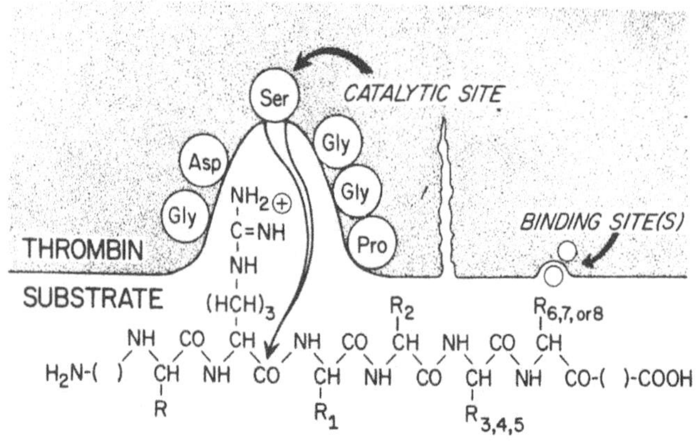

Fig. 7. Illustration in which an arginyl bond of a protein is broken by thrombin when secondary binding site(s) are 6, 7, or 8 amino acid residues to the right of arginine. Serine is at active center of thrombin.

activity was indeed decreased, but it was first increased under special conditions (Fig. 8). This esterase enhancement also occurred with classical thrombin (27). It is amazing that a single molecule can modify its own function so extensively in terms of its environment.

A derivative of prothrombin also modifies the function of thrombin. This was found recently when the 0 fragment of prothrombin was isolated. It is one of the fragments that can arise from prothrombin when thrombin activity generates. In the presence of the 0 fragment, the proteolytic function of thrombin was strongly depressed (37). (Fig. 9). This latter effect could partially or perhaps even fully represent the antithrombin IV effect. It is an instance of inhibition by means of product formation.

Reflection at Halfway Station

On this journey o'er the road travel'd slowly the past 35 years, I notice several facts. There were no road signs or rules, and the destination was really not known. I kept going without interruption. Eager associates contributed important ideas and technical help. Wars and political oppression did not stop the work. Except when funds were requested, no one asked about the cost of the work. No one asked whether other demands on my energy were made or whether the work was done in a palace or in crowded quarters. In this connection,

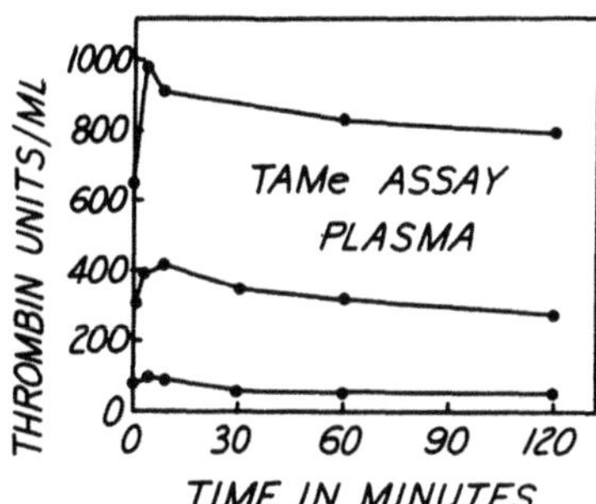

Fig. 8. One volume of defibrinated plasma was mixed with an equal volume of purified thrombin. Solutions stood at room temperature, and activity was measured by TAMe assay. Initial tendency was for activity to increase this phenomenon is called esterase enhancement.

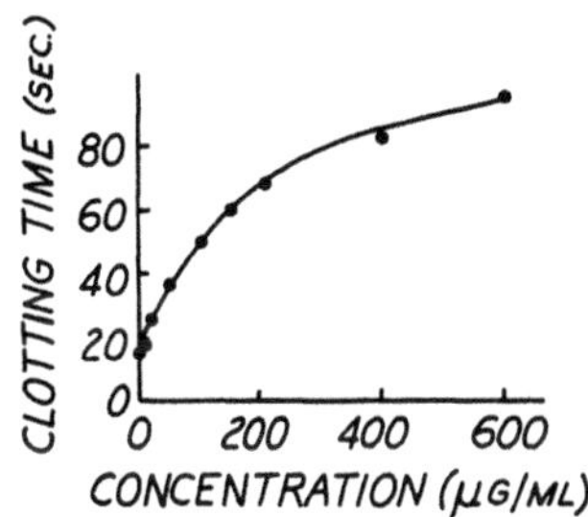

Fig. 9. Fibrin formation by thrombin inhibited by purified 0 fragment of prothrombin. Fibrinogen concentration, 10 mg/ml.

a Canadian once made me feel uncomfortable at a dinner by saying that I would have had a better press if I had done my work at a big-name place. Unlike the Canadian, former President Harry Truman would have said Ha-vud. I grew with my university, which is now the sixth largest in the United States, and I have an endowed chair in the second largest medical school in the United States. Unexpected findings proved to be the most important. How can one say in a grant application that one intends to find that which one does not know he is going to find? I believe the only honest answer is that one cannot do it. If the result of the proposed work can be forecast, why do the work? I can, however, propose to build an instrument or to sequence a protein and the like. Such enterprises, in themselves, are not likely to uncover the unexpected.

Quantitative measurements were essential. Likewise, the latest instruments were needed; and each time there was purified material, information was forthcoming. Protein purification is arduous, and the meaning of that statement is probably comprehended only by someone who has tried purification himself. The recent revolution in purification technology, with the invention and manufacture of instruments to measure product components, has changed that aspect very appreciably. It was difficult to prepare enough thrombin for the experiments. During World War II, I set up an industrial operation for the manufacture of thrombin (Thrombin Topical, Parke, Davis and Co.), but I did not use this resource, because I wanted the highest-quality thrombin for theoretical work. There were substantial differences between one preparation and another.

The blood coagulation system has feedback components. It is a cybernetic system (21). The procoagulants are balanced by anticoagulant forces. During my time, infatuation with procoagulants was a dominant characteristic of the popular perspective. I take satisfaction in my work with an important inhibitor. In the same spirit of enthusiasm, I devoted energy to the purification of thrombin, autoprothrombin C, and their precursors. Whoever had the purified enzymes had a monopoly of a peculiar kind for the study of antithrombin III.

Antithrombin III and Autoprothrombin C

The next task undertaken was the isolation of autoprothrombin C (factor Xa), a "new" clotting factor. Our laboratory was the first to have sufficient quantities for chemical studies (23). But there was not enough for all we wished to do. We also had thrombin and the knowledge that thrombin would combine with antithrombin III. The working hypothesis was that antithrombin III inactivated autoprothrombin C. Evidence to the contrary would have been disliked intensely. We deployed to go directly to the crucial question, with minimal use of autoprothrombin C.

Autoprothrombin C was placed in the role of a proteolytic enzyme in the following system (28):

$$\text{prothrombin} \xrightarrow[]{\substack{\text{Ca}^{+2} \\ \text{phospholipid} \\ \text{Ac-globulin} \\ \text{autoprothrombin C}}} \text{thrombin}$$

At that time, this equation still needed further verification. Only one person, Haskel Milstone at Yale University (perhaps also Lenggenhager in Switzerland), had documented its use for thrombin formation. One main point was stated as quoted below (Fig. 10) (28):

> Our experiments involved quantitative work in which plasma was first mixed with purified thrombin. In such a reaction, antithrombin III and thrombin are neutralized simultaneously. When autoprothrombin C was then added to the thrombin-treated plasma, the amount of autoprothrombin C destroyed by the plasma was much less than for an equivalent amount of plasma that had not inactivated thrombin. We think this was due to the fact that antithrombin III, which neutralized thrombin by combining with it in an enzyme-substrate complex, could then not also neutralize autoprothrombin C.

A similar experiment with serum as the source of antithrombin III was performed (18) (Fig. 11).

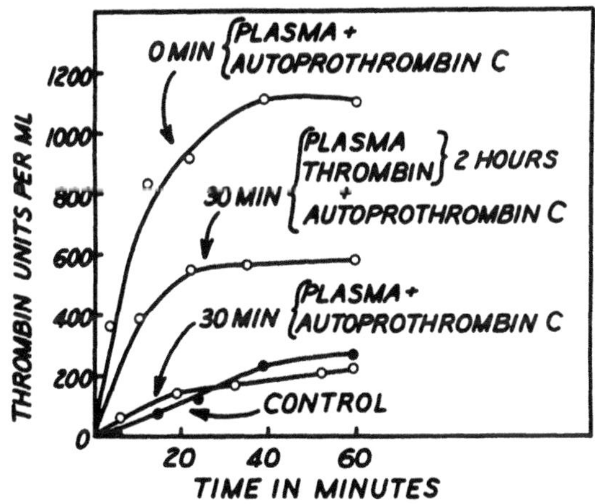

Fig. 10. Thrombin generation from purified prothrombin complex in presence of Ac-globulin, platelet extract, and calcium ions (control). No autoprothrombin C added. Top, control materials with autoprothrombin C and defibrinated plasma. Bottom, control materials with autoprothrombin C after standing with plasma for 30 min. Antithrombin inactivated autoprothrombin C. Middle, materials with autoprothrombin C after standing with thrombin-treated plasma for 30 min. Thrombin bound to antithrombin, and remaining antithrombin, was not sufficient to inactivate all the autoprothrombin C.

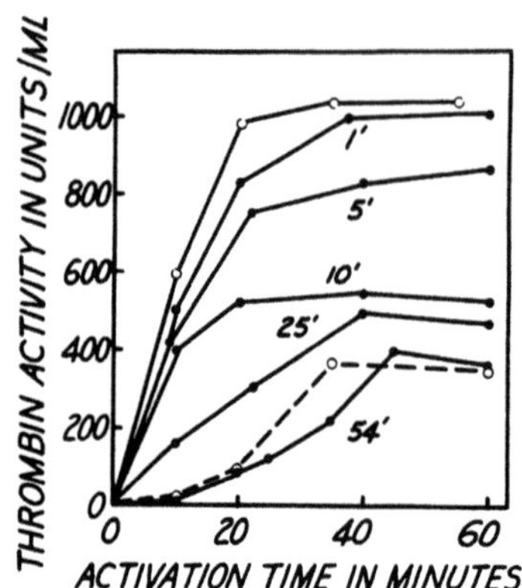

Fig. 11. Reaction mixture consisted of purified prothrombin complex, calcium ions, and platelet concentrate. An aliquot of autoprothrombin C that was free of thrombin was mixed with an equal volume of barium carbonate-adsorbed bovine serum and allowed to stand at 28°. After 1, 5, 10, 25, and 54 min, the procoagulant power was tested in the above reaction mixture. Curve (o---o) represents serum alone in the reaction mixture; curve (o——o) with complete conversion of prothrombin to thrombin represents autoprothrombin C alone in the reaction mixture. Experiment features three main points: autoprothrombin C functions with platelets, autoprothrombin C activity is inactivated by serum, and prothrombin is only partially converted to thrombin with limited amounts of autoprothrombin C.

In the first experiments, it was assumed that antithrombin III in plasma or serum was the inactivating agent. It was natural to ask whether purified antithrombin III would inactivate autoprothrombin C. For that purpose, purified antithrombin III was needed, and I asked Frank C. Monkhouse at the University of Toronto for some material. Without asking what we wanted it for, he supplied it promptly. It was then possible to establish antithrombin III as the inactivating agent of autoprothrombin C (24). I regarded this as an important advance in our knowledge related to antithrombin III. It is now 10 years later, and a popular book on blood coagulation has appeared without even taking notice of the work. We thus see another way in which a person gets lost in scientific work - even the work gets lost or neglected.

It remains to be seen how heparin and antithrombin prevent the activation of autoprothrombin III (factor X). Autoprothrombin C promotes its own formation by autocatalysis. Thus, it is likely that the neutralization of the enzyme prevents any more from forming

(26). In a similar way, soybean trypsin inhibitor prevents coagulation by neutralizing autoprothrombin C.

The autoprothrombin C used in the experiments had been derived from the prothrombin complex with the use of thromboplastin. The enzyme was also prepared by activation with autoprothrombin C. This produced another form of the enzyme, which also was inactivated by antithrombin III (33). There was, however, a difference between the two enzymes: 1 ml of plasma neutralized more autoprothrombin C (w/w basis) when formed with thromboplastin than when the enzyme was produced by using autoprothrombin C for the activation (19).

Other studies, such as the purification of autoprothrombin III now got priority, although I hoped for an opportunity to do more with equations to describe the function of antithrombin III. My need was for someone proficient in mathematics. When this need was fulfilled, more experiments were done with antithrombin III. Several equations were applied to describe fundamental relations. A paper was written and accepted for publication, and my associate on the project returned to his own country. When the galley proof arrived as usual, my graduate student, Frederick A. Dombrose, whose name was not on the paper, read it and was disappointed. He wished that the data might have been considered in a "better way" and pressed hard during the dicussion. He left, but returned within the hour to renew the discussion because he sensed my predicament. I took the position that a person should not tear anything down without willingness and ability to add something better. I asked whether he was ready to go to work on this with me and postpone the purification of Ac-globulin (2, 4). He declared himself ready. The galley proof was then scrapped, and I paid the publisher of _Thrombosis et Diathesis Haemorrhagica_ the cost of printing the first manuscript, which was substantial. I then asked James A. Sedensky of my Department of Physiology to help because of his special qualifications for the work before us. Furthermore, under the circumstances, two mathematics experts constituted a fail-safe combination.

We worked diligently and grew rapidly. After several months, a paper was written and published. Among other features, the paper highlights two considerations (3): a working hypothesis, including equations, for a mutual depletion system was presented to account for the known characteristics of enzyme inhibition by antithrombin III; and we compared the neutralization of thrombin with the neutralization of autoprothrombin C and found that each enzyme responds to antithrombin III according to similar kinetic relations. The paper is informative and concentrated, and hence not easily summarized.

Antithrombin III in 1974

My work with antithrombin III continues. For example, there is autoprothrombin II-A. This protein was discovered and isolated in this laboratory (11, 29). It is a competitive inhibitor of autoprothrombin C and under suitable condit ons induces fibrinolysis. Antithrombin III neutralizes the activity of autoprothrombin II-A.

Prethrombin-E and thrombin-E are esterase enzymes in the prothrombin activation sequence (37). They are neutralized by antithrombin III (Table I). Prethrombin-E is the immediate precursor of thrombin. It has the same amino acid composition as thrombin, but, in contrast with thrombin, the B chain of thrombin is still attached to the A chain (Fig. 12). Although the enzyme has esterase activity, it is not proteolytic. Thrombin-E is also not proteolytic, but has higher esterase activity than the thrombin from which it is derived by autolysis. During autolysis, the B1 chain, with about 77 amino acid residues, is removed from the B chain (32). With it goes an active histidine site. Whether this B1 chain, which has NH_2-terminal isoleucine, is gone, as in the case of thrombin-E, or held to A chain, as in the case of prethrombin-E, the enzyme is not protected from inactivation by antithrombin III. In other words, the active histidine residue is not needed for binding to antithrombin III. The serine active center is sufficient. The B1 chain is doubtless needed for the digestion of antithrombin III, and I predict that prethrombin-E and thrombin-E cannot free themselves from antithrombin III, as occurs with classical thrombin.

TABLE I

Inactivation of Enzymes with Purified Antithrombin, Tosyllysinechloroketone (TLCK) and Diisopropylfluorophosphate (DFP)

Enzyme	Antithrombin	TLCK	DFP
Prethrombin-E	Inhibited	Inhibited	Inhibited
Thrombin	Inhibited	Inhibited[a]	Inhibited
Thrombin-E	Inhibited	Active	Inhibited

[a]Esterase and clotting activity inactivated.

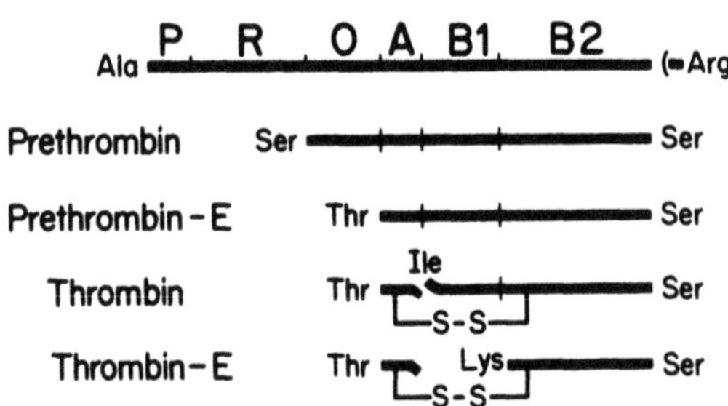

Fig. 12. Purified prothrombin viewed as a single chain after unfolding. In the model, the molecular weight, with carbohydrate included, is proportional to the length assigned to each fragment. Carbohydrate is found mainly in the R fragment and B1 chain. Each bond broken by autoprothrombin C or thrombin is an arginyl. The NH_2-terminal amino acid for R fragment is threonine. Thrombin, as well as autoprothrombin C, can remove PR fragment. From prethrombin to prethrombin-E requires autoprothrombin C. From thrombin to thrombin-E requires thrombin. Prethrombin-E has esterase activity. By breaking its Arg-Ile bond with autoprothrombin C, the B1 portion of the B chain is free and the structure is that of classical thrombin. By autolysis of thrombin, the B1 chain is broken from the B chain, and the active histidine site that it contains can no longer function. Thrombin-E has only esterase activity. There are uncertainties about the COOH-terminal end of prothrombin. In the prothrombin activation sequence, there is one pathway: prothrombin ⟶ prethrombin ⟶ prethrombin-E ⟶ thrombin ⟶ thrombin-E. This involves esterase ⟶ esterase + proteolytic ⟶ esterase functions. Esterase activity is inactivated by antithrombin in prethrombin-E, thrombin, and thrombin-E.

Importance of Antithrombin III

Thrombin and autoprothrombin C are the dominant powerful enzymes in blood coagulation and, as a corollary, antithrombin III is the dominant inhibitor. Blood coagulation can be comprehended as a cybernetic system in which there are three basic reactions:

1. Formation of autoprothrombin C

2. Formation of thrombin

3. Formation of fibrin

If the first and second reactions are canceled out or regulated with antithrombin III, then antithrombin III has an impressive role. Additionally, if heparin is present, other properties of antithrombin III are brought into action, including interference with the thrombin-fibrinogen reaction.

Summary

By devising and applying quantitative methods for the assay of thrombin and autoprothrombin C and by developing techniques for their purification, it was possible to obtain information about the function and properties of antithrombin. The inhibitor is a protein for which the initial purification steps consist of removing fibrinogen from plasma by heating to 56° for 3 min, removing prothrombin complex by adsorption on barium carbonate, adsorbing the antithrombin on aluminum hydroxide, and eluting with phosphate buffer. Antithrombin is limited in its capacity to neutralize thrombin activity, and, under some conditions, the rate of inhibition was accelerated, but equivocal results were involved. Heparin cofactor was found to be essential for retarding the formation of thrombin, and, by inference, it is essential for retarding the formation of autoprothrombin C. Heparin cofactor and antithrombin III are the same.

Thrombin adsorbs on fibrin, and this has been referred to as the "antithrombin I effect." Interference with the thrombin-fibrinogen reaction by mixtures of antithrombin III and heparin is called the "antithrombin II phenomenon." The acceleration of thrombin inactivation at the time thrombin forms is called the "antithrombin IV effect." It was discovered that antithrombin III neutralizes thrombin, as well as autoprothrombin C. The inhibitor and the enzyme form a mutual depletion system.

To assay for antithrombin III, a standard quantity of thrombin (about 1,100 U/ml) was reacted with antithrombin III for 2 hr. The percent thrombin inactivated was then measured. In random samples of human blood, a wide range of antithrombin III concentration was found. The inhibitor is relatively stable in plasma and serum. It is not changed in concentration when Dicumarol therapy is instituted. Ether extraction of plasma reduces antithrombin III activity. Seitz filtration of plasma did not remove activity.

Under special conditions, antithrombin III enhances esterase activity of thrombin. Under special conditions, thrombin regenerates from the thrombin-antithrombin III complex. Antithrombin III neutralizes the activity of prethrombin-E and thrombin-E; consequently, an active histidine center found in the B1 chain of thrombin is not essential for the binding of antithrombin. Autoprothrombin II-A activity was neutralized by antithrombin III. Autoprothrombin C was found to be neutralized by antithrombin III; the amounts required varied with the molecular forms of autoprothrombin C. Thrombin and autoprothrombin C apparently occupy the same binding sites on antithrombin III. An equation was developed to account for all the known characteristics of antithrombin III functions. The kinetic aspects of thrombin neutralization were found to correspond exactly with those of autoprothrombin C. Antithrombin III is a high-capacity inhibitor of the two most powerful enzymes in blood coagulation.

This paper has described one way in which scientific endeavor functions. I have recorded a personal account and analyzed progression in the creative art known as laboratory experimentation. Situations conducive to repression were pointed out. I presented evidence for the opinion that the political orientation of my United States, as well as some other countries, is "more" enlightened than the world scientific perspective.

Journey's Meaning

Not I, not any one else can travel that road for you,
You must travel it for yourself.

It is not far, it is within reach,
Perhaps you have been on it since you were born and did
 not know,
Perhaps it is everywhere on water and on land.

-Walt Whitman

ACKNOWLEDGEMENTS

Preparation of this report was supported by research grant HL 14142-04 from the National Heart and Lung Institute, National Institutes of Health, U.S. Public Health Service. What I have written about I learned on the basis of experiments, and, because this is an unusual report, references are exclusively to papers from my laboratory.

REFERENCES

1. BRINKHOUS, K.M., SMITH, H.P., WARNER, E.D. and SEEGERS, W.H., Amer. J. Physiol., 125 (1939) 683.

2. DOMBROSE, F.A. and W.H. SEEGERS, Thromb. Diath. Haemorrh. Suppl. (1973) in press.
3. DOMBROSE, F.A., SEEGERS, W.H. and SEDENSKY, J.A., Thromb. Diath. Haemorrh., 26 (1971) 103.
4. DOMBROSE, F.A., YASUI, T., ROUBAL, Z., ROUBAL, A. and SEEGERS, W.H., Prep. Biochem., 2 (1972) 381.
5. FELL, C., IVANOVIC, N., JOHNSON, S.A. and SEEGERS, W.H., Proc. Soc. Exp. Biol. Med., 85 (1954) 199.
6. FELL, C., JOHNSON, J.F. and SEEGERS, W.H., Amer. J. Clin. Path., 24 (1954) 153.
7. IVANOVIC, N., JOHNSON, J.F., OLWIN, J.H. and SEEGERS, W.H., Proc. Soc. Exp. Biol. Med., 85 (1954) 496.
8. KLEIN, P.D. and SEEGERS, W.H., Blood, 5 (1950) 742.
9. LANDABURU, R.H. and SEEGERS, W.H., Proc. Soc. Exp. Biol. Med., 94 (1957) 708.
10. LANDABURU, R.H. and SEEGERS, W.H., Canad. J. Biochem. Physiol., 37 (1959) 1361.
11. MAMMEN, E.F., THOMAS, W.R. and SEEGERS, W.H., Thromb. Diath. Haemorrh., 5 (1960) 218.
12. MONKHOUSE, F.C., FRANCE, E.S. and SEEGERS, W.H., Circ. Res., 3 (1955) 397.
13. SEEGERS, W.H., Proc. Soc. Exp. Biol. Med., 51 (1942) 172.
14. SEEGERS, W.H., J. Phys. Colloid Chem., 51 (1947) 198.
15. SEEGERS, W.H., Arch. Biochem. Biophys., 36 (1952) 484.
16. SEEGERS, W.H., Coagulation of the Blood, The Harvey Lectures, Series XLVII, Academic Press, New York, 1952, p. 180.
17. SEEGERS, W.H., J. Mich. State Med. Soc., 51 (1952) 597.
18. SEEGERS, W.H., Prothrombin, Harvard University Press, Cambridge, (1962).
19. SEEGERS, W.H., Prothrombin in Enzymology, Thrombosis and Hemophilia, Thomas, Springfield, (1967).
20. SEEGERS, W.H., Ann. N. Y. Acad. Sci., 146 (1968) 593.
21. SEEGERS, W.H., Ser. Haemat., (1974) in press.
22. SEEGERS, W.H., BRINKHOUS, K.M., SMITH, H.P. and WARNER, E.D., J. Biol. Chem., 126 (1938) 91.
23. SEEGERS, W.H., COLE, E.R., HARMISON, C.R. and MARCINIAK, E., Canad. J. Biochem. Physiol., 41 (1963) 1047.
24. SEEGERS, W.H., COLE, E.R., HARMISON, C.R. and MONKHOUSE, F.C., Canad. J. Biochem., 42 (1964) 359.
25. SEEGERS, W.H., JOHNSON, J.F. and FELL, C., Amer. J. Physiol., 176 (1954) 97.
26. SEEGERS, W.H. and LANDABURU, R.H., Proc. Soc. Exp. Biol. Med., 95 (1957) 710.
27. SEEGERS, W.H., LANDABURU, R.H. and JOHNSON, J.F., Science, 131 (1960) 726.
28. SEEGERS, W.H. and MARCINIAK, E., Nature, 193 (1962) 1188.
29. SEEGERS, W.H., McCOY, L.E., GROBEN, H.D., SAKURAGAWA, N. and AGRAWAL, B.B.L., Thrombosis Res., 1 (1972) 443.
30. SEEGERS, W.H., MILLER, K.D., ANDREWS, E.B. and MURPHY, R.C., Amer. J. Physiol., 169 (1952) 700.

31. SEEGERS, W.H., NIEFT, M. and LOOMIS, E.C., Science, 101 (1945) 520.
32. SEEGERS, W.H., REUTERBY, J., MURANO, G., McCOY, L.E. and AGRAWAL, B.B.L., Thromb. Diath. Haemorrh. Suppl., 47 (1971) 325.
33. SEEGERS, W.H., SCHRÖER, H. and KAGAMI, M., Canad. J. Biochem., 42 (1964) 1425.
34. SEEGERS, W.H. and SMITH, H.P., Amer. J. Physiol., 137 (1942) 348.
35. SEEGERS, W.H. and SMITH, H.P., Proc. Soc. Exp. Biol. Med., 52 (1943) 159.
36. SEEGERS, W.H., SMITH, H.P., WARNER, E.D. and BRINKHOUS, K.M., J. Biol. Chem., 123 (1938) 751.
37. SEEGERS, W.H., WALZ, D.A., REUTERBY, J. and McCOY, L.E., Thrombosis Res., 1974, in press.
38. SEEGERS, W.H. and WARE, A.G., in Blood Clotting and Allied Problems, (Ed. Flynn J.E.), Josiah Macy, Jr. Foundation, New York, 1948, p. 64.
39. SEEGERS, W.H., WARNER, E.D., BRINKHOUS, K.M. and SMITH, H.P., Science, 96 (1942) 300.
40. SEEGERS, W.H., YOSHINARI, M. and LANDABURU, R.H., Thromb. Diath. Haemorrh., 4 (1960) 293.
41. WALZ, D.A., SEEGERS, W.H., REUTERBY, J. and McCOY, L.E., Thrombosis Res., (1974) in press.
42. WARE, A.G. and SEEGERS, W.H., Amer. J. Clin. Path., 19 (1949) 471.

THE COAGULATION-FIBRINOLYTIC MECHANISM AND THE ACTION OF HEPARIN

Robert D. ROSENBERG

Department of Medicine, Harvard Medical School and
Beth Israel Hospital, Boston, Massachusetts (USA)

During the coagulation of blood, thrombin is produced from its precursor, prothrombin, by the concerted action of factor V and activated factor X (factor Xa) in the presence of a lipid surface and calcium ions. Thrombin consists of a heavy and a light polypeptide chain connected by disulfide crosslinks (18). Once this proteolytic enzyme is generated, it specifically cleaves two pairs of unique arginine-glycine peptide bonds in fibrinogen (5) and initiates the development of a fibrin clot. The clotting activity of thrombin is critically dependent upon a group of amino acid residues that constitute its active center region, and predominant among these is a highly reactive serine on the heavy polypeptide chain (13).

Thrombin gradually loses its enzymatic activity when added to defibrinated plasma or serum. At the beginning of this century, Contejean (7), Morowitz (21), Howell (15), and others recognized this phenomenon and postulated that antithrombin, a specific inhibitor of thrombin, must be present in plasma under normal physiologic conditions. By 1916, McLean (19) had isolated heparin from the liver, as well as the heart, and demonstrated its potent anticoagulant action. Confusion about its inhibitory effects on purified procoagulants was resolved by Brinkhous *et al*. (6), who showed in 1939 that heparin was effective as an anticoagulant only in the presence of a plasma component, which they termed "heparin cofactor."

In the 1950's, the work of Lyttleton (17), Waugh and Fitzgerald (30), and Monkhous *et al*. (20) indicated that plasma antithrombin activity and plasma heparin cofactor activity are intimately related.

These investigators suggested that heparin acts to accelerate, by a factor of 50-100, the rate at which antithrombin neutralizes thrombin. In 1968, this hypothesis was substantiated by Abildgaard (1) who obtained small amounts of a purified human plasma protein that functions in both capacities.

Our own investigations advance the most convincing evidence to date that both plasma antithrombin and plasma heparin cofactor activities reside in the same molecular species. In addition, we present the first molecular analysis of the thrombin-inhibitor interaction. This mechanism has allowed us to predict heretofore unsuspected actions of antithrombin and heparin-activated antithrombin on serine proteases of the coagulation (8) and fibrinolytic (14) systems.

Human Antithrombin-Heparin Cofactor

We have purified human antithrombin by aluminum hydroxide adsorption-elution, Sephadex G-200 filtration, DEAE-Sephadex fractionation, DEAE-cellulose chromatography, and isoelectric focusing in sucrose density gradients. The final yield of this procedure averages 12%, and the final product is homogeneous by disc gel electrophoresis, sodium dodecyl sulfate gel electrophoresis, and immunoelectrophoresis (Fig. 1) (9, 26).

This purified protein progressively inhibits thrombin in the absence of heparin (antithrombin activity), but instantaneously neutralizes the enzyme when the acidic mucopolysaccharide is present (heparin cofactor activity) (Fig. 2).

Monospecific antisera directed against the purified protein were produced in rabbits. The γ-globulin fraction of these antisera, or 0.15 M NaCl in 0.01 M Tris-HCl (pH, 7.5), was mixed in ratios of 1:2, 1:1, 2:1, and 3:1 with heat-defibrinated pooled human plasma. These mixtures were incubated at 37° for 1 hr and at 4° for 16 hr. They were centrifuged at 5,000 g for 20 min and assayed for antithrombin as well as heparin cofactor activity. Compared with their buffer controls, the activities of antithrombin and heparin cofactor remaining in the mixtures were 43% versus 43%, 24% versus 32%, 23% versus 24%, and 10% versus 8%, respectively. Furthermore, the γ-globulin fraction of normal rabbit serum was equivalent to the buffer control when employed in the place of the γ-globulin directed against the inhibitor. In addition, 75 - 80% of the antithrombin and heparin cofactor activities could be immunoprecipitated from pooled human plasma defibrinated with small amounts of purified thrombin (5 NIH units/ml). Therefore, our inhibitor is responsible for the major portion of both these plasma activities.

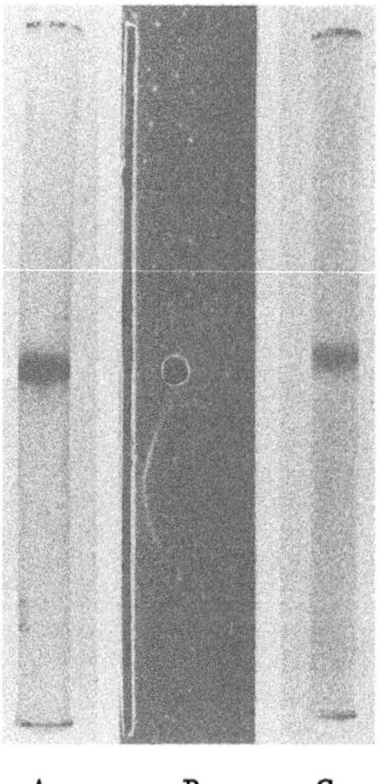

Fig. 1. Electrophoresis and immunoelectrophoresis of the purified inhibitor. The migration of protein is toward the anode (bottom). A, disc gel electrophoretic pattern obtained with 40 μg of protein; B, immunoelectrophoretic pattern obtained with 10 μg of protein in the circular well and 50 μl of concentrated γ-globulin prepared from rabbit antihuman sera; C, sodium dodecyl sulfate gel electrophoretic pattern obtained with 40 μg of protein.

Mechanism of Action of Antithrombin-Heparin Cofactor

Interaction of the inhibitor with the active center serine of thrombin. - Human thrombin was purified by the method of Rosenberg and Waugh (17). In a highly specific fashion, DFP alkylphosphorylates a unique serine residue in thrombin (DIP-thrombin) (3). Employing the native enzyme and this specific modification, we can show that the active center serine of thrombin is essential for interaction with antithrombin. Antithrombin and heparin were incubated for 30 sec. This incubation mixture was added to an equal volume of DIP-thrombin, native thrombin, or buffer. The concentration of native thrombin carefully chosen to neutralize all the antithrombin without any free thrombin being detected under these experimental conditions. After 5 min of incubation at 37°, thrombin was added, in a volume equivalent to the intermediate incubation mixture, and the esterolytic activity of residual thrombin was measured at 10 sec. Five separate incubation mixtures were used for each combination

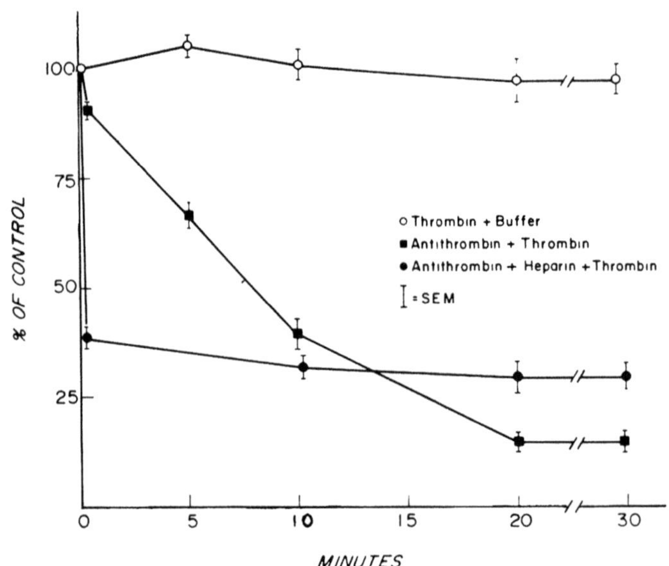

Fig. 2. Inhibition of thrombin by antithrombin in presence and absence of heparin. Final concentrations of enzyme and inhibitor were 0.162 and 0.164 absorbance units/ml, respectively. When heparin was added to incubation mixtures, its final concentration was 12 units/ml. Esterolytic activity of thrombin was measured with tosyl-L-lysine methyl ester (Cyclo Chemical Company, Los Angeles, Calif.) by continuously recording uptake of 0.01 M NaOH in a pH-stat fitted with automatic burette and chart recorder. Buffer used for this assay was 0.0005 M Tris-HCl in 0.15 M NaCl (pH, 8.3). All protein preparations used in these studies were extensively dialyzed against this assay buffer. Concentration of substance was 0.015 M, and reaction velocity at 37° ±0.1° was recorded for first 3 min. Composition of each incubation mixture is indicated in text and consisted of 0.15 ml, of which 0.1 ml was added to 0.9 ml of substrate in the assay buffer. Each assay sequence was randomized and included a series of standard thrombin dilutions. Plots of thrombin concentration versus reaction velocity were linear, with remarkably constant slope and intercept.

studied. As expected, if native thrombin were initially added to almost saturated antithrombin, all the final thrombin addition was present at the time of the esterolytic assay. However, if DIP-thrombin or buffer were intially admixed, smaller quantities of the final thrombin addition remained at 10 sec by esterolytic assays (Fig. 3). A statistical analysis demonstrated that DIP-thrombin could not be distinguished from buffer. However, both these bland additives significantly differ from the initial native thrombin addition. Since the concentration of DIP-thrombin was approximately four times that of the native thrombin initially added, we could easily have detected binding, provided that 25% or more of the alkylphosphorylated enzyme had successfully competed for sites on antithrombin. Furthermore, under identical conditions but in the absence of heparin, the addition of DIP-thrombin was also indistinguishable from that of buffer. Therefore, the active center serine of thrombin is of major importance for interaction with the inhibitor in the presence or absence of heparin.

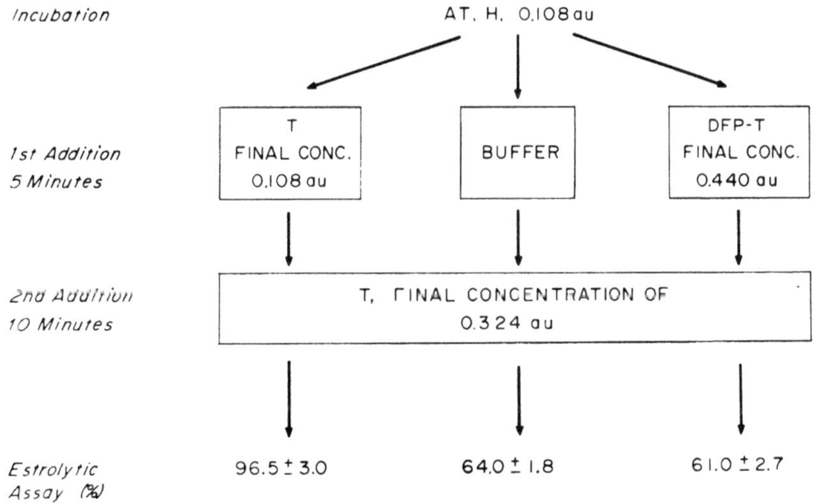

Fig. 3. Behavior of DIP-thrombin toward antithrombin. To alkylphosphorylate active serine of thrombin, 10 ml of enzyme solution at concentration of 0.165 absorbance units/ml were treated with 10 μl of 1 M DFP in 2-propanol. This inactivation was conducted at ionic strength of 0.15 - 0.35, pH of 7.5 or 8.3, and 24°. The 350-fold molar excess of DFP produced time-dependent inhibition of enzyme, so that 1% or less of esterolytic and proteolytic activity is present after 4 hr of incubation. Under these conditions, approximately 1 mole of DIP is incorporated per mole of thrombin.

This inability of the alkylphosphorylated enzyme to interact with the inhibitor was confirmed by filtering the inhibitor, with and without heparin, through columns of thrombin-Sepharose, DIP-thrombin-Sepharose, and ethanolamine-Sepharose (9, 26). When disc gel electrophoresis was used to study the thrombin-antithrombin interaction, the development of a complex was observed. This complex was stable after treatment with various combinations of 8 M urea, 2% sodium dodecyl sulfate, 0.1 M β-mercaptoethanol, and 6 M guanidine hydrochloride. These findings suggested that the interaction could be analyzed by serially denaturing samples and observing the molecular species with sodium dodecyl sulfate gel electrophoresis.

Therefore, antithrombin and thrombin were mixed, aliquots were sequentially removed, immediately denatured, and analyzed by sodium dodecyl sulfate gel electrophoresis (Fig. 4).

Both antithrombin and thrombin migrate as single bands in this electrophoretic system, with apparent molecular weights of 62,300 ± 1,500 and 33,800 ± 900, respectively. When these proteins are incubated together, both bands disappear over 5 min as a complex of apparent molecular weight 88,700 ± 700 is formed (Fig. 5).

The interaction of antithrombin and thrombin was also studied in the presence of heparin (5 units/ml). The protein concentrations, conditions of incubation, and analytic methods were identical with those described above. Figure 5 shows the immediate formation of

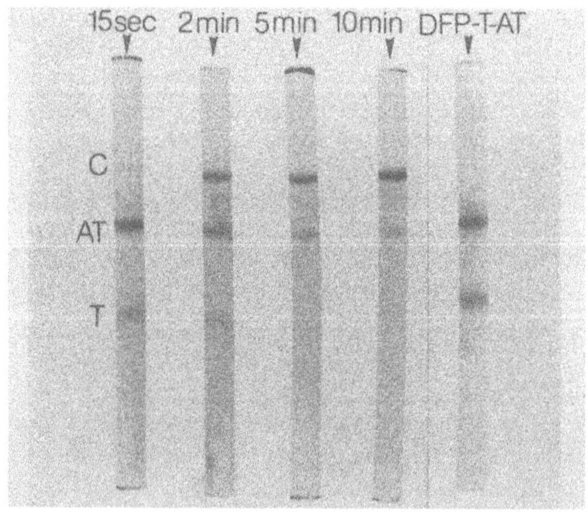

Fig. 4. Sodium dodecyl sulfate gel electrophoretic analysis of thrombin-antithrombin and DFP-treated thrombin-antithrombin interactions. C, complex; AT, antithrombin; T, thrombin.

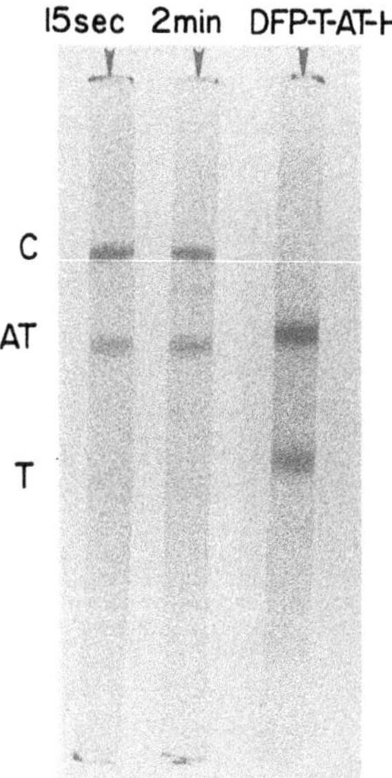

Fig. 5. Sodium dodecyl sulfate gel electrophoretic analysis of thrombin-antithrombin-heparin and DFP-treated thrombin-antithrombin-heparin interactions. C, complex; AT, antithrombin, T, thrombin.

complex. Densitometric analysis of these and other sodium dodecyl sulfate gels (not shown) suggested that 15 sec of incubation in the presence of heparin yielded an amount of complex equivalent to that observed at 3.5 min of incubation without heparin.

Sodium dodecyl sulfate gel electrophoresis has been used to study the interaction of DIP-thrombin with antithrombin. In the presence or absence of heparin, DIP-thrombin and antithrombin were incubated for 10 min. Sodium dodecyl sulfate gel electrophoretic analysis of the reaction products demonstrated only DIP-thrombin and antithrombin bands, without formation of complex (Fig. 4 and 5).

<u>Residues on the inhibitor critical for complex formation with thrombin and interaction with heparin</u>. - Given the highly acidic nature of heparin, one would expect that positive groups on antithrombin, such as ε-amino lysyl residues, form the binding site for this negatively charged anticoagulant.

We therefore guanidinated the lysyl residues of the inhibitor and examined the interaction of the modified protein with heparin and thrombin.

Antithrombin was dialyzed at 24° against either 0.5 M O-methylisourea in 0.01 M EDTA (pH 9.5) or a control solution at the same ionic strength and pH. The results of four separate experiments are presented in Table I and compared with samples of the inhibitor stored at 4°.

Dialysis against the control solution had a relatively small effect on the antithrombin and heparin cofactor activity. Dialysis against 0.5 M O-methylisourea virtually eliminated heparin cofactor activity, whereas antithrombin activity was only slightly reduced. In addition, sodium dodecyl sulfate gel electrophoretic analysis of thrombin-guanidinated inhibitor interactions revealed a quali-

TABLE I

Guanidination of Inhibitor with O-Methylisourea

Samples[a]	Assay, %	
	Antithrombin	Heparin Cofactor
Inhibitor stored at 4°	100[b]	100[b]
Inhibitor dialyzed against the control solution at 24°	91 ± 4	84 ± 6
Inhibitor dialyzed against the modification solution at 24°	75 ± 6	10 ± 0.8

[a]Aliquots of antithrombin (0.6 - 1.5 ml) were dialyzed at 24° against either 200 ml of a solution composed of 0.5 M O-methylisourea in 0.01 M EDTA (pH, 9.5) or 200 ml of a control solution of identical conductivity composed of 0.55 M NaCl in 0.01 M EDTA (pH, 9.5). After 62 hr, all samples were further dialyzed against 1 l of 0.15 M NaCl in 0.01 M Tris-HCl (pH, 8.3) for 16 hr at 4°, and assayed for antithrombin and heparin cofactor activity. The values given represent the means and standard errors of four separate experiments with different preparations of inhibitor. In each experiment, single samples were dialyzed against the appropriate buffer.

[b]The inhibitor stored at 4° is arbitrarily set at 100% activity.

tatively similar result. Compared with inhibitor dialyzed against the control solution, the guanidinated species formed little protease-protease inhibitor complex at 10 sec in the presence of heparin, but an equivalent amount of protease-protease inhibitor complex at 10 min in the presence or absence of heparin.

When the O-methylisourea-modified inhibitor and its control sample were subjected to amino acid analysis, the only changes observed in the modified inhibitor were a 63% decrease in the lysine content and the appearance of an equivalent amount of homoarginine. Other basic amino acids, as well as neutral and acidic amino acids, were unchanged.

The physical interaction of heparin with native and guanidinated inhibitor was examined to determine whether a loss of heparin cofactor activity was due to a reduced ability of the protein to bind this acidic mucopolysaccharide. A dilute heparin-Sepharose slurry was prepared so that it was equivalent to 10% by weight of packed gel and buffer. In each experiment, three sets of guanidinated and control samples were mixed in 0.2-ml aliquots with 0.2 ml of the heparin-Sepharose slurry. The solutions were incubated at 37° for 3 min, centrifuged at 5000 g for 5 min, and the supernatants were examined immunologically for inhibitor. In two separate experiments, 37% ± 3% of the control sample and 100% ± 6% of the guanidinated sample were found in the supernatant solution.

Thus, under conditions in which 63% of the native inhibitor bound to heparin-Sepharose, little of the guanidinated species was adsorbed. To demonstrate that heparin was responsible for this phenomenon, the experiments were repeated with Sepharose 4B in place of heparin-Sepharose. Now, 100% ± 5% of both inhibitor samples were found in the supernatant solution.

Of course, one might question whether guanidinated ε-amino groups are at the binding site and thus directly involved in the interaction with heparin, or whether the inhibition of heparin binding results from the modification of ε-amino groups elsewhere in the molecule, with a resulting change in the conformation of the inhibitor. Inasmuch as the guanidinated inhibitor loses little, if any, antithrombin activity and can still form an undissociable complex with thrombin, a major conformational alteration of the inhibitor seems unlikely. But, in an attempt to determine if heparin binds to lysine residues, we tested whether this acidic mucopolysaccharide could protect these critical residues of the inhibitor against modification by O-methlyisourea. Antithrombin was treated with 0.3 M O-methylisourea or a control solution at the same ionic strength and pH in both the presence and the absence of heparin (Table II). The data clearly indicate that little or no loss of activity occurred during dialysis at 24° in the presence and absence of heparin.

TABLE II

Guanidination of Inhibitor in Presence of Heparin

Samples[a]	Heparin Cofactor Assay, %
Inhibitor stored at 4°	100[b]
Inhibitor dialyzed against the control solution at 24°	86 ± 0.7
Inhibitor dialyzed against the modification solution at 24°	25 ± 1
Inhibitor and heparin dialyzed against the control solution at 24°	98 ± 3
Inhibitor and heparin dialyzed against the modification solution at 24°	74 ± 3

[a]These experiments employed antithrombin and heparin at final concentrations of 0.3 and 500 units/ml, respectively. Samples of the inhibitor, with and without heparin, were dialyzed for 62 hr at 24° against either 200 ml of solution consisting of 0.3 M O-methylisourea in 0.01 M EDTA (pH, 9.5) or 200 ml of control solution of matched conductivity with 0.39 M NaCl in 0.01 M EDTA (pH, 9.5). After 16-hr dialysis at 4° against 1 l of 0.15 M NaCl in 0.01 M Tris-HCl (pH, 8.3), aliquots were assayed for heparin cofactor activity. The values given represent the means and standard errors of four separate experiments with different preparations of inhibitor. In each experiment, single samples were dialyzed appropriately.
[b]The inhibitor stored at 4° is arbitrarily set at 100% activity.

Furthermore, treatment with 0.3 M O-methylisourea inactivated 71% of the inhibitor, compared with the control dialysate in the absence of heparin, and 24% compared with the control dialysate in the presence of heparin. Therefore, heparin protected 65% of the inhibitor against inactivation due to guanidination.

The protection afforded by heparin against inactivation by O-methylisourea suggests that lysyl groups are directly involved in heparin binding. However, a detailed knowledge of the mechanism of heparin binding will be required to substantiate this tentative conclusion.

Since thrombin is known to have a remarkably narrow specificity for unique arginine-x bonds (5), we suspected that our inhibitor employed this residue as a reactive site. We therefore analyzed the interaction of thrombin with an arginine-modified inhibitor. Antithrombin was dialyzed at 24° against either 0.015 M 1,2-cyclohexanedione with 0.01 M EDTA in 0.1 M triethylamine (pH, 11.0) or a control solution at the same ionic strength and pH. The results of four separate experiments are given in Table III.

The data indicate that dialysis against the control solution resulted in only small losses of antithrombin and heparin cofactor activity. However, the addition of 1,2-cyclohexanedione to the

TABLE III

Modification of Inhibitor with 1,2-Cyclohexanedione

Samples[a]	Assay, %	
	Antithrombin	Heparin Cofactor
Inhibitor stored at 4°	100[b]	100[b]
Inhibitor dialyzed against the control solution at 24°	86 ± 5	88 ± 5
Inhibitor dialyzed against the modification solution at 24°	5 ± 3	7 ± 3

[a]Aliquots of antithrombin (0.6 - 1.5 ml) were dialyzed in the dark at 24° against either 200 ml of a solution composed of 0.015 M 1,2-cyclohexanedione with 0.01 M EDTA in 0.10 M triethylamine (pH, 11.0) or 200 ml of a control solution that did not contain 1,2-cyclohexanedione. After 6 hr of contact with either solution, the samples were dialyzed against 0.15 M NaCl in 0.01 M Tris-HCl (pH, 8.3) at 4° for 16 hr and assayed for antithrombin and heparin cofactor activity. The values given represent the means and standard errors of four separate experiments with different preparations of inhibitor. In each experiment, single samples were dialyzed against the appropriate buffer.
[b]The inhibitor stored at 4° is arbitrarily set at 100% activity.

solution virtually eliminated both activities. In addition, a sodium dodecyl sulfate gel electrophoretic analysis of thrombin-1,2-cyclohexanedione-treated inhibitor interactions revealed a qualitatively similar result. When compared with inhibitor dialyzed against the control solution, the 1,2-cyclohexanedione-treated inhibitor formed little protease-protease inhibitor complex at 10 - 15 sec and 10 min with and without optimal amounts of heparin.

When the 1,2-cyclohexanedione-modified inhibitor and the control sample were subjected to amino acid analysis, the only change noted was a 21% reduction in the arginine content of the modified inhibitor. Other basic amino acid residues, as well as neutral and acidic amino acid residues, were unchanged. However, measurements at 440 nm demonstrated that the modified inhibitor had appreciable absorbance in this spectral region, whereas the control sample had hardly any absorbance at this wavelength.

Liu _et al_. (16) have studied the inactivation of many protease inhibitors with 1,2-cyclohexanedione. They have observed the emergence of this absorbance band when some degree of lysine modification has occurred. However, these products appear unstable during acid hydrolysis. Therefore, the loss of heparin cofactor activity observed with 1,2-cyclohexanedione could be due partially to lysine alterations undetected by amino acid analysis, rather than arginine modification. However, the loss of antithrombin activity noted with this reagent can be ascribed to arginine modification, in that guanidination has no effect on this inhibitory function.

To demonstrate that arginine residues are essential for heparin cofactor function, a more stringently controlled modification was used. Heparin has already been shown to protect critical lysine residues against guanidination. Therefore, this sulfated mucopolysaccharide was used to protect these lysine groups during arginine modification. Since 1,2-cyclohexanedione had little effect below a pH of 10.5, whereas heparin-dependent protection of lysine residues is minimal at or above this pH, we employed an alternate method for modifying arginine residues. The reagent 2,3-butanedione was chosen, because it has a specificity identical with that of 1,2-cyclohexanedione and is active at a lower pH. The conditions of pH and ionic strength were chosen to correspond to those previously employed for the protection of critical lysyl residues by heparin. In these modification studies, the inhibitor was dialyzed at 24° against either 0.115 M 2,3-butanedione and 0.39 M NaCl in 0.01 M EDTA (pH, 9.5) or a control solution at the same pH and ionic strength. The results of three separate experiments are given in Table IV. Dialysis of the inhibitor against the control solution, with or without heparin, resulted in little or no loss of inhibitor activity. However, exposure of the inhibitor to 2,3-butanedione produced the same significant loss of heparin cofactor activity in the presence or absence of heparin.

TABLE IV

Modification of Inhibitor with 2,3-Butanedione in the Presence and Absence of Heparin

Samples[a]	Heparin Cofactor Assay, %
Inhibitor stored at 4°	100[b]
Inhibitor dialyzed against the control solution at 24°	93 ± 8
Inhibitor dialyzed against the modification solution at 24°	14 ± 7
Inhibitor and heparin dialyzed against the control solution at 24°	107 ± 8
Inhibitor and heparin dialyzed against the modification solution at 24°	12 ± 8

[a]The inhibitor and heparin were employed at final concentrations of 0.3 absorbance unit/ml and 500 units/ml, respectively. Aliquots of inhibitor (0.6 - 1.0 ml), with and without heparin, were dialyzed for 6 hr at 24° against either 200 ml of a solution consisting of 0.115 M 2,3-butanedione and 0.39 M NaCl in 0.01 M EDTA (pH, 9.5) without 2,3-butanedione. After a 16-hr dialysis at 4° against 0.15 M NaCl in 0.01 M Tris-HCl (pH, 8.3), aliquots of the samples, if indicated, were assayed for heparin cofactor activity. The values given represent the means and standard errors of three separate experiments with different preparations of inhibitor. In each experiment, single samples were dialyzed appropriately.

[b]The inhibitor stored at 4° is arbitrarily set at 100% activity.

Therefore, arginine residue(s) are of critical importance for this inhibitor function. This conclusion is strengthened by the known specificity of the thrombin active site for arginine-x bonds and our demonstration that this active site is required for the formation of a protease-protease inhibitor complex in the presence of heparin.

Whether one or more arginine residue(s) are required for

inhibitor function and whether the same residue(s) are critical for both antithrombin and heparin cofactor activity remain undetermined. However, in analogy with other protease inhibitors, we suggest that a unique arginine residue forms the reactive site of the inhibitor. Furthermore, we belive that the simplest and most attractive hypothesis of inhibitor function would require this reactive site to be employed for both antithrombin and heparin cofactor action.

Summary

Antithrombin neutralizes the activity of thrombin by the formation of a 1:1 stoichiometric complex of enzyme and inhibitor via an interaction of a reactive site (arginine) and an active center (serine). Heparin binds to the lysyl residues of antithrombin and accelerates this interaction. The dramatic increase in the rate of complex formation most probably depends on a heparin-induced conformational alteration of the inhibitor that renders the reactive site arginine* more accessible to the active center serine of thrombin. Indirect evidence of this heparin-dependent conformational event has been obtained.

Multiple Actions of Heparin on the Coagulation and Fibrinolytic Mechanisms

The coagulation cascade is composed of a series of linked proteolytic reactions. At each stage of this mechanism, a parent zymogen is converted to a corresponding serine protease, which catalyzes a zymogen-serine protease transition. Thus, of the seven proteins directly or indirectly involved in the conversion of prothrombin to thrombin, four are ultimately activated to serine proteases (factors XII_a, XI_a, IX_a, and X_a), and two are cofactors for these proteolytic events (factors V and VIII)** (10, 22, 31).

Previous investigators have shown that factor X_a is neutralized by antithrombin and heparin (4, 28, 32). On the basis of this evidence and our knowledge of the biochemical mechanism of antithrombin action, we predicted that most serine proteases produced within the coagulation cascade would be neutralized by this inhibitor and that heparin would accelerate each of these interactions.

---

*Alternately, the binding of heparin to the inhibitor could lead to the exposure of a new reactive site that allows a more rapid interaction with serine proteases. Thus, different arginine residues might be employed for antithrombin and heparin cofactor functions.

**The characterization of human factor VII and its activation remains undetermined.

To test this concept, we have examined the interaction of antithrombin and heparin with partially purified factor XI_a, a serine protease that is generated at an early stage of the coagulation cascade. The physical properties, and substrate specificity of factor XI_a differ markedly from those of factor X_a and thrombin.

Factor XI was purified from human plasma by DEAE-cellulose chromatography, celite adsorption-elution, hydroxyapatite chromatography, and polyacrylamide P-150 gel filtration, according to the method of Amir _et al_. (2), with minor modifications. The final product is free of factors I, II, V, VII, VIII, IX, and X, as judged by fibrinogen determinations and one-stage assays using congenitally deficient plasmas as substrates.

To ascertain whether antithrombin inactivates factor XI_a, antithrombin was incubated with factor XI_a, and residual factor XI_a activity was sequentially measured as a function of time. As shown in Fig. 6, there was a progressive decline in factor XI_a activity. No significant reduction in this activity was observed, however, when buffer was substituted for antithrombin.

Although methodologic problems make such measurements difficult, we wish to test whether heparin accelerates the factor XI_a-antithrombin interaction. Heparin is known to inhibit (through antithrombin) the action of factor X_a on prothrombin (12), as well as that of thrombin on fibrinogen, and a valid determination of the proteolytic activity of factor XI_a depends on the degree to which this species triggers these terminal reactions of the coagulation cascade. Therefore, heparin and antithrombin-heparin complexes must be removed from incubation mixtures before they are assayed for factor XI_a. The addition of 9 mg of DEAE-cellulose to test incubations, centrifugation of these mixtures, and use of the supernatants proved, however, to be a rapid means of selectively eliminating free or bound heparin without altering the concentration of factor XI_a.

Several experiments demonstrated the validity of this approach. First, when heparin and factor XI_a were incubated together and processed in the above faction, no significant decrease in factor XI_a activity was detected (Fig. 7). Second, when antithrombin and heparin were incubated together for 5 min, the mixtures treated with DEAE-cellulose, and the resulting supernatants incubated with factor XI_a, only a minimal reduction in factor XI_a activity was seen (Fig. 7). Third, when factor XI_a was incubated with buffer and treated with DEAE-cellulose, the supernatants obtained showed no significant reduction in the initial factor XI_a activity (Fig. 7).

To test whether heparin accelerates the interaction between factor XI_a and antithrombin, mixtures of antithrombin, heparin, and factor XI_a at final concentrations of 36, 0.7, and 100 units/ml,

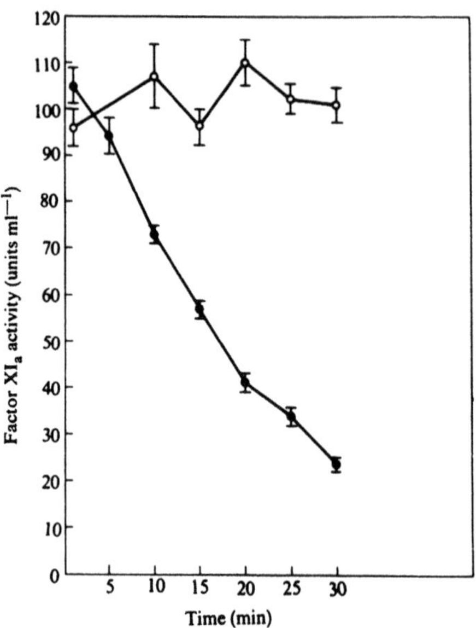

Fig. 6. Inhibition of factor XI_a by antithrombin. Factor XI_a was incubated with buffer or antithrombin in albuminized plastic tubes at 37°. 0.02 ml of factor XI_a was added to either 0.38 ml of 0.15 M NaCl in 0.01 M Tris (pH, 7.5) or 0.18 ml of this buffer and 0.2 ml antithrombin. Final concentrations of factor XI_a and antithrombin used were 100 units/ml (0.002 AU/ml) and 64 units/ml (0.060 AU/ml), respectively. Mixtures were separately incubated for various periods and assayed in triplicate for residual factor XI_a. Each point represents mean obtained for two separate incubation mixtures ±s.e.m. o--o, factor XI_a +buffer; •--•, factor XI_a+antithrombin. Factor XI_a activity was measured by modification of method of Nossel (23), using cephalin, and initial concentrations of factor XI_a preparations were estimated by comparison with factor XI content of pooled plasma obtained from 25 normal subjects. Assays of factor XI in plasma were performed by method of Rapaport _et al._ (25) with mean value of this plasma component arbitrarily set at 100 units/ml. Antithrombin-heparin cofactor activity was determined by modification of method of Abildgaard (1) in presence of heparin or in its absence.

respectively, were incubated together for 1, 3, or 5 min before the addition of DEAE-cellulose. After centrifugation, the supernatants were analyzed, and factor XI_a activity was found to be virtually absent in all incubation mixtures (Fig. 7). In other experiments (not shown), shorter periods of incubation (0.1 - 0.5 min) revealed

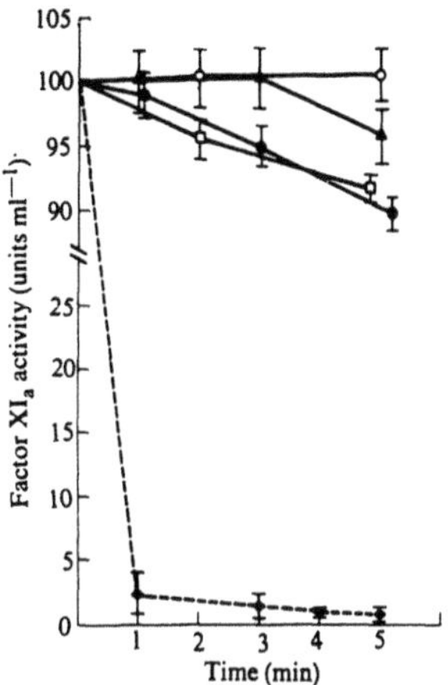

Fig. 7. Inhibition of factor XI_a by antithrombin and heparin. Two distinct sequences of addition of factor XI_a to other components were used. The first incubation sequence used duplicate samples of 1.6 ml composed of factor XI_a added to either buffer, heparin, antithrombin, or antithrombin and heparin. In this instance, factor XI_a was employed at final concentration of 100 units/ml, while heparin and antithrombin, if present, were used at final concentrations of 0.7 and 36 units/ml, respectively. All samples were incubated for 1-5 min at 37°, dried DEAE-cellulose (9 mg) was added with gentle agitation, mixtures were immediately centrifuged at 15,000 g for 30 sec and resulting supernatants were assayed in triplicate for residual factor XI_a activity. The second incubation sequence used duplicate samples of 1.2 ml consisting of antithrombin and heparin at final concentrations of 48 and 0.93 units/ml, respectively. Both samples were incubated at 37° for 5 min, dried DEAE-cellulose (9 mg) was added with gentle agitation, and mixtures were centrifuged at 15,000 g for 30 sec. Factor XI_a activities were determined in triplicate. Values given represent mean factor XI_a activity ±s.e.m. In all cases, components have been either dialyzed against or extensively diluted in buffer (0.15 M NaCl) in 0.01 M Tris (pH, 7.5); o--o, factor XI_a+buffer; ▲--▲, factor XI_a+heparin; □--□, factor XI_a+heparin-antithrombin-treated supernatant; ●--●, factor XI_a+antithrombin; ◆--◆, factor XI_a+heparin+antithrombin.

the same degree of factor XI_a inhibition. These results are to be compared with those obtained when the inactivation was examined in identical conditions, but with the omission of heparin. In these mixtures, only a minimal time-dependent reduction in factor XI_a activity was observed (Fig. 7). Therefore, heparin dramatically accelerates the formation of factor XI_a-antithrombin complex. On the basis of the neutralization of factor XI_a by antithrombin in the absence of heparin, as well as our knowledge of the mechanism of inhibitor action, we may confidently presume that inactivation of factor XI_a follows this interaction.

Thus, factor XI_a, like factor X_a and thrombin, is a serine protease that is slowly but progressively inhibited by antithrombin in the absence of heparin and almost instantaneously neutralized by antithrombin in the presence of this anticoagulant. On the basis of these data and the known mechanism of action of antithrombin, we predicted that antithrombin ought to inactivate the other remaining serine proteases of the coagulation cascade and that heparin will accelerate each of these interactions. Indeed, Aronson (personal communication), using antithrombin prepared by us, has already confirmed this hypothesis for factor IX_a. As pure human coagulation factor proteins become available, we hope to test this hypothesis rigorously for each step of the coagulation mechanism.

As a logical extension of antithrombin's pattern of inhibitor specificity, we have attempted to determine whether other serine proteases produced in systems separate from but linked to the coagulation cascade would be inactivated by this inhibitor and its acidic mucopolysaccharide cofactor.

The conversion of plasminogen to plasmin represents the central event of the fibrinolytic mechanism. Once this potent serine protease is evolved, it hydrolyzes fibrinogen and fibrin, as well as a variety of other biologically important proteins. Normal blood is capable of localizing and restricting the activity of plasmin. This is accomplished partially by the presence of several plasma protease inhibitors. Although the quantitative significance of each of these antiplasmins has not been evaluated, most authors have assumed that α_2-macroglobulin is of primary importance.

The degree of inhibition of the proteolytic activity of plasmin by antithrombin was studied in the presence and absence of heparin. Plasminogen was purified by the affinity chromatographic technique of Deutsch and Mertz (11) and was converted to plasmin with urokinase (14). To determine the stability of our plasmin preparations, this enzyme was incubated for various periods at 37°. Curve A Fig. 8 shows that there was no significant loss in plasmin activity over a 30-min incubation period. To examine the antithrombin-plasmin interaction, the inhibitor was incubated with buffer for 3 min at 37°; then plasmin was added for various periods. Curve B of Fig. 8 illustrates the resulting time-dependent decay in activity during 30 min of incubation. To analyze the interaction of plasmin with antithrombin and heparin, the inhibitor was preincubated with the mucopolysaccharide for 3 min at 37° before the addition of plasmin. The final concentrations of the enzyme and inhibitor were maintained at their previous values, but heparin was added at 10 units/ml.

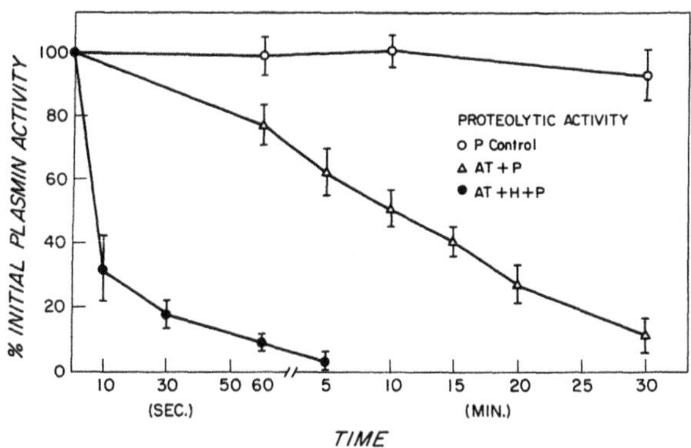

Fig. 8. Proteolytic activity of plasmin (14), plasmin with antithrombin (14), and plasmin with antithrombin and heparin (14) were followed for various periods. To assay proteolytic activity of plasmin, 0.2 ml of a test sample was added to 0.1 ml of ^{125}I-labeled α-casein. Before use in the assay, labeled substrate was dissolved in 0.09 M NaCl in 0.06 M Tris-HCl (pH, 7.5) at a final concentration of 1 mg/ml. Resulting assay mixtures were incubated at 37° for 1 min. Proteolysis of the substrate was quenched by addition of 0.2 ml of unlabeled α-casein (1.4%) followed by 1.0 ml of 0.5 M perchloric acid. Reaction mixture was placed at 4° for 10 min and subsequently centrifuged at 5,000 g and 4° for 10 min. Supernatants were decanted, and 1 ml aliquots were counted for 1 min in a Nuclear-Chicago gamma counter. "Background" control was measured by replacing plasmin sample in reaction mixture with 0.2 ml of 0.09 M NaCl in 0.06 M Tris-HCl (pH, 7.5) and processing solutions in manner outlined above. To quantitate peptides released from α-casein by plasmin, background counts were subtracted from those of the test sample. A standard titration curve of 1 - 20 μg of this enzyme was employed to determine the amount of plasmin in the test sample. When counts from perchloric acid supernatant minus those of the control were plotted as a function of plasmin concentration, a linear relation (r=0.985) was obtained. Plasmin activities were determined in triplicate. Values given represent mean plasmin activity ±s.e.m.

Curve C of Fig. 8 reveals that the rate of neutralization of plasmin by antithrombin was accelerated in the presence of heparin, without any appreciable alteration in the final amount of enzyme inactivated by the inhibitor.

In other experiments conducted at a fixed plasmin concentration (15 μg/ml), the quantity of antithrombin in the reaction mixture was varied from 2.5 to 15 μg/ml, and a proportional increase was noted in the final amount of enzyme neutralized in the presence or absence of heparin. Using these results and the known molecular weights of the enzyme and inhibitor, the equivalence point of this reaction was estimated to occur at a 1.15:1 molar ratio of plasmin to antithrombin.

The kinetics of the plasmin-antithrombin interaction in the presence and absence of heparin has also been examined with esterolytic assays. The results obtained were identical with those depicted above, employing a proteolytic assay of plasmin's activity (14).

Initial experiments with the plasmin-antithrombin complex indicated that it, too, was stable with respect to denaturing agents. Therefore, plasmin and antithrombin were mixed at 37°, aliquots were sequentially removed, denatured, and analyzed by sodium dodecyl sulfate gel electrophoresis in which reducing agents had been omitted.

As shown in Fig. 9, both antithrombin and plasmin migrated as single bands in this electrophoretic system with apparent molecular weights of 62,300 ± 1,600 and 81,400 ±1,800, respectively. When these proteins were incubated together, both bands decreased in intensity over 10 min, as a complex of apparent molecular weight

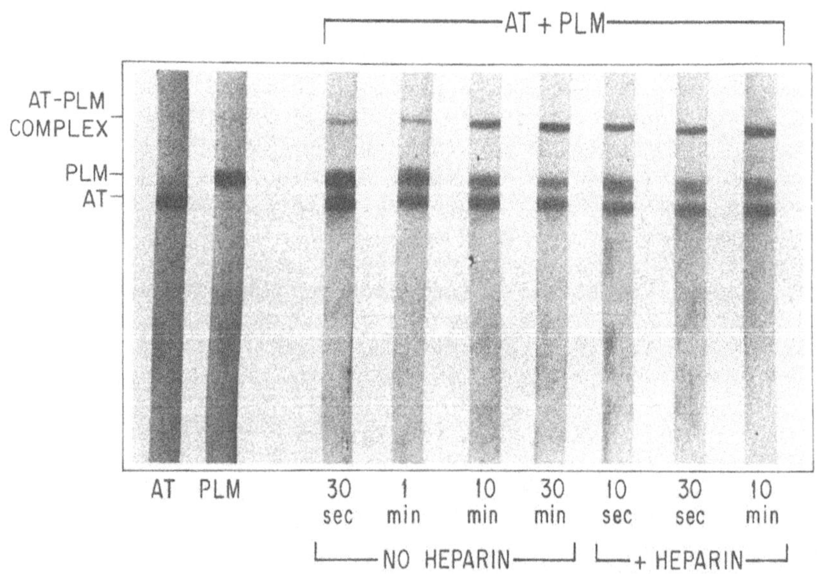

Fig. 9. Nonreduced sodium dodecyl sulfate gel electrophoretic analysis of plasmin-antithrombin interactions in presence and absence of heparin. AT, antithrombin; PLM, plasmin.

142,000 ± 3,500 was gradually formed. The failure of the plasmin band to disappear resulted from the presence of some biologically inactive enzyme in our preparations.

In the presence of heparin, the antithrombin-plasmin complex was formed within 10 sec. (Fig. 9).

Sodium dodecyl sulfate gel electrophoretic studies of the plasmin-antithrombin interaction were also conducted in which samples were denatured in the presence of reducing agents. Under these conditions, antithrombin migrated as a single band with apparent molecular weight of 61,700 ±1,300, whereas plasmin was observed to consist of heavy and light polypeptide chains of apparent molecular weights 55,600 ± 2,200 and 24,200 ± 1,800, respectively. Both with and without heparin, incubation of the enzyme with the inhibitor resulted in the appearance of a new band of apparent molecular weight 83,900 ± 3,100 and a corresponding reduction in the intensity of the light chain of plasmin and of antithrombin. The intensity of the heavy chain of plasmin did not appear to be altered by this interaction. Inasmuch as the active serine center of this enzyme is on the light polypeptide chain (29), these results indicate that this portion of the enzyme may be critically important in its interaction with antithrombin. In fact, preliminary results obtained with specific chemical modifications of the enzyme and inhibitor indicate that this supposition is correct and that plasmin is neutralized by antithrombin via a mechanism similar to but not identical with that proposed for the inactivation of thrombin. These data evoke the intriguing possibility that antithrombin-heparin cofactor may regulate the hemostatic balance between coagulation and fibrinolysis by modulating the activities of both systems. However, the importance of this inhibitor as an antiplasmin must await a comparative analysis of the inactivation of plasmin by all protease inhibitors normally present in plasma.

ACKNOWLEDGEMENTS

This work was supported by National Institutes of Health grants HL 17533 and HL 13955.

REFERENCES

1. ABILDGAARD, U., Scand. J. Clin. Lab. Invest., 21 (1968) 89.
2. AMIR, J., PENSKY, J. and RATNOFF, O.D., J. Lab. Clin. Med., 79 (1972) 106.
3. BELL, W.N. and ALTEN, H.G., Nature, 174 (1954) 880.

4. BIGGS, R., DENSON, K.W.E., AKMAN, N., BORRETT, R. and HADDEN, M., Brit. J. Haematol., 19 (1970) 283.
5. BLOMBACK, B., BLOMBACK, M., HESSEL, B. and IWANAGA, S., Nature, 215 (1967) 1445.
6. BRINKHOUS, K., SMITH, H.P., WARNER, E.D. and SEEGERS, W.H., Amer. J. Physiol., 125 (1939) 683.
7. CONTEJEAN, C., Arch. Physiol. Norm. Pathol., 7 (1895) 45.
8. DAMUS, P.S., HICKS, M. and ROSENBERG, R.D., Nature, 246 (1973) 355.
9. DAMUS, P.S. and ROSENBERG, R.D., Proc. of Soc. of Acad. Surgeons, New Orleans Meeting, (1972) p. 34.
10. DAVIE, E.W. and RATNOFF, O.D., The Proteins, Academic Press, New York, 1965, vol. 3, chap. 16.
11. DEUTSCH, D. and MERTZ, E.T., Science, 170 (1970) 1095.
12. DOMBROSE, F.A., SEEGERS, W.H. and SEDENSKY, J.A., Thromb. Diath. Haemorrh., 26 (1971) 1.
13. GLADNER, J.A. and LAKI, K., J. Amer. Chem. Soc., 80 (1958) 1263.
14. HIGHSMITH, R.F. and ROSENBERG, R.D., J. Biol. Chem., 249 (1974) 4.
15. HOWELL, W.H., Harvey Lect., 2 (1916) 272.
16. LIU, W.H., FEINGSTEIN, G., OSUGA, D.J., HAYNES, R. and FEENEY, R.E., Biochemistry, 7 (1968) 2886.
17. LYTTLETON, J.W., Biochem. J., 58 (1954) 15.
18. MAGNUSSON, S., Proc. Biochem. Soc., 110 (1968) 25.
19. McLEAN, J., Amer. J. Physiol., 41 (1916) 250.
20. MONKHOUS, F.C., FRANCE, E.S. and SEEGERS, W.H., Circ. Res., 3 (1955) 397.
21. MOROWITZ, P., Springfield, Illinois, Charles C. Thomas, 1968.
22. NEMERSON, Y. and ESNOUF, M.P., Proc. Nat. Acad. Sci., 70 (1973) 310.
23. NOSSEL, H.G., The Contact Phase of Blood Coagulation, Blackwell, (Oxford), 1964.
24. PLAPP, B.Z., MOORE, S. and STEIN, W.H., J. Biol. Chem., 246 (1971) 939.
25. RAPAPORT, S.I., SCHIFFMAN, S. and PATCH, M.J., J. Lab. Clin. Med., 59 (1961) 771.
26. ROSENBERG, R.D. and DAMUS, P.S., J. Biol. Chem., 248 (1973) 6490.
27. ROSENBERG, R.D. and WAUGH, D.F., J. Biol. Chem., 245 (1970) 5049.
28. SEEGERS, W.H., COLE, E.R., HARMISON, C.R. and MONKHOUSE, F.C., Can. J. Biochem., 42 (1964) 359.
29. SUMMARIA, L., HSIEN, B., GROSKOPF, W.R., ROBBINS, K.C. and BARLOW, G.H., J. Biol. Chem., 242 (1967) 5046.
30. WAUGH, D.F. and FITZGERALD, M.A., Amer. J. Physiol., 184 (1956) 627.
31. WUEPPER, K.D. and COCHRANE, C.G., J. Exp. Med., 135 (1972) 1.
32. YIN, E.T., WESSLER, S. and STOLL, P.J., J. Biol. Chem., 246 (1971) 3694.

HEPARIN INTERACTION WITH ACTIVATED FACTOR X AND ITS INHIBITOR

E. Thye YIN, Linda EISENKRAMER and Janet V. BUTLER

Department of Medicine, the Jewish Hospital of St. Louis, St. Louis, Missouri 63110 (USA)

We would like to present an improved method for the purification of the inhibitor of activated factor X (heparin cofactor, antithrombin III) from various species of plasma, and some observations on the *in vitro* response of plasma of various species to heparin.

Purification of inhibitor of activated factor X. - Most of the steps in the purification of the plasma inhibitor of activated factor X (XaI) have already been described (5). The first step of the purification procedure is carried out on a Sephadex G-200 column. The XaI activity peak is pooled, concentrated by lyophilization, reconstituted to 10% of the original plasma volume, and dialyzed against 0.15 M NaCl. The dialyzed material is chromatographed on a heparin-agarose column (Fig. 1). The water-insoluble heparin is prepared as described elsewhere (1). The XaI eluted from the heparin column is then concentrated by dialysis against 50% polyvinylpyrrolidone at a pH of 7.0 and dialyzed against 0.05 M NaCl and 0.1 M Tris HCl at a pH of 8.32. This fraction is then chromatographed on a DEAE-Sephadex column (5).

Figure 2 shows the status of the XaI fractions (human), as revealed by analytical disc electrophoresis on polyacrylamide gels. Using this improved method, samples of plasma XaI from human, baboon, cow, and rabbit have been isolated in homogeneous form. Their amino acid compositions and molecular weights have also been determined. The antibody to human XaI developed in the rabbit fully cross-reacts immunologically with the baboon XaI and minimally with the bovine and rat XaI.

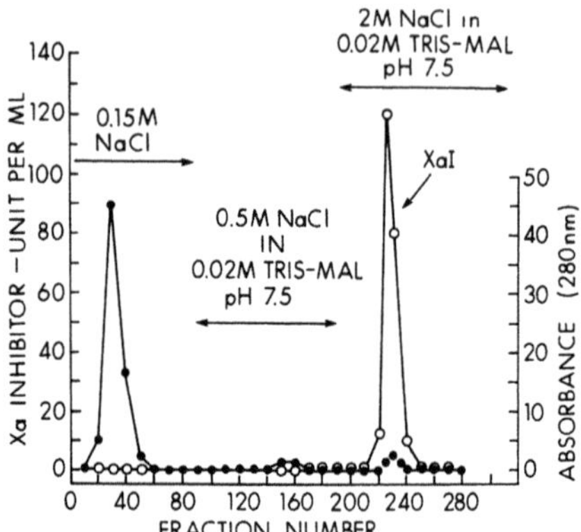

Fig. 1. Affinity-column chromatography of plasma XaI on heparin-agarose.

The XaI isolated by this procedure is highly stable during storage at -20°, compared with that obtained by the previous method (5). The overall recovery is at least 25%; and from 1 liter of plasma, one can obtain at least 60 mg of homogeneous material.

Plasma response to heparin in different species. - During our attempts to determine the antithrombotic effect of a low-dose heparin regimen in animals, we encountered several problems that led us to suspect that different species respond differently to low doses of heparin *in vivo*, as well as *in vitro*. This finding is most readily demonstrated through the assay for the activity of plasma XaI (E.T. Yin, J.V. Butler and J. Singer, unpublished data). When normal human plasma is serially diluted in 0.15 M NaCl and tested in the XaI activity assay, a curve represented by X - - - X in Fig. 3 is obtained in the absence of added heparin. On the basis of our previous findings (2-4), we anticipated that, when a constant amount of heparin was added to each plasma dilution, the residual factor Xa clotting time in the test system would have a linear relation similar to that in the absence of heparin. However, an unexpected plasma dilution response to the added heparin was obtained, as shown in Fig. 3. An unexpected higher sensitivity of the test system to heparin was observed when the plasma was diluted to below 6%. In contrast, when the same tests were performed on the rabbit plasma, the response to added heparin varied almost linearly with plasma concentration. Although it is not shown in Fig. 3, baboon plasma behaved like the human plasma; and rat plasma behaved like the rabbit plasma. Furthermore, purified XaI from the plasmas of human, baboon, rat, and rabbit all behaved alike in the presence of heparin. Thus, in evaluating the antithrombotic property of heparin preparations, one must be very cautious whenever animal models are used.

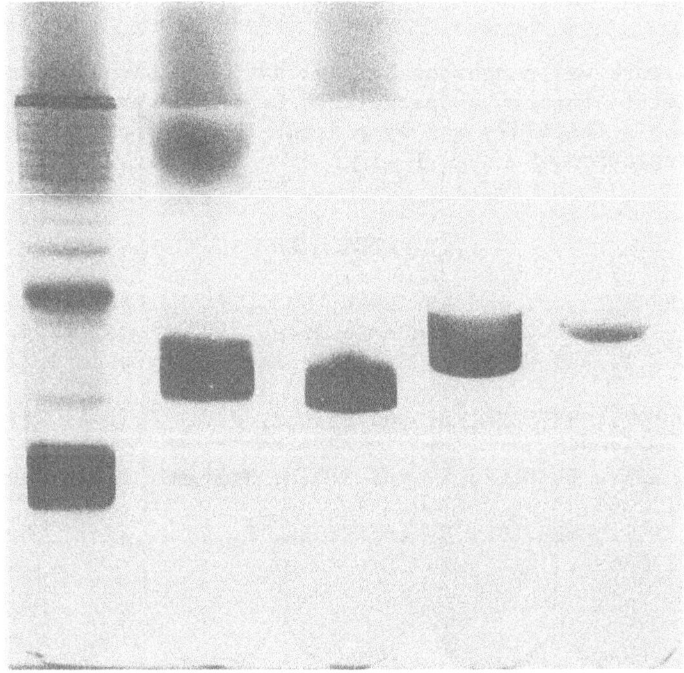

Fig. 2. Left to right: disc electrophoresis on polyacrylamide gel (pH 9.5): whole plasma, postheparin-agarose chromatography, and final product after DEAE-Sephadex chromatography (150 μg of protein used); disc electrophoresis on polyacrylamide gel in the presence of sodium dodecylsulfate: nonreduced (150 μg of protein used), and nonreduced (15 μg of protein used).

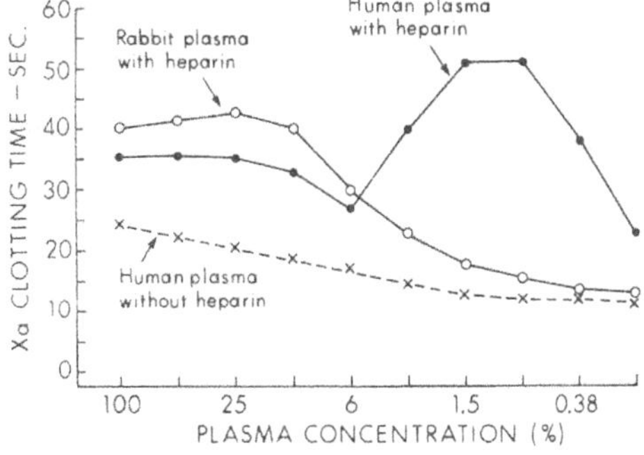

Fig. 3. Response of plasma in various dilutions to heparin.

ACKNOWLEDGEMENTS

This work was supported in part by the National Heart and Lung Institute through a Specialized Center of Research (grant) in Thrombosis (HL14147) and by a grant-in-aid from the American Heart Association.

REFERENCES

1. CUATRECASAS, P. and ANFINSEN, C.B., in Methods in Enzymology, (Ed. Jakoby, W.B.), Academic Press, 1971, vol. 22 p. 345.
2. YIN, E.T. and WESSLER, S., Biochim. Biophys. Acta, 201 (1970) 387.
3. YIN, E.T., WESSLER, S. and BUTLER, J.V., J. Lab. Clin. Med., 81 (1973) 298.
4. YIN, E.T., WESSLER, S. and STOLL, P.J., J. Biol. Chem., 246 (1971) 246.
5. YIN, E.T., WESSLER, S. and STOLL, P.J., J. Biol. Chem., 246 (1971) 3694.

EVIDENCE OF A CATALYTIC ROLE OF HEPARIN IN ANTICOAGULATION REACTIONS

Sanford N. GITEL

Department of Medicine, Jewish Hospital of St. Louis, and Washington University School of Medicine, St. Louis, Missouri 63110 (USA)

The nature of heparin involvement in the inhibition of factor Xa or thrombin by their natural plasma inhibitor has not been completely elucidated. This report describes some of the properties of heparin that may be important in this inhibition reaction.

Materials and Methods

Bovine serum albumin (BSA) and plasma deficient in factor VII and X were obtained from Sigma Chemical Company. Heparin sodium (Liquaemin 10) - 1,000 U.S.P. units/ml - was obtained from Organon.

Thrombin was prepared according to the procedure of Yin and Wessler (14) and further purified using sulfopropyl Sephadex. Factor X was purified by the procedure of Duckert *et al.*(5) and activated with Russell's viper venom, and the venom was removed by chromatography on DEAE cellulose. Tritiated fragment 1 was isolated as described by my co-worker and me (7) from ^{3}H -prothrombin, tritiated by the procedure of Van Lenten and Ashwell (12). Factor Xa inhibitor (XaI) was isolated by the procedure of Yin (described elsewhere in this volume).

Factor X was assayed as described previously (16) by a modification of the method of Bachmann *et al.* (1). Prothrombin was assayed by the procedure of Hjort *et al.* (8). Thrombin activity was derived from a standard calibration curve with NIH human thrombin (lot 3B) as a reference. Heparin and XaI were assayed as described by Yin *et al.* (15).

TABLE I

Heparin Conservation During Xa Inhibition

Heparin Concentration	Total Heparin	Total XaI, ug	Ratio of XaI to Heparin [a]	Final Heparin Concentration, u/ml Predicted	Observed
0.1	0.7	20	5	0.09	0.095
0.1	0.7	40	10	0.09	0.1
0.05	0.35	40	20	0.045	0.05
0.05	0.35	50	24	0.04	0.035

[a] On the basis of a molecular weight of 60,000 for XaI and an average molecular weight of 10,000 for Heparin.

Agarose-bound heparin was prepared by the method of Cuatrecasas (3), as described by Yin.

Heparin Conservation During Inhibition Reactions

Table I shows the fate of heparin as it catalyzes the inhibition of Xa by XaI. Purified inhibitor in the presence of the indicated amount of heparin was titrated with Xa so that no inhibitor activity remained and only a small amount of Xa activity could be detected. The amount of heparin remaining was then determined by the method of Yin et al. (15). The final two columns in Table I show that the recovery of heparin is quantitative, with ratios of inhibitor to heparin ranging from 5 to 24. These data suggest that heparin is acting as a true catalyst and need not remain associated with the XaI-Xa complex.

Protein Binding to Heparin Affinity Columns

Data by Danishefsky presented elsewhere in these proceedings show that the carboxyl and aminosulfate groups of heparin are essential for heparin anticoagulant action. It would be expected, in the formation of any immobilized heparin, that the utilization of the carboxyl and aminosulfate groups may alter the reactivity or protein binding characteristics of the heparin. Thus, attaching heparin to agarose with soluble-carbodimide coupling reactions will remove the free carboxyl groups, altering the reactivity of the heparin. An example of this could be in Danishefsky and Tzeng's inability to observe strong binding of XaI to their heparin-agarose gel (4), whereas we and others (13) have observed excellent binding to heparin-agarose prepared by the method of Cuatrecasas (3). The latter method would not affect free carboxyls, but rather takes advantage of free amino groups present on the heparin used to prepare the gel. The somewhat increased pH of the reaction may contribute to the hydrolysis of some sulfates, however. Therefore, not all data concerning immobilized heparin can be separated from the method of immobilization.

Table II lists the interactions of several proteins with a 0.6 x 7 cm analytic column packed with heparin bound to agarose. The columns were eluted with a linear gradient of NaCl containing BSA at 1 mg/ml. The last column of the Table shows the elution position of each protein in terms of ionic strength. The data presented in this column show that the XaI is very tightly bound to the heparin-agarose affinity column. Thrombin, as reported by Gentry and Alexander (13), and Xa exhibit strong binding to the heparin-agarose gel, whereas prothrombin and fragment 1 (the lipid binding fragment of prothrombin that cannot be converted to thrombin)

TABLE II

Binding of Proteins to Heparin-Agarose

Protein	Solvent	Concentration of Salt Required for Elution[a]
Fragment 1	H_2O with BSA	0.1 M NaCl
Prothrombin	Plasma	0.25 M NaCl
Xa	H_2O with BSA	0.65 M NaCl
Xa	H_2O with BSA, 2/mM Ca^{+2}	0.8 M NaCl
Thrombin	H_2O with BSA	0.9 M NaCl
XaI	Plasma	1.5-1. 6 M NaCl

[a] All solutions contained BSA at a concentration of 1 mg/ml.

exhibit little or no binding to the heparin. A comparison of the charge characteristics of some of the proteins in this Table may give a better understanding of the mode of binding of thrombin and Xa to the heparin-agarose gel. Thrombin, which has a net positive charge of about 15 (11), could be expected to bind to the negatively charged heparin; thus its binding could be a site-specific interaction or a nonspecific ionic binding. Fragment 1 has a net negative charge of 7 (R.A. Henriksen and C.M. Jackson, unpublished data), and it is not surprising that it shows no binding to the heparin column. However, Xa has a net negative charge of about 15-20 (6, 9), and it still shows considerable binding to the heparin gel; this indicates a possible specific heparin binding site. Factor Xa binding to heparin is enhanced by the addition of Ca^{+2} in the physiologic range. This might be due to the decrease in negative charge of both the heparin and the Xa.

Inasmuch as XaI, thrombin, and Xa are each tightly bound to heparin, we suggest that the role of heparin in promoting XaI activity is similar to the role of phospholipid in prothrombin activation (2, 10) - i.e., the heparin binds both the XaI and the Xa, thus allowing their reaction to proceed at an increased rate. Once the XaI-Xa complex is formed, the heparin would be released, and it could then catalyze further the inhibition of Xa or thrombin.

ACKNOWLEDGEMENTS

This work was supported by the National Heart and Lung Institute through a Specialized Center of Research in Thrombosis grant HL-14147 to the Washington University Medical Center.

REFERENCES

1. BACHMANN, F., DUCKERT, F. and KOLLER, F., Throm. Diath. Haemorrh., 2 (1958) 24.
2. BARTON, P.G., JACKSON, C.M. and HANAHAN, D.J., Nature, 214 (1967) 923.
3. CUATRECASAS, P., J. Biol. Chem., 245 (1970) 3059.
4. DANISHEFSKY, I. and TZENG, F., Thomb. Research, 4 (1974) 237.
5. DUCKERT, F., YIN, E.T. and STRAUB, W., Prot. Biol. Fluids Proc. Colloq. Bruges, 41 (1960).
6. FUJIKAWA, K., LEGAZ, M.E. and DAVIE, E.W., Biochemistry, 11 (1972) 4892.
7. GITEL, S.W., OWEN, W.G., ESMON, C.T. and JACKSON, C.M., Proc. Natl. Acad. Sci. US, 70 (1973) 1344.
8. HJORT, P., RAPAPORT, S.I. and OWREN, P.A., J. Lab. Clin. Med., 46 (1955) 89.
9. JACKSON, C.M., Biochemistry, 11 (1972) 4873
10. JOBIN, F. and ESNOUF, M.P., Biochem. J., 102 (1967) 666.
11. MAGNUSSON, S., in The Enzymes (Ed. BOYER) Academic Press, New York, 3 (1971) p. 277.
12. VAN LENTEN, V. and ASHWELL, G., J. Biol. Chem., 246 (1970) 1889.
13. WILSON-GENTRY, P. and ALEXANDER, B., Fed. Proc. Abstracts, 57th Meet., 1973
14. YIN, E.T. and WESSLER S., J. Biol. Chem., 243 (1968) 112.
15. YIN, E.T., WESSLER, S. and BUTLER, J.V., J. Lab. Clin. Med., 81 (1973) 298.
16. YIN, E.T., WESSLER, S. and STOLL, P.J., J. Biol. Chem., 246 (1971) 3694.

STRUCTURAL AND FUNCTIONAL RELATIONSHIPS OF HUMAN ANTITHROMBIN III AND ALPHA$_1$-ANTITRYPSIN

Anna D. BORSODI[+], Ralph A. BRADSHAW[+], Indulal K. RUGHANI and Robert M. BRUCE
Department of Biological Chemistry[+] and Pulmonary Disease Division, Department of Medicine, Washington University, St. Louis, Missouri 63110 (USA)

One of the principal mechanisms for the control of blood coagulation in man is the interaction of serum inhibitors with proteolytically active clotting factors, with concomitant losses in thrombogenic activity. One of the inhibitors, antithrombin III, which migrates as an α_2-globulin on gel electrophoresis (1, 9, 13, 20), has received considerable attention. It appears to be identical with heparin cofactor and factor x_a inhibitor (21) and provides the major site of action of heparin in inhibiting thrombin formation. However, it should be noted that the inhibition of thrombin is not limited to antithrombin III. Two other plasma proteins, an α_2-macroglobulin (1, 3, 4, 11, 17), and the α_1-proteolytic inhibitor (2, 14, 17) also show such activity. The latter inhibitor is probably identical with α_1-antitrypsin (14), the protein responsible for most of the inhibition of trypsin activity in human serum.

Several similarities in structure and function have led us to examine in greater detail the relation between antithrombin III and α_1-antitrypsin. They are isolated by similar procedures and, as shown in Table I, have similar isoelectric points and carbohydrate contents. Unfortunately, the reported molecular-weight data (1, 6, 14-16) for these proteins are still ambiguous enough to prevent careful comparison of this property.

The functional properties of these proteins are obviously similar. Both are capable of specifically interacting with various proteases of the "serine" family to effect inhibition of the enzymatic activity. In fact, there is a clear crossover in specificity, in that α_1-antitrypsin inhibits trypsin as well as chymotrypsin, elastase and plasmin (8, 10, 14), and antithrombin III inhibits trypsin and

TABLE I

Properties of Human α_1-Antitrypsin and Antithrombin III

	α_1-antitrypsin	antithrombin III
Molecular weight	45,000[a]	47,000[b]
	50,000[c]	64,000[d]
	60,000[e]	
Isoelectric point	5.10[c]	5.11[f]
Carbohydrate content, %	11[c]	9[g]
	14[e]	

[a]Bundy *et al.* (5) and Tan and Gans (19).
[b]Rimon *et al.* (14).
[c]Crawford (6).
[d]Abilgaard (1).
[e]Schultz *et al.* (16).
[f]Rosenberg and Damus (15).
[g]Yin *et al.* (20).

chymotrypsin (4). It is well established that all members of the "serine" protease family from mammalian sources share a common three-dimensional structure, in terms of the general conformation of the polypeptide backbone (7). Therefore, it seems probable that antithrombin III and α_1-antitrypsin also share common structural features, possibly including a common evolutionary precursor.

The possibilities of such structural similarities are intriguing, but the observations that we wish to report here deal more with functional aspects. In particular, we have examined the effect of heparin on antithrombin III and α_1-antitrypsin in the plasma of patients with a genetic α_1-antitrypsin deficiency. The serum of heterozygous α_1-antitrypsin-deficient patients (Pi MZ) contains 35-60% of the normal trypsin inhibitory capacity, whereas homozygotes (Pi ZZ) contain about 15%, probably owing to the action of other serum proteins (12). Patients with this deficiency have already been observed to be deficient in antithrombin III activity (8, 18). In fact, Gans and Tan (8) suggested that more than 75% of the antithrombin activity was found in the α_1-antitrypsin fraction. Figure 1 shows the average results of studies on plasma from five patients with α_1-antitrypsin deficiency (phenotype Pi ZZ). We also found that these samples contain only 30% antithrombin III activity, measured as factor Xa inhibitor activity (20) (E.T. Yin, personal communication). However, when both the normal (Pi MM) plasma and homozygous α_1-antitrypsin-deficient plasma are treated with heparin, the expected rise in Xa inhibition activity is observed not only in the normal sample, but also in the genetically deficient one, reaching values in the latter approaching that shown by normal serum in the presence of heparin. The trypsin inhibitory capacity of the

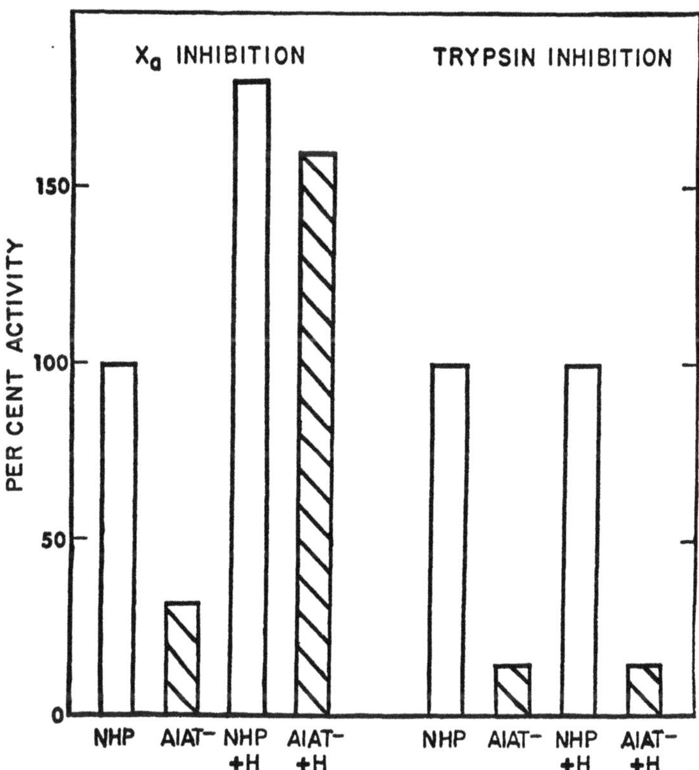

Fig. 1. Factor Xa inhibitor activity and trypsin inhibitory capacity of normal and α_1-antitrypsin-deficient plasma. NHP, normal human plasma; AIAT$^-$, α_1-antitrypsin-deficient plasma; H, heparin.

same samples is unaffected by heparin in either the normal or the α_1-antitrypsin-deficient patients.

Immunologic assays for both antithrombin III and α_1-antitrypsin have also been performed. The results can be summarized as follows: Purified human antithrombin III does not cross-react with antisera to α_1-antitrypsin; and plasma of patients genetically deficient in α_1-antitrypsin shows greatly decreased content, but normal immunoreactive antithrombin III (N. Alkjaersig, personal communication).

Thus, we conclude that the residual antithrombin III activity measured either as antithrombin or as factor Xa inhibitor activity, which is present in homozygous α_1-antitrypsin-deficient patients, is attributable to the antithrombin activity that can be stimulated by heparin and that this molecule does not significantly contribute to the antitrypsin activity of normal serum. In addition, in agreement with the observations of others a considerable portion of the antithrombin III or factor Xa inhibitor activity of normal serum is due to α_1-antitrypsin, which is not affected by heparin.

Although these results clearly indicate different structure-function relations for antithrombin III and α_1-antitrypsin, they do not answer the question of possible structural relation. The molecular characterization necessary to answer this question are in progress.

ACKNOWLEDGEMENTS

This work was supported by the Specialized Center of Research for Thrombosis grant HL 14147 and a Research grant from the National Institutes of Health, HL 16118.

R.A.B. is the recipient of a U.S. Public Health Service research career development award AM 23968.

A.D.B. is on leave of absence from the Institute of Biochemistry, Biological Research Center of The Hungarian Academy of Sciences, Szeged.

REFERENCES

1. ABILGAARD, U., Scand. J. Clin. Lab. Invest., 19 (1967) 190.
2. ABILGAARD, U., Scand. J. Clin. Lab. Invest., 20 (1967) 207.
3. ABILGAARD, U., Scand. J. Clin. Lab. Invest., 21 (1968) 89.
4. ABILGAARD, U. and EGEBERG, O., Scand. J. Haemat., 5 (1968) 155.
5. BUNDY, H.F. and MEHL, J.W., J. Biochem., 234 (1959) 1124.
6. CRAWFORD, I.P., Arch. Biochem. Biophys., 156 (1973) 215.

7. DESNUELLE, P., NEURATH, H. and OTTESEN, M., (Eds) Structure-Function Relationship of Proteolytic Enzymes, Internatl. Symposium, (Copenhagen) 1969, Academic Press.
8. GANS, H. and TAN, B.H., Clin. Chim. Acta, 17 (1967) 111.
9. HENSEN, A. and LOELIGER, E.A., Thromb. Diath. Haemorrh. (Suppl. 1) 9:1 (1963).
10. KUEPPERS, F. and BEARN, A., Proc. Soc. Exp. Biol. Med., 121 (1966) 1207.
11. LACHANTIN, G.F., PLESSET, M.L., FRIEDMAN, J.A. and HART, D.W., Proc. Soc. Expl. Biol. Med., 121 (1966) 444.
12. LAURELL, C.B. and ERIKSSON, S., Clin. Chim. Acta, 11 (1965) 395.
13. LYTTLETON, J.W., Biochem. J., 58 (1954) 8.
14. RIMON, A., SHAMASH, Y. and SHAPIRO, B., J. Biol. Chem., 241 (1966) 5102.
15. ROSENBERG, R.D. and DAMUS, P., J. Biol. Chem., 248 (1973) 6490.
16. SCHULTZE, H.E., HEIDE, U. and HAUPT, H., Klin. Wochenschr., 40 (1962) 427.
17. STEINBUCH, M., BLATRIX, C. and JOSSO, F.L., Proc. XI Congr. Intl. Soc. Haemat., (Sydney) 1966, p. 7.
18. SZCZEKLIK, A. and TERESIAK, T., Dolaczkowska, Pol. Arch. Med. Wewn., 47 (1971) 407.
19. TAN, B.H. and GANS, H., Pharm. Weekblad, 104 (1969) 717.
20. YIN, E.T., WESSLER, S. and STOLL, P.J., J. Biol. Chem., 246 (1971) 3694.
21. YIN, E.T., WESSLER, S. and STOLL, P.J., J. Biol. Chem., 246 (1971) 3712.

HEPARIN, LYSOLECITHIN, AND PLATELET FUNCTION

J. H. JOIST[+] and J. F. MUSTARD[++]
[+]Division of Laboratory Medicine, Washington University School of Medicine, Barnes Hospital, St. Louis, Missouri (USA) and [++]Department of Pathology, McMaster University, Hamilton, Ontario (Canada)

There is substantial evidence that heparin affects platelet function. Several investigators have demonstrated inhibition by heparin of platelet aggregation and the platelet-release reaction, both *in vitro* (18) and *in vivo* (3). Others, however, have reported data indicating that heparin added to blood or platelet-rich plasma *in vitro* potentiates platelet aggregation and the platelet-release reaction induced by a variety of stimuli (6, 7, 25). The latter contention appears to be supported by recent data on platelet function and concentration in heparinized patients during extracorporeal circulation (24) and several reports on thrombocytopenia induced by administration of heparin in experimental animals (8, 19) and man (10, 11). In other studies (4, 5, 21, 22), however, heparin administration was found to have little or no detectable effect on the platelet count in circulating blood. It seems clear from the results of some of the *in vitro* studies cited that heparin or, perhaps, some heparin preparations may exert a direct effect on platelets.

There is recent evidence that the effect of heparin on platelets *in vivo* may be more complex than previously realized, i.e., that it is mediated through effects of heparin on plasma lipoprotein metabolism. Heparin is known to induce the release of lipoprotein lipase from tissue sites into the bloodstream (20) in association with the liberation of free fatty acids (FFA) from triglycerides. Both induction of platelet aggregation and the platelet-release reaction and potentiation of these reactions induced by ADP, collagen, or thrombin by long-chain, saturated fatty acids have been reported (1, 12-13, 16, 23). Although FFA can cause platelet aggregation and potentiation of the effect of ADP, collagen or thrombin, it seems well established that these effects on platelets occur only under conditions in which the FFA concentration exceeds the capacity of plasma albumin to bind FFA (13). More recently, Besterman and Gillett (3) have reported a 60-70% increase in the rate of formation of lysolecithin (L-PC) in plasma obtained from patients after the intravenous administration of as little as 1,000 units of heparin. This increase in L-PC formation associated with a respective increase in plasma lecithin degradation appeared to be due to the heparin-induced release or activation of a phospholipase distinct from lecithin-cholesterol-acyl transferase. These investigators also demonstrated that both the rate and the extent of irreversible platelet aggregation induced by collagen or epinephrine were reduced in post-heparin-platelet-rich plasma.

Besterman and Gillett (3) found that L-PC added to citrated platelet-rich plasma _in vitro_ caused instantaneous and dose-dependent inhibition of only the second phase of ADP- or epinephrine-induced aggregation and suggested that the effect of L-PC on platelet aggregation was due to inhibition of the platelet release reaction. In studies reported elsewhere (14), we found that L-PC affected the so-called "release" phase of aggregation induced by ADP in citrated human platelet-rich plasma and in higher concentrations inhibited the primary phase of ADP-induced aggregation (Fig. 1). However, L-PC at similar concentrations also inhibited ADP-induced aggregation of rabbit platelets in citrated platelet-rich plasma and ADP-induced

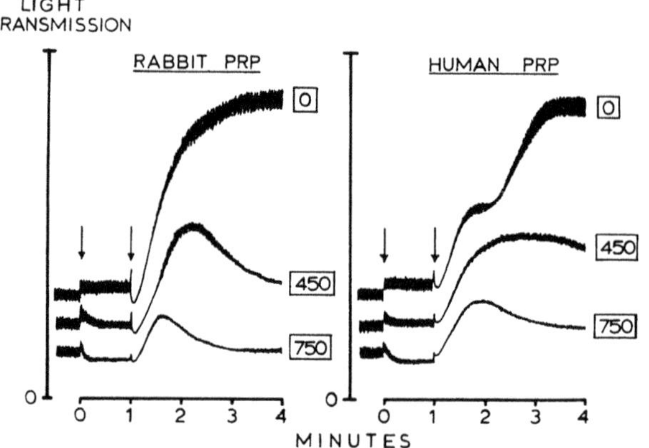

Fig. 1. Effect of L-PC on ADP-induced platelet aggregation in citrated rabbit or human platelet-rich plasma (PRP). PRP (300,000/ mm^3) was stored at room temperature, and 0.45-ml aliquots were pre-warmed at 37^{0} and transferred to an aggregometer. 0.05-ml aliquots of L-PC (to give the final concentrations indicated in the boxes after the aggregation tracings) were added at the first arrow, and 0.05-ml aliquots of ADP (to give the final concentration of 5 μM) were added at the second arrow.

aggregation in suspensions of washed rabbit or human platelets. Inasmuch as ADP-induced aggregation in these systems is not associated with release from platelets of granular constituents, L-PC appears to affect both primary ADP-induced aggregation and the release reaction. ADP-, collagen-, and thrombin-induced platelet aggregation (Fig. 2) in citrated platelet-rich plasma were shown to be partially or completely reversible over a period of 60-90 min. In contrast, when L-PC was added to suspensions of washed rabbit (2) or human (17) platelets in Tyrode solution, containing 0.35% bovine serum albumin, inhibition of ADP-, collagen-, and, particularly

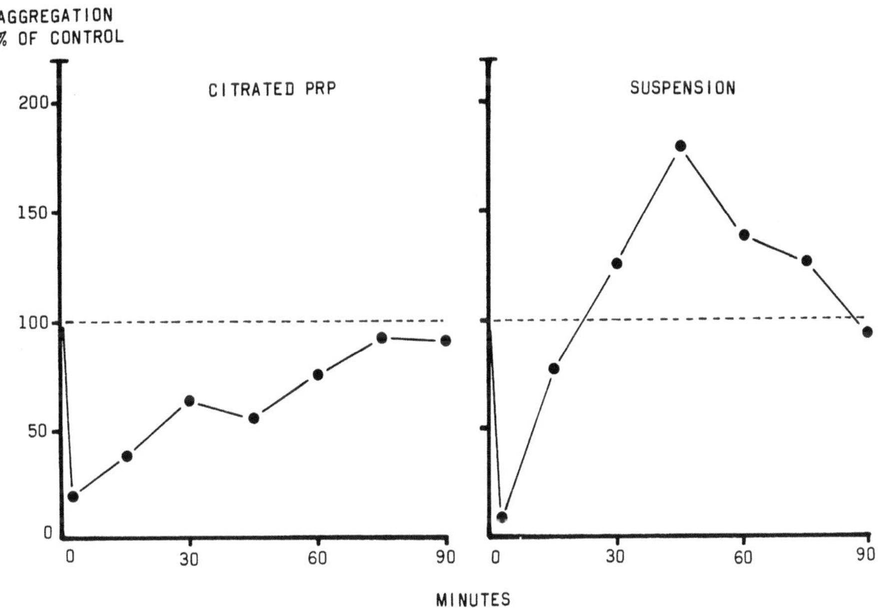

Fig. 2. Thrombin-induced aggregation of human platelets in citrated PRP or Tyrode albumin solution during prolonged incubation with L-PC. PRP or platelet suspensions ($300{,}000/mm^3$) were divided into two aliquots and incubated at 37°. L-PC was added to one aliquot of PRP or platelet suspension to give the final concentrations of 150 μM and 75 μM, respectively. Saline as control was added to the other portion. 0.45 ml aliquots were removed from the incubation mixtures at specified intervals and transferred to an aggregometer. The aggregation response to the addition of standard concentrations of thrombin (0.2-0.3 units/ml of PRP and 0.01-0.03 units/ml of suspension) was recorded, and the results were measured in chart paper units. The measurements obtained with the samples to which L-PC had been added were expressed as percent of those obtained with the samples to which saline had been added.

thrombin-induced aggregation was quickly reversible and was followed by transient potentiation of aggregation. These effects could be observed at concentrations that did not cause appreciable changes in platelet shape or platelet lysis, as indicated by loss from platelets of lactate dehydrogenase.

Elsbach et al. (9) recently demonstrated that platelets can metabolize L-PC that they take up from the medium by both acylation in the presence of CoA and ATP and deacylation to yeild water-soluble glycerophosphorylcholine (GPC) and FFA (Fig. 3). We obtained similar results by using ^{14}C choline-labeled L-PC and a suspension of washed rabbit platelets. Thrombin-induced aggregation, studied simultaneously, was markedly inhibited initially, but returned to normal within 15-20 min and was subsequently potentiated for 15-20 min. Furthermore, we obtained evidence (15) that rabbit platelets take up and metabolize endogenous plasma L-PC in a manner similar to that in which they take up and metabolize L-PC added to platelet-rich plasma or platelet suspensions. In these experiments, $^{32}PO_4$ was injected intraperitoneally into rabbits; this resulted in labeling of all the major plasma phospholipids. Blood was collected after 21 hr in 3.8% sodium citrate, and the platelet-free plasma prepared by differential centrifugation was extensively dialyzed to remove non-bound ^{32}P. Unlabeled washed rabbit platelets were then resuspended in this labeled plasma, and aliquots of the incubation mixture were removed at various intervals for platelet isolation, lipid extraction, and phospholipid analysis. As shown in Fig. 4, rapid labeling of a

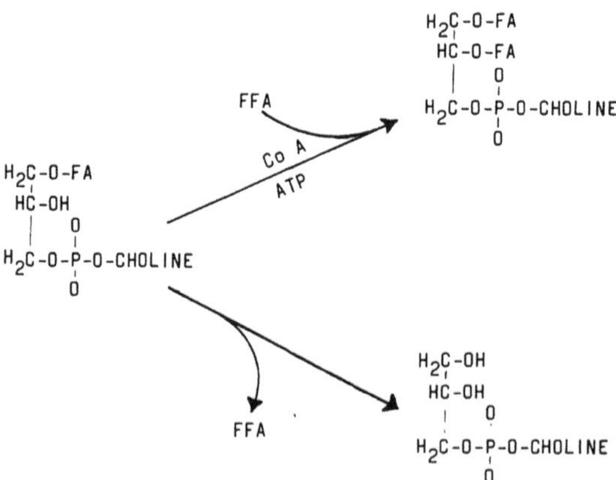

Fig. 3. Pathways of L-PC metabolism by human platelets, according to Elsbach et al. (9).

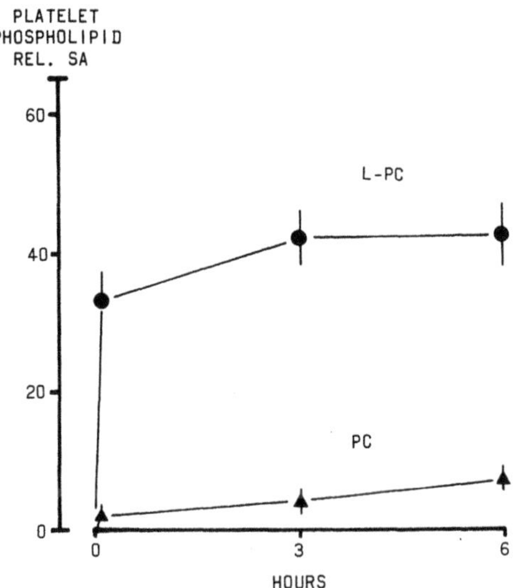

Fig. 4. Labeling of rabbit platelet phospholipids during incubation with dialyzed platelet-free plasma in which all the major phospholipids had been labeled _in vivo_ with $^{32}PO_4$. The platelet phospholipid relative specific activity (Rel. SA) denotes the fraction percent of the platelet phospholipid containing label derived from plasma phospholipids that showed significant labeling. The means and standard errors of the means of three experiments are shown.

substantial part of platelet L-PC and a slow, continuous increase in platelet phosphatidylcholine (PC) occurred, whereas little or no radioactivity was detected in the other platelet phospholipids. Substantial radioactivity was found in the water phase of the platelet extracts associated with GPC (not shown). The finding of a close association between uptake of L-PC by platelets and inhibition of platelet aggregation may be explained by assuming that L-PC covers or alters receptor sites on the platelet membrane. As L-PC is metabolized, this inhibition disappears. The association of rapid metabolism of L-PC by platelets and potentiation of platelet aggregation observed with suspensions of washed platelets lead us to investigate the effect on thrombin-induced platelet aggregation of the products of platelet metabolism of L-PC. Neither PC nor GPC up to 0.5 mM had measurable effects on thrombin-induced platelet aggregation. However, as mentioned earlier, long-chain FFA, the third product, is known to potentiate ADP-, collagen-, or thrombin-induced platelet aggregation, and it seems possible that accumulation of FFA on the platelet membrane is responsible for the potentiation. Our failure, so far, to observe potentiation of ADP- or collagen-induced aggregation in citrated or heparinized rabbit or human plasma may be due to the fact that FFA formed during L-PC hydrolysis by platelets is rapidly removed from the platelet surface by plasma albumin. Indirect support for this assumption is derived from our finding that L-PC did not cause potentiation of thrombin-induced platelet aggregation of washed rabbit platelets suspended in Tyrode solution containing 2 or 4%, instead of the standard 0.35%, albumin.

An obviously somewhat speculative hypothesis of the effects of heparin on platelet function is depicted in Fig. 5. Heparin may induce the release from vascular endothelium of phospholipase and lipoprotein lipase, which in turn may lead to liberation of both L-PC and FFA from plasma lipoproteins. Inasmuch as both substances compete for binding sites on the albumin molecule, it is conceivable that conditions could arise under which the available binding sites on plasma albumin are occupied by L-PC and FFA, perhaps allowing additional FFA liberated from lipoproteins to accumulate on the platelet membrane. This could result in an increase in platelet reactivity to surface or chemical stimuli. Conversely, saturation of the albumin binding sites by FFA may lead to increased uptake and metabolism by platelets of L-PC resulting in inhibition of aggregation and possibly (as a result of accumulation of L-PC-derived FFA on the platelet surface) to inpotentiation of platelet function. Although the evidence reviewed here seems to indicate that L-PC may be an important factor in the complex interactions among heparin, plasma lipoproteins, proteins, and platelets, its significance with respect to the regulation of platelet function _in vivo_, particularly with regard to hyperlipidemia and vascular disease, remains to be established.

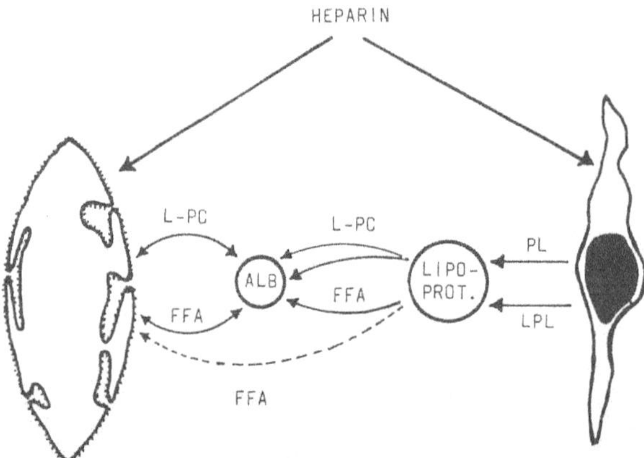

Fig. 5. Mechanisms of action of heparin on blood platelets. The structure on the left side of the figure represents an endothelial cell. PL, phospholipase; LPL, lipoprotein lipase; ALB, albumin.

REFERENCES

1. ARDLIE, N.G., KINLOUGH, R.A., GLEW, C. and SCHWARTZ, C.J., Aust. J. Exp. Biol. Med. Sci., 44 (1966) 105.
2. ARDLIE, N.G., PACKHAM, M.A. and MUSTARD, J.F., Brit. J. Haemat., 19 (1970) 7.
3. BESTERMAN, E.M.M. and GILLETT, M.P.T., Atherosclerosis, 17 (1973) 503.
4. BORCHGREVINCK, C.F., Acta. Med. Scand., 170 (1961) 365.
5. DAVEY, M.G. and LANDER, H., J. Clin. Path., 21 (1968) 55.
6. EIKA, C., Scand. J. Haemat., 9 (1972) 748.
7. EIKA, C., Scand. J. Haemat., 9 (1972) 665.
8. EIKA, C. and GODAL, H.C., Scand. J. Haemat., 8 (1971) 481.
9. ELSBACH, P., PETTIS, P. and MARCUS, A., Blood, 37 (1971) 675.
10. FIDLAR, E. and JAQUES, L.B., J. Lab. Clin. Med., 33 (1948) 1410.
11. GOLLUB, S. and ULIN, A.W., J. Lab. Clin. Med., 59 (1962) 430.
12. HASLAM, R.J., Nature, 202 (1964) 765.
13. HOAK, J.C., SPECTOR, A.A., FRY, G.L. and WARNER, E.D., Nature, 228 (1970) 1330.
14. JOIST, J.H., DOLEZEL, G., CUCUIANU, M.P., NISHIZAWA, E.E. and MUSTARD, J.F., Proc., IV. Congr., Int. Soc. Thromb. Hemostasis, Vienna, 1973.
15. JOIST, J.H., DOLEZEL, G., LLOYD, J.V. and MUSTARD, J.F., Thrombosis Res. (Suppl. 1) 4 (1974) 81.
16. MAHADEVAN, V., SINGH, H. and LUNDBERG, W.O., Proc. Soc. Exp. Biol. Med., 121 (1966) 82.
17. MUSTARD, J.F., PERRY, D.W., ARDLIE, N.G. and PACKHAM, M.A., Brit. J. Haemat., 22 (1972) 193.
18. O'BRIEN, J.R., SHOOBRIDGE, S.M. and FINCH, W.J., J. Clin. Path., 22 (1969) 28.
19. QUICK, A.J., SHANBERGE, J.N. and STEFANINI, M., J. Lab. Clin. Med., 33 (1948) 1424.
20. ROBINSON, D.S., In Comprehensive Biochemistry, (Eds. Florkin, M. and Stotz, E.H.) Elsevier, Amsterdam, 18 (1970) 51.
21. ROWSELL, H.C., GLYNN, M.F., MUSTARD, J.F. and MURPHY, E.A., Amer. J. Physiol., 213 (1967) 915.
22. SALZMAN, E.W., Transfusion, 3 (1963) 274.
23. SHORE, P.A. and ALPERS, H.S., Nature, 200 (1963) 1331.
24. THOMPSON, C., FORBES, C.D., MARTIN, E. and PRENTICE, C.R.M., Clin. Sci., 44 (1973) 21.
25. ZUCKER, M.B., Thrombos. Diathes. Haemorrh., 28 (1972) 393.

EFFECT OF LUNG AND GUT HEPARIN ON EXPERIMENTAL ARTERIAL THROMBOSIS

Hau C. KWAAN[+] and Ali HATEM[++]

[+]Hematology Section, Veterans' Administration Research Hospital, and [++]Department of Medicine, Northwestern University Medical School, Chicago, Illinois (USA)

The most widely used form of heparin for the treatment of thromboembolism has for many years been a preparation derived from porcine intestinal mucosa (hereafter referred to as gut heparin). Recently, a preparation of heparin extracted from bovine lung (hereafter referred to as lung heparin) was made available commercially for therapeutic purposes. Novak *et al*. (2), in 1972, reported that blood samples obtained after intravenous administration of gut heparin to human volunteers showed a tendency toward shorter platelet aggregation time and longer aggregate dispersion time in response to ADP, whereas volunteers given lung heparin exhibited changes in the opposition direction. They also demonstrated that platelet thrombus formation *in vitro* using Chandler's loop technique was accelerated in the blood samples obtained after injection of gut heparin, compared with that after injection of lung heparin. Goldberg *et al*. (1) reported that the incidence of loss of radial arterial pulse that followed brachial arterial catheterization with gut heparin as an anticoagulant was significantly reduced when lung heparin was used. However, comparison of the anticoagulant activity of the two heparin preparations - measuring clotting and not platelet function, such as using clotting time and partial thromboplastin time (PTT) - revealed no significant differences between the two preparation.

We report here an *in vivo* comparison of the two forms of heparin, based on an experimental arterial thrombosis model in dogs. In the thrombus obtained in this model, the platelet was shown histologically and immunologically to be the major component.

Model

The studies on the two forms of heparin have been carried out on mongrel dogs weighing 15-25 kg. Screening of the dogs consisted of the following tests: platelet count, hematocrit, PTT, and platelet aggregation. These tests were also repeated 15 and 30 min. after administration of heparin.

In each experimental animal, the vessel to be used was dissected out for about 10 cm of its length, and a 4 mm incision was made after placement of bulldog clamps at both ends of the vessel segment. A polyethylene tube 13 mm long, 3 mm in outside diameter, and 1.2 mm in inside diameter was inserted atraumatically and fixed in place with suture ties around the vessel and the tube.

Flow measurements were recorded on a chart recorder with an electromagnetic flow meter and external probes (Statham Instruments, Inc., and Carolina Medical Electronics) after dissection of the vessel, immediately on release of the clamps after insertion of the polyethylene tube, and continuously thereafter until total occlusion took place, or for an arbitrary period of 1-3 hr if the tube remained patent to any extent.

At the end of the experiment, the vessel was again clamped and a segment of the vessel and the contained tube removed. The segment of vessel was then opened, and the tube, adherent or contained thrombus, and intima were examined.

The thrombi were studied with hematoxylin and eosin, Lendrum's stain, and Carstairs' modification of Lendrum's stain. Immunofluorescent studies were also done using rabbit antidog-platelet and antidog-fibrinogen serum.

Results

The Thrombus

In almost every instance, the thrombus or plug that formed was at the distal end of the tube, relative to blood flow. It was round to oval, and it appeared firmly adherent to the outer lip of the tube. The plug was mainly gray with brown and red parts, depending on the length of time before the tube was removed after total occlusion. Sometimes, a circular extension would be found growing into the tube. The plug would vary from minute to 4 mm in diameter, depending on whether parts of it had broken off.

Histologic sections demonstrated masses of tightly adherent platelets of typical morphology and staining properties and fibrin that varied from minimal to substantial, depending on the segment of the thrombus studied and, again, the length of time before the tube and thrombus were removed from the vessel after total occlusion. White cells and mononuclear cells were evident within the masses of platelets, especially in areas where platelets appeared degranulated or contracted; this suggested a more direct role for white cells in the dynamics of thrombus formation. Closer examination of the platelet masses showed them to be distributed lamellarily.

Immunofluorescent staining with labeled antidog-platelet serum showed strong fluorescence in the thrombi, and with antidog-fibrinogen serum, weakly positive. The latter staining is compatible with the picture of staining of platelet fibrinogen.

Blood Flow

Control animals. - After a definite lag of 3-5 min during which flow continued at the initial rate, significant platelet deposition occurred, as indicated by a reduction in flow. This progressed until complete occlusion took place. The times required to achieve a flow rate of 75%, 50%, 25%, and 0% of the initial rate are shown in Fig. 1 to have averaged 13, 20, 25, and 34 min respectively, in 13 animals. The times of individual animals are shown in Fig. 2.

Animals given gut heparin. - Compared with the controls, the flow rate in dogs given gut heparin showed an earlier initiation of the occlusive process, as indicated by the shorter time taken to

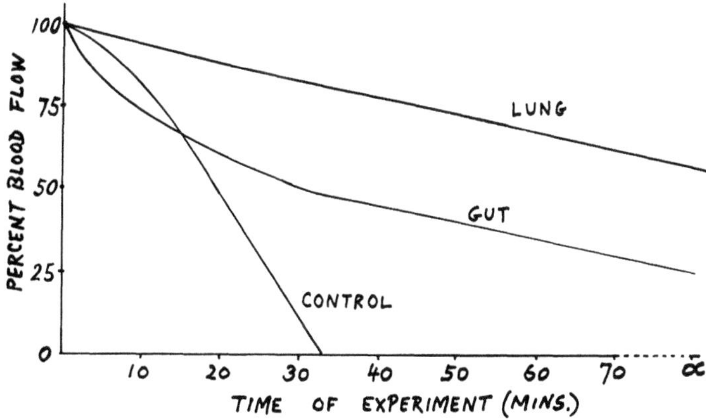

Fig. 1. Mean reduction in blood flow as percent of initial flow in arteries after insertion of polyethylene tubing, in control dogs, dogs on gut heparin and dogs on lung heparin.

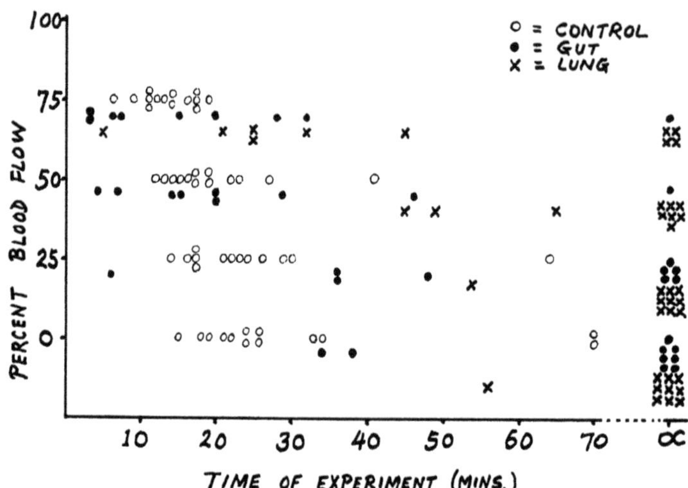

Fig. 2. Changes in flow rate expressed as percent of initial flow in control dogs, dogs on gut heparin and dogs on lung heparin.

achieve a 75% flow rate, with eight of the nine animals reaching this degree of flow reduction (Fig. 1 and 2). However, the reduction in flow rate was not consistently progressive, owing to breakage and embolization of parts of the thrombus. This became evident as resurgence of flow to varying degrees up to the initial rate. Five of the nine animals in this group did not achieve reduction in flow to 25% of the initial rate.

Animals given lung heparin. - Of the 10 dogs given lung heparin, decrease in flow to 75% of the initial rate could not be achieved in four. In the other six, this flow rate was reached much later than in the dogs given gut heparin. Only one dog underwent a reduction in flow to 25% of the initial rate.

Discussion

A thrombus formed primarily from platelet deposition is produced in the present model of arterial thrombosis. Our observations suggested platelet deposition in two phases. A primary adhesion and aggregation of platelets was the first step of platelet deposition onto the polyethylene tubing. Release reactions of platelets in the primary thrombus resulted in secondary aggregation of circulating platelets. This led to additional deposition and growth of the thrombus. Fibrin formation took place on the surface, thus, after some period, a layering effect was seen.

Apparently, heparin by its anticoagulant action, interfered with the second phase of platelet thrombus formation. Both heparin preparations were able to prevent complete occlusion of the blood flow in our experimental model. However, in the early or primary phase of platelet thrombus formation, the gut heparin accelerated the platelet deposition, whereas the lung heparin did not. The number of experiments is too small to make firm conclusions, but such finding of *in vivo* effects on platelet thrombus formation support previous *in vitro* observations of increased platelet aggregation by gut heparin. The present arterial thrombus model is much more closely akin to the thrombus seen clinically on prosthetic heart valves, on arteriovenous shunts, and in reconstructive arterial bypasses. Thus, a comparison of the efficacy of the two forms of heparin is warranted for these clinical conditions.

REFERENCES

1. GOLDBERG, E., O'REILLU, M. and CHAITHIRAPHAN, S., Lancet, 1 (1972) 789.
2. NOVAK, E., SEKHAR, N.C., DUNHAM, N.W. and COLEMAN, L.L., Clin. Med., (1972) 22.

DISCUSSION OF THE PAPERS OF JOIST AND MUSTARD AND KWAAN AND HATEM

BRINKHOUS

Dr. Jaques presented a paper in the *Journal of Laboratory and Clinical Medicine* about 1948 on this subject and perhaps he'd like to comment on this.

JAQUES

Perhaps the only thing I should say about that paper is that Dr. Quick never forgave us for the fact that the editor received our paper 24 hours before the paper by Quick, Shanberge and Stefanini. I will make one correction; I think it is important to have the story straight. We, like every group, have our fair number of legends and stories that ain't so. Actually the first commercial heparin for clinical use, that was produced by a number of manufacturers starting in 1936, was heparin from beef lung. As far as I could determine and I was in pretty close contact with manufacturers all through those years, this was the sole source of heparin to well into the 1950's. Then there was this gradual change which was due, as pointed out yesterday, to the fact that pork intestinal mucosa was a waste product from the manufacture of sausage casing. This provided a cheap source for heparin and that is the reason for the shift.

BRINKHOUS

I might comment that Dr. Mason and I several years ago looked at heparin and platelet aggregation and came to the conclusion that heparin acted in the later phases of aggregation, the stabilization phase, in which thrombin and fibrin are important but it did not have any action earlier so we just assumed it was the heparin cofactor action that inhibited the maturation phase of the platelet aggregate.

LIPOPROTEIN LIPASE

Thomas OLIVECRONA, Olle HERNELL, and Torbjörn EGELRUD

Department of Chemistry, Umeå University, Umeå (Sweden)

Intravenous injection of heparin releases lipase activity from tissue sites to the circulating blood (8, 20, 30, 31, 34). One major enzyme activity released is the so-called lipoprotein lipase (LPL). This enzyme is inhibited by high salt concentration, and its activity against long-chain triglycerides is greatly enhanced by a serum component (31) identified as the apolipoprotein CII (also called R-Glu or apo-LP-Glu) of the very low-density lipoproteins (9, 23, 36). Enzymes with these characteristics are present in several tissues - e.g., adipose tissue (34), skeletal and heart muscle (31), and lactating mammary gland (47) - and, in some species, in milk (24, 33). They probably have an important role in the metabolism of the triglyceride-rich plasma lipoproteins (42). Electron microscopic and other studies indicate that in some tissues these lipoproteins can attach to the capillary endothelium, where they are then acted on by LPL (5). The triglycerides are hydrolyzed, and fatty acids and possibly also partial glycerides are transferred into the tissue through the capillary endothelium, leaving a remnant lipoprotein, a particle that has lost most of its triglyceride. This remnant lipoprotein is further metabolized by other, less well-known systems.

LPL activity of some tissues varies with nutritional and hormonal state. LPL activity is high in adipose tissue in the fed state, but it decreases on fasting (25, 41, 46). In heart and skeletal muscle, the opposite may be true (6, 26, 46). The tissue uptake of fatty acids from circulating lipoprotein triglycerides varies in parallel with enzyme activity (4, 6, 17). Variations in enzyme activity in individual tissues may determine not primarily the total rate at which triglyceride is cleared from the plasma, but rather the distribution of what is taken up by the tissues.

Injection of heparin thus releases into the blood enzymes that normally act at the capillary endothelium in some tissues. This accelerates the clearing of lipoprotein triglyceride from the blood, probably by hydrolysis in the circulating blood, and as a result the plasma triglyceride concentration decreases, whereas that of free fatty acids may increase (42). If the uptake by, for instance, adipose tissue depends on the LPL activity in that tissue, one would expect the uptake to decrease as the enzyme activity is washed out from its tissue sites by heparin. The data in Table I show that heparin does indeed cause a dramatic redistribution of the tissue uptake of triglyceride fatty acids of chylomicron. Chyle was obtained from a donar rat fed labeled (with 9, 10-[^{3}H]palmitic acid) fatty acids and injected intravenously into two groups of five anesthetized rats each. Each rat received 0.5 ml of chyle. Blood and tissue samples were obtained 10 min later, and their radioactivity determined. One groups of rats had received heparin at 2 mg/kg of body weight 15 min before chyle injection; the other group served as controls and received saline, instead of heparin. It can be seen that heparin injection accelerated the removal of radioactive lipid from the blood, decreased the fraction in skin (used as a sample of adipose tissue), and increased that in the liver.

TABLE I

Effect of Heparin on Metabolism of Labeled Chylomicra in rats[a]

Group	Radioactivity 10 min after Injection, %		
	In Blood	In Liver	In Skin
Saline (control)	20.3 $\pm$ 9.5	18.8 $\pm$ 6.1	10.6 $\pm$ 2.6
Heparin	5.1 $\pm$ 1.5	42.4 $\pm$ 3.4	4.7 $\pm$ 0.6
Significance	$p < 0.02$	$p < 0.001$	$p < 0.005$

[a]Data from Brown and Olivecrona (7). Data were obtained in an investigation designed to study some effects of insulin on lipid metabolism. All rats were insulin-treated alloxan diabetic rats. There were no significant difference in body weight or blood sugar content between the groups.

The release of the enzyme occurs rapidly after the injection of heparin (43, 44). Because heparin does not leave the blood rapidly, this suggests that the enzyme is close to the circulating blood. Histochemical studies with perfused adipose tissue (5) and with mammary gland (45) also suggest that one site of action of the enzyme is at the luminal surface of the capillary endothelium. There is a considerable excess of LPL in the tissue, compared with the amounts necessary for the uptake that occurs. Probably, only a fraction of the tissue enzyme is in such a location that it can act on the plasma triglycerides. Direct support of this hypothesis has been obtained in experiments with perfused rat hearts (6). These hearts will take up a considerable amount of triglyceride from the perfusing medium. If heparin is added to the medium, some but not all of the LPL will be released into the perfusate. Later, the heart will no longer take up much triglyceride from a perfusing medium; this suggests that the fraction of LPL active in facilitating the uptake of triglyceride has been washed out with the heparin perfusion. In studies with isolated adipocytes, Schotz and associates have whown that the enzyme is actively secreted by these cells (51). This process may be stimulated by heparin (49). They also have shown that adipocytes can secrete more enzyme activity in vitro than can be demonstrated in the cells at the start of the experiment, even in the presence of inhibitors of protein synthesis (50). The inference from these and other studies is that the cells may contain inactive precursors of the enzyme, which (by mechanisms as yet unknown) can be transformed into the active enzyme.

LPL-Heparin Complexes

Does heparin form a complex with the enzyme under physiologic conditions? To study this question, we used an agarose gel with heparin covalently linked to it (27) - i.e., an insolubilized preparation of heparin - and bovine milk as a source of enzyme. Skim milk was obtained by centrifugation and then dialyzed to remove the calcium and therby break the casein micelles to which the enzyme is bound. It was then further dialyzed against a buffer with a pH of 7.4 and ionic strength of 0.16 - i.e., a pH and ionic strength about those of blood plasma. The gels were preequilibrated with 0.157 M NaCl in 5 mM Na-barbital at a pH of 7.4 and at 4°. In test tubes, 1 ml of the gel was mixed with 1 ml of the dialyzed skim milk, the system was allowed to equilibrate at 4° for 1 hr with intermittent shaking and was then centrifuged, and the amounts of enzyme and of total protein remaining in the supernatant were measured (39). After equilibration of the dialyzed skim milk with the heparin-Sepharose, most of the enzyme activity had disappeared from the supernatant (Table II). After equilibration with a number of other ion exchangers, most of the enzyme activity remained in the supernatant. The binding to heparin thus implies a specific affinity. It should be

TABLE II

Binding of Lipoprotein Lipase to Heparin-Sepharose[a]

Gel	Lipoprotein Lipase in Supernatant, units/ml[b]
Heparin-Sepharose	9.5
Ethanolamine-Sepharose	93.5
CM-Sephadex C 50	87
SE-Sephadex C 50	90
DEAE-Sephadex A 50	102.5

[a]Data from Olivecrona et al.[39]
[b]One unit of enzyme catalyzes the release of 1 micromole of fatty acid per minute from triglyceride under optimal conditions at 37°.

noted that the experiments were carried out with a crude enzyme preparation containing large amounts of other milk proteins and at an ionic strength of 0.16.

Purification of LPL

This affinity suggested a method for purifiying the enzyme. Because of its instability and its tendency to aggregate, LPL has thus far resisted purification by the classic methods of protein purification. The enzyme in postheparin plasma has been purified by Fielding (15, 16), who utilized the binding of the enzyme to its lipoprotein substrate as the main purification step. In our hands, this method does not work well with the milk enzyme. We believe, however, that bovine milk is the most advantageous source of enzyme (33). To obtain the same amount of enzyme activity as is present in 1 liter of skim milk, one would need at least 2 liters of rat postheparin plasma, or 60 kg of hen adipose tissue. Furthermore, the enzyme in milk is more stable than the tissue enzymes we have tried and is therefore easier to work with. The LPL is the main and perhaps the only lipolytic enzyme normally present in the milk (H. Castberg, T. Egelrud, and T. Olivecrona, unpublished observations).

Presumably, it occurs in the milk as a result of leakage from the mammary tissue, where it is present in high activity during lactation (47); it probably has no function in the milk.

The main purification is obtained by affinity chromatography on heparin-Sepharose (13). The enzyme can be eluted from the column by increased salt concentration at a pH of 7.4. This experiment shows that the binding to heparin is reversible and involves ionic interactions. The method yielded a preparation that was more than 90% pure, as judged by gel electrophoresis in several systems. The constituent polypeptide chain was estimated by SDS-gel electrophoresis to have a molecular weight of about 62,000 and to have bound carbohydrate, as evidenced by a positive PAS reaction. The preparation has been valuable for studies of the substrate specificity of the enzyme (12, 38) and its interaction with apolipoproteins from human plasma (22). However, the amounts obtained thus far have not been sufficient for studies of the protein chemistry of the enzyme. Attempts are now being made to scale up the procedure to obtain material for more detailed studies of the enzyme and its interactions.

The amount of enzyme estimated to be present in human tissues is too small to encourage attempts to purify them for studies of their chemical properties. However, we have recently adapted our method of enzyme purification to human milk (24). In bovine milk, most of the enzyme is in the skim milk, where it is bound to the casein micelles (Table III). In contrast, most of the enzyme in human milk is bound to the lipid droplets and is thus found in the cream fraction. Therefore, although the initial enrichment steps were different, the main purification was again by affinity chromatography on heparin-Sepharose. We obtained a preparation that was about 50% pure, as judged by gel electrophoresis, and that showed many similarities to the bovine milk enzyme. Thus, by SDS-polyacrylamide gel electrophoresis, the constituent polypeptide chain was estimated to have a molecular weight of about 63,000, and it gave a positive PAS reaction. Furthermore, as far as we have studied, the two enzymes have the same subtrate specificity and probably about the same specific activity. However, there is less enzyme activity in human milk than in bovine milk, and the enzyme activity varies widely between different mothers (24), so it would be difficult to obtain sufficient quantities of the human-milk enzyme for final purification and chemical studies.

We have raised a rabbit antibody to the purified bovine-milk enzyme (T. Egelrud, O. Hernell, and T. Olivecrona, unpublished data). The antibody completely inhibits the activity of the bovine-milk enzyme and it inhibits, in a similar fashion, the activity of the human-milk LPL. We have found that the antibody can also inhibit the activity of human postheparin-plasma LPL; the stoichiometry of this inhibition was very similar to that for the human-milk LPL, which suggests a relation between the enzyme in the milk and in

TABLE III

Lipoprotein Lipase in Bovine and Human Milk

Milk	Total Enzyme Activity, units/ml[a]	Fraction of Enzyme Activity in Cream, % of total	Fraction of Enzyme Activity in Skim Milk, % of total
Bovine	100-200	5-10	90-95
Human	0-90	> 85	< 15

[a]One unit of enzyme catalyzes the release of 1 micromole of fatty acid per minute from triglyceride under optimal conditions at 37°.

postheparin-plasma. These studies indicate that there are many similarities between the properties of the bovine-milk enzyme and those of the human-milk and postheparin-plasma enzyme, which encourages us to continue our efforts to obtain quantities of the purified bovine-milk enzyme for further studies.

Effect of Heparin on the Stability and Activity of LPL

The binding experiments and the affinity chromatography demonstrate that the enzyme and heparin can form a complex, if they are together at physiologic pH and ionic strength. This has now been shown for LPL from several origins - e.g., bovine milk (13, 39), human milk (24), human postheparin plasma (14; Egelrud *et al*., unpublished data), pig adipose tissue (2), and hen adipose tissue (11). Thus, affinity for heparin may be a general property of enzymes of this class. This raises the question of whether the enzyme is active when in a complex with heparin - and perhaps only when in a complex with heparin (29). After purification on heparin-Sepharose, the enzyme preparation contains some heparin-like material. It is difficult to quantitate these small amounts of heparin accurately, but they probably correspond to less than 2% of the weight of the enzyme (28). It is reasonable, but it has not been proved, that this heparin represents bleeding from the column-bed material. If the enzyme activity is measured under otherwise optimal conditions, addition of small amounts of exogenous heparin does not markedly influence it. Thus, added heparin does not appear to be a necessary cofactor for the enzyme activity under optimal conditions.

To test whether the small amounts of heparin-like material in the preparation might be necessary, the enzyme preparation was exposed to a heparin eliminase, which under identical conditions was shown to degrade added heparin. In this experiment (28), enzyme was preincubated with the heparin eliminase, and the remaining activity was then measured. As a control, the preincubation was also carried out with boiled heparin eliminase. The active heparin eliminase did not promote the inactivation of the enzyme. Thus, it appears unlikely that a heparin-like material is a necessary cofactor for the enzyme activity. However, addition of heparin at 1 μg/ml at an ionic strength of 0.16 did not markedly change the enzyme activity (28). This corresponded to a molar ratio of heparin to enzyme of about 60:1 and suggests that, under otherwise optimal conditions, the enzyme activity is the same when it is alone as when it is a complex with heparin. Large amounts of heparin inhibit the enzyme activity.

A characteristic property of these enzymes is that their activity is inhibited by high salt concentration in the assay medium (30). Heparin impedes this inhibition at some salt concentrations (28). The low activity observed at high salt concentration could be due to either irreversible loss of enzyme activity or reversible inhibition of the enzyme. To differentiate these alternatives, we studied the time course of the release of fatty acid (28). In the absence of heparin, the release was linear with time at ionic strengths of 0.05 and 0.16. At an ionic strength of 0.40, the time course was not linear, suggesting a time-dependent progressive loss of enzyme activity. At the same ionic strength and a heparin concentration of 1 μg/ml, the release process was again approximately linear, demonstrating that heparin stabilizes the enzyme under these conditions. Similar results were obtained in experiments on the heat inactivation of enzyme (28). Stabilization of LPL activity by heparin had been shown by others (16, 40). However, in addition to its effect in stabilizing the enzyme, heparin also has a direct effect on the inhibition of enzyme activity by salt. When the enzyme was incubated under the same conditions but at 20°, instead of 37° no progressive loss of enzyme activity was observed, even in the absence of heparin (28). Yet, in the presence of heparin, the enzyme activity was increased about twofold. Our studies suggest that the enzyme has about the same catalytic activity under optimal conditions, when alone and with heparin. However, the enzyme-heparin complex has several properties distinguished from those of the enzyme alone. For instance, heparin stabilizes the enzyme activity and makes it less sensitive to inhibition by high salt concentration in the assay medium. An additional effect of heparin is to increase the solubility of the enzyme; when the eluate from the heparin-Sepharose column is dialyzed to decrease the salt concentration, the enzyme precipitates out of solution, but no such precipitation occurs in the presence of heparin.

Previous studies using crude enzyme preparations had suggested a variety of effects of heparin on LPL. Many of these observations need to be reevaluated with more purified systems. Lipase activity is released by intravenous injection of heparin (8, 14-16, 18, 20, 29, 35, 37, 48), but also by injection of a large variety of other polyanions (for references, see Bernfeld and Kelley (3)). Because at least two different lipase activities are released, we do not know whether these other substances release both or only one of these enzyme activities. LPL activity in crude systems is often enhanced by a suitable concentration of heparin (3, 30). This activation has been reported to require a polysaccharide containing N-sulfate groups and thus may be more specific than the release of lipase activity (3). The activation is not observed with all enzyme preparations and may also vary with the substrate. At higher concentrations, heparin usually inhibits the enzyme activity (3, 28, 32). In crude assay systems, complex effects may be observed. For instance, when crude skim milk was incubated in an assay in which whole serum was the source of activator, the addition of heparin at 1 μg/ml gave about a tenfold stimulation of the enzyme activity (28). However, little or no effect of heparin was observed when purified lipoproteins or purified enzyme was used. The effect of heparin apparently was removal of an inhibition, rather than a true stimulation (28). The many conflicting results in the literature on effects of heparin on enzyme activity in crude assay systems should thus be interpreted with great caution. The possibility that heparin exerts its effect through interaction with other components of the system - e.g., with the substrate lipoproteins - should always be considered.

Other Enzymes Released into the Blood by Heparin

LPL is not the only enzyme activity released into the circulating blood by heparin injection. Another triglyceride lipase is also released (14, 18, 35, 37, 48), probably from the liver (1, 35, 37, 53). This enzyme differs markedly from the LPL. It is inhibited by serum, whereas its activity is not decreased by high salt concentration. This enzyme also binds to heparin-Sepharose and has been purified by affinity chromatography on this gel (1, 14, 18), from which it is eluted before and well separated from the LPL activity. It has about the same minimal molecular weight as the LPL and is also a glycoprotein (14). However, it does not react with the antibody to the purified bovine-milk enzyme (Egelrud et al., unpublished data), as the postheparin-plasma LPL does. The time course of its appearance in the blood is also somewhat different from that of LPL. There are no indications that this "liver lipase" requires a heparin-like substance for activity; on the contrary, low concentrations of heparin inhibit it. This inhibition is seen at low, but not at high salt concentrations and is increased by serum.

A diamine oxidase is also released into the blood by heparin

(52). This occurs in all vertebrate species that have been studied (for references, see Hansson (21)). This enzyme is passed into the bloodstream from the liver in the guinea pig, but from the intestine in the rat and in the rabbit. A considerable proportion of the enzyme reaches the blood via the lymph in contrast with the triglyceride lipase activities, which are low in lymph (10). The rise of diamine oxidase in the blood occurs rapidly (10), suggesting that it, like the lipases, is released from sites close to the circulating blood. The biologic importance of the diamine oxidase released into the blood is not settled, but there have been suggestions that it plays a role in plasma histamine clearance in anaphylaxis (19).

Conclusions

Several enzymes are released into the bloodstream by heparin injection. The release probably occurs from tissue sites close to the circulating blood, as evidenced by the rapid appearance of the enzymes in the blood. At least two triglyceride lipases are released. These enzymes probably have important roles in the metabolism of the circulating plasma lipoproteins. We have studied mainly lipoprotein lipase, which is a biochemically interesting enzyme with a high affinity for heparin and whose optimal activity against long-chain triglycerides requires interaction with a specific apolipoprotein from the very low-density plasma lipoproteins. Through affinity chromatogrpahy, we hope to obtain sufficient quantities of this enzyme from bovine milk for more detailed studies of its chemistry and interactions. Release of the enzyme from its tissue sites into the circulating blood by heparin injection causes a profound derangement of the metabolism of circulating plasma lipoproteins, which are cleared more rapidly from the blood and with a different tissue distribution from normal.

ACKNOWLEDGEMENTS

The original work reported in this paper was supported by the Swedish Medical Research Council (B 03X-727). The experiments on the binding of enzyme to heparin and on the effects of heparin on the stability and activity of the enzyme were carried out in collaboration with Drs. Per-Henrik Iverius and Ulf Lindahl at the Institute of Medical Chemistry in Uppsala.

REFERENCES

1. ASSMAN, G., DRAUSS, R.M., FREDRICKSON, D. and LEVY, R.I., J. Biol. Chem., 248 (1973) 1992.

2. BENSADOUN, A., EHNHOLM, C., STEINBERG, D. and BROWN, W.V., J. Biol. Chem., 249 (1974) 220.
3. BERNFELD, P. and KELLEY, T.F., J. Biol. Chem., 238 (1963) 1236.
4. BEZMAN, A., FELTS, J.M. and HAVEL, R.J., J. Lipid Res., 3 (1962) 427.
5. BLANCHETTE-MACKIE, E.J. and SCOW, R.O., J. Cell Biol., 51 (1971) 1.
6. BORENSZTAJN, J. and ROBINSON, D.S., J. Lipid Res., 11 (1970) 111.
7. BROWN, D.F. and OLIVECRONA, T., Acta Physiol. Scand., 66 (1966) 9.
8. BROWN, R.K., BOYLE, E. and ANFINSEN, C.B., J. Biol. Chem., 204 (1953) 423.
9. BROWN, W.V. and BAGINSKY, M.L., Biochem. Biophys. Res. Comm., 46 (1972) 375.
10. DAHLBÄCK, O., HANSSON, R., TIBBLING, G. and TRYDING, N., Scand. J. Clin. Lab. Invest., 21 (1968) 17.
11. EGELRUD, T., Biochim. Biophys. Acta, 296 (1973) 124.
12. EGELRUD, T. and OLIVECRONA, T., Biochim. Biophys. Acta, 306 (1973) 115.
13. EGELRUD, T. and OLIVECRONA, T., J. Biol. Chem., 247 (1972) 6212.
14. EHNHOLM, C., SHAW, W. and BROWN, W.V., Report and abstract No. 58 at the Third International Sympsoium on Atherosclerosis in West-Berlin, 1973.
15. FIELDING, C.J., Biochim. Biophys. Acta, 206 (1970) 109.
16. FIELDING, C.J., Biochim. Biophys. Acta, 178 (1969) 499.
17. GARFINKEL, A.S., BAKER, N. and SCHOTZ, M.C., J. Lipid Res., 8 (1967) 274.
18. GRETEN, H., WALTER, B. and BROWN, W.V., FEBS letters, 27 (1972) 306.
19. HAHN, F., KRETZSCHMAR, R., TESCHENDORF, H. and MITZE, R., Int. Arch. Allergy, 39 (1970) 339.
20. HAHN, P.F., Science, 98 (1943) 119.
21. HANSSON, R., Scand. J. Clin. Lab. Invest., 31 (1973) 129.
22. HAVEL, R.J., FIELDING, C.J., OLIVECRONA, T., SHORE, V.G., FIELDING, P.E. and EGELRUD, T., Biochemistry, 12 (1973) 1828.
23. HAVEL, R.J., SHORE, V.G., SHORE, B. and BIER, D.M., Circ. Res., 27 (1970) 595.
24. HERNELL, O. and OLIVECRONA, T., J. Lipid Res., 15 (1974) 367.
25. HOLLENBERG, C.H., Am. J. Physiol., 197 (1959) 667.
26. HOLLENBERG, C.H., J. Clin. Invest., 39 (1960) 1282.
27. IVERIUS, P.-H., Biochem. J. (London), 124 (1971) 677.
28. IVERIUS, P.-H., LINDAHL, U., EGELRUD, T. and OLIVECRONA, T., J. Biol. Chem., 247 (1972) 6610.
29. KORN, E.D., in Colloq. Intern. Centre nat. recherche sci. (Paris), Paris, 99 (1961) 139.
30. KORN, E.D., J. Biol. Chem., 215 (1955) 1.
31. KORN, E.D., J. Biol. Chem., 215 (1955) 15.
32 KORN, E.D., J. Biol. Chem., 237 (1962) 3423.
33. KORN, E.D., J. Lipid Res. 3 (1962) 246.

34. KORN, E.D. and QUIGLEY, T.W., J. Biol. Chem., 226 (1957) 833.
35. KRAUSS, R.M., WINDMUELLER, H.G., LEVY, R.I. and FREDRICKSON, D.S., J. Lipid Res., 14 (1973) 286.
36. LaROSA, J.C., LEVY, R.I., HERBERT, P., LUX, S.E. and FREDRIKSON, D.S., Biochem. Biophys. Res. Comm., 41 (1970) 57.
37. LaROSA, J.C., LEVY, R.I., WINDMUELLER, H.G. and FREDRICKSON, D.S., J. Lipid Res., 13 (1972) 356.
38. NILSSON-EHLE, P., EGELRUD, T., BELFRAGE, P., OLIVECRONA, T. and BORGSTRÖM, B., J. Biol. Chem., 248 (1973) 6734.
39. OLIVECRONA, T., EGELRUD, T., IVERIUS, P.-H. and LINDAHL, U., Biochem. Biophys. Res. Comm., 43 (1971) 524.
40. ROBINSON, D.S., Quart. J. Exp. Physiol., 41 (1956) 195.
41. ROBINSON, D.S., J. Lipid Res., 1 (1960) 332.
42. ROBINSON, D.S., in Compr. Biochem. (Ed. Florkin, M. and Schotz, E.H.) Elsevier, Amsterdam, vol. 18, 1970, p. 51.
43. ROBINSON, D.S. and HARRIS, P.M., Quart. J. Exptl. Physiol., 44 (1959) 80.
44. ROBINSON, D.S. and JENNINGS, M.A., J. Lipid Res., 6 (1965) 222.
45. SCHOEFL, G.I. and FRENCH, J.E., Proc. Roy. Soc. B., 169 (1968) 153.
46. SCHOTZ, M. and GARFINKEL, A.S., Biochim. Biophys. Acta, 270 (1972) 472.
47. SCOW, R.O., MENDELSON, C.R., ZINDLER, O., HAMOSH, M. and BLANCHETTE-MACKIE, E.J., in Dietary Lipids and Postnatal Development (Ed. Galli, C., Jacini, G. and Pecile, A) Raven Press, New York, 1973, p. 77.
48. SHORE, B. and SHORE, V., Amer. J. Physiol., 201 (1961) 915.
49. STEWART, J.E. and SCHOTZ, M.C., J. Biol. Chem., 249 (1974) 904.
50. STEWART, J.E. and SCHOTZ, M.C., J. Biol. Chem., 246 (1971) 5749.
51. STEWART, J.E., WHELAN, C.F. and SCHOTZ, M.C., Biochem. Biophys. Res. Comm., 34 (1969) 376.
52. TRYDING, N., Scand. J. Clin. Lab. Invest., Suppl. 17, 86 (1965) 196.
53. WAITE, M. and SISSON, P., J. Biol. Chem., 248 (1973) 7985.

"LIPOLIPIN": A GLYCOPROTEIN INHIBITOR OF POSTHEPARIN PLASMA LIPOPROTEIN LIPASE

Premanand V. WAGH
Connective Tissue Laboratory, Veterans Administration Hospital, Little Rock, Arkansas 72206, and Department of Biochemistry, University of Arkansas Medical Center, Little Rock, Arkansas 72201 (USA)

It has been established that the uptake of circulating blood triglyceride fatty acids by extrahepatic tissues is mediated through the hydrolysis of triglycerides by lipoprotein lipase (LPL). LPL hydrolyzes triglycerides only in the presence of lipoproteins, and its activity is considered essential for a normal rate of clearance of triglycerides from plasma (15).

The inhibition of *in vitro* activity of LPL from postheparin plasma or tissues by a wide variety of substances - such as 1 M NaCl, divalent cations, and protamine sulfate - has been reviewed elsewhere (10, 11, 16). Nikkila first proposed the presence of an endogenous inhibitor of LPL on the basis of an observation that partially purified LPL was more active than the original postheparin plasma (13). The presence of LPL inhibitors in serum and plasma (2, 5, 7), in platelets and white blood cells (3, 12), and in other tissues (8, 9) has since been reported, but, because the inhibition of LPL was measured by using either crude extracts or partially purified fractions derived therefrom, the exact nature of the inhibition could not be ascertained. Nevertheless, studies of Hollett (5, 6) and of Ishii (8) indicate that crude glycoprotein preparations obtained from plasma, serum, and various organs inhibited postheparin plasma LPL activity both *in vitro* and *in vivo*. It was therefore essential to learn whether their crude preparations contained a pure glycoprotein or some macromolecule other than a glycoprotein that inhibited LPL. The kind of evidence necessary to evaluate the physiologic role of the inhibitor, if it were found to be glycoprotein, would come from isolating such a species in a highly purified state and studying its properties with respect to the mechanism of inhibition.

We recently isolated a purified glycoprotein from the intimal region of the porcine aorta (18). Because this glycoprotein is unique and is an inhibitor of postheparin plasma LPL, I propose to call it "lipolipin" (from lipoprotein lipase inhibitor). The evidence of its inhibition of LPL is presented in this report. A preliminary discussion of some of these data has been presented elsewhere (19).

Materials and Methods

Lipolipin (porcine aortic intimal glycoprotein) was obtained essentially by a procedure described earlier (18).

Source of LPL. - Healthy mongrel dogs, fed *ad libitum* and weighing 15 kg, were anesthesized with pentobarbital (25 mg/kg of body weight). Heparin (lipo-hepin, 1000 u/ml) was administered intravenously (70 u/kg of body weight), and blood was drawn from the femoral artery 5 min later and collected in precooled centrifuge tubes containing enough sodium citrate to make the final concentration of the salt 0.02 M. Postheparin plasma was collected by centrifuging at 1980 *g* for 20 min in a refrigerated centrifuge and stored at -20° This plasma was used as the enzyme source. There was no loss of activity for 4 months under the storage conditions.

Assay. - The extent of lipolysis was measured by the release of free fatty acids (FFA) in the incubation mixture. The final volume and pH for each assay were 0.8 ml and 8.2, respectively, throughout the experiments. The incubation mixture contained postheparin plasma (0.2 ml), Ediol (stabilized coconut oil emulsion, Calbiochem) diluted with water and containing various concentrations of triglycerides (0.1 ml), and 0.2 M NH_4Cl-NH_4OH buffer at a pH of 8.6 (0.5 ml). Whenever the materials were to be tested for LPL inhibition, they were included in the assay system after dissolution in the buffer and substituted for buffer without added ingredients. All incubations were carried out at 30° on a Dubnoff metabolic shaker at 160 cycles/min. Duplicate samples (0.1 ml each) were removed after 0 and 30 min of incubation and used for FFA determination. The velocity was linear up to 30 min. Ediol was added to zero time after preincubation of enzyme and buffer with or without added materials for 10 min at 37°. The reaction velocities were expressed as micromoles of FFA liberated per milliliter of postheparin plasma per minute.

Analysis. - The triglyceride content of Ediol was estimated by the method of Van Handel and Zilversmit (17). The determination of FFA was carried out by the colbalt soap procedure of Novak (14), with palmitic acid as the standard. The volume of each reagent used was twice the quantity cited in the original procedure.

Results

The plot of the reciprocal velocity against reciprocal triglyceride substrate (Ediol) was linear between concentrations of 0.75 and 5.0 mM (Fig. 1). Concentrations greater than 5 mM were not suitable for kinetic studies, because Ediol tended to form two layers during incubation (4), and that resulted in erroneous measurements of FFA. Therefore, triglyceride concentrations within

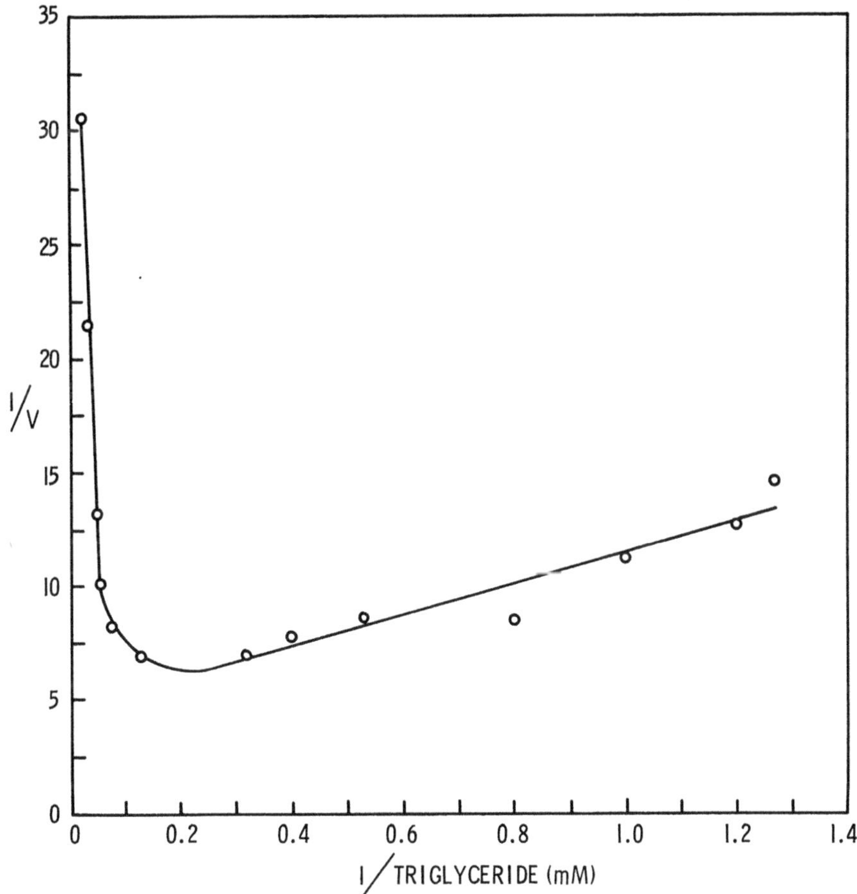

Fig. 1. Lineweaver-Burke plot of the reciprocal of velocity against the reciprocal of Ediol-triglyceride concentration for postheparin plasma LPL. Incubation mixture contained 0.2 ml of postheparin plasma 0.5 ml of buffer (0.2 M NH_4OH-NH_4Cl at a pH of 8.6) and 0.1 M Ediol (with various concentrations of triglycerides). Velocity is expressed as micromoles of FFA liberated per milliliter of postheparin plasma per minute.

the linear portion of the double-reciprocal plot were used for inhibition studies.

The inhibition of LPL activity at various lipolipin concentrations and a triglyceride concentration of 2.5 mM in Fig. 2. Inhibition of LPL increased exponentially with lipolipin concentration. At the triglyceride concentration used, the reaction velocity was reduced by 50% when the concentration of lipolipin was 1.5×10^{-5} M (molecular weight, 71,000). The inhibition was 95% with 1 M NaCl at alkaline pH in the assay system; this indicated that lipolysis

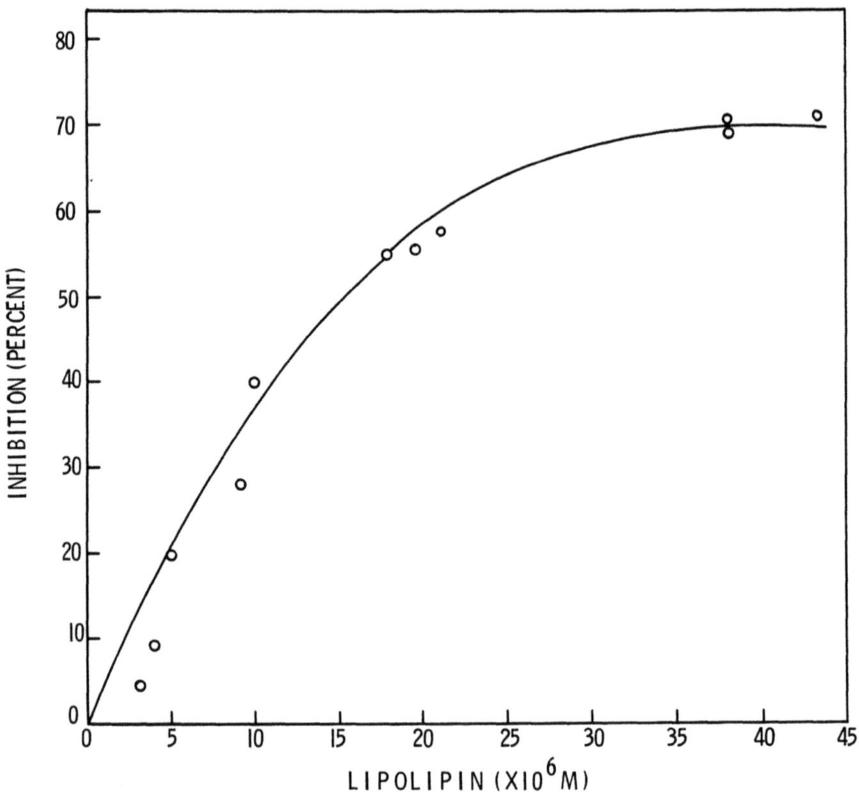

Fig. 2. Inhibition of LPL as a function of lipolipin concentration. Incubation mixture contained 0.2 ml of postheparin plasma, 0.5 ml of buffer (0.2 M NH_4OH-NH_4Cl at a pH of 8.6) containing various amounts of lipolipin and 0.1 ml of Ediol (20 micromoles triglyceride per milliliter). Final concentration of triglyceride was 2.5 mM. Percent inhibition was calculated as the decrease in LPL activity as related to assay mixtures without added lipolipin.

was due to LPL, and not to other lipases.

To determine the nature of inhibition, we measured velocity of lipolysis at various concentrations of triglyceride and the inhibitor. It is clear from Fig. 3 that the inhibition is noncompetitive. The value of K_i determined from these data was 1.39 x 10^{-5}. Because the enzyme was not purified and the nature of the true substrate in the complex assay system used is not known, the significance of K_i cannot yet be evaluated.

The stability of lipolipin under various conditions was examined by electrophoresis (1) on polyacrylamide gel. Lipolipin is unstable

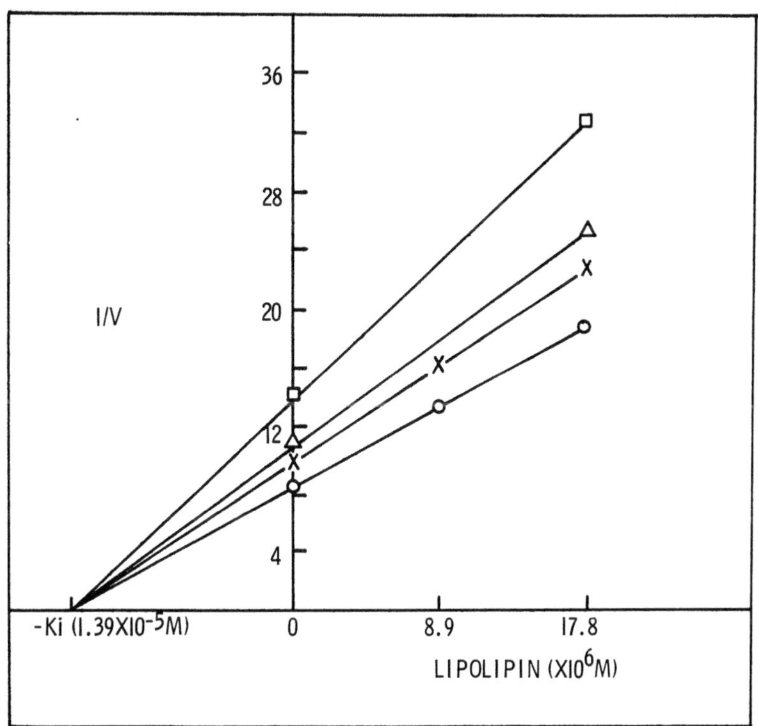

Fig. 3. Dixon plot of LPL activity. Incubation mixture contained 0.2 ml of postheparin plasma, 0.5 ml of buffer (0.2 M NH_4OH-NH_4Cl at a pH of 8.6) containing various concentrations of lipolipin and 0.1 ml Ediol (with various amounts of triglycerides). Velocity is expressed as micromoles of FFA liberated per milliliter of postheparin plasma per minute. Final concentrations of triglyceride were: □----□, 0.7 mM; Δ----Δ, 1.6 mM; x----x, 2.5 mM; and o----o, 4.3 mM.

when stored in neutral tris-HCl buffer or as freeze-dried powder after approximately 2 weeks. The instability is indicated by additional bands in the gel, and the lipolipin loses its LPL-inhibiting property very rapidly.

Discussion

A number of studies have suggested the presence of an endogenous inhibitor of LPL. The reported study here provides the first clear evidence that a highly purified glycoprotein isolated from the intimal region of porcine aorta is an inhibitor of LPL. The inhibition is of a linear noncompetitive type and is similar to that observed by Hollett (5), who used a crude preparation of the inhibitor from postheparin dog plasma. Although the kinetic data conclusively favor the argument that a unique glycoprotein is an inhibitor of LPL, we could not evaluate the stoichiometry between the enzyme and the inhibitor because of the complex nature of the reaction system.

A glycoprotein isolated from porcine mitral heart valves is essentially identical in chemical composition with that obtained from the aortic intimal region (M.M. Baig, personal communication, 1974). The mechanism by which lipolipin inhibits heparin-induced LPL activity is under investigation.

ACKNOWLEDGMENTS

The author wishes to thank Mr. Bruce I. Roberts for the glycoprotein preparation and Mrs. Janine Jones, Miss Ann P. Leverich, and Mr. Robert Apple for enzyme assays. This work was supported by Veterans Administration Research Funds (project 9166-01).

REFERENCES

1. DAVIS, B.J., Ann. N.Y. Acad. Sci., 121 (1964) 404.
2. ENGLEBERG, H., Amer. J. Physiol., 181 (1955) 309.
3. FEKETE, L.L., LEVER, W.F. and KLEIN, E., J. Lab. Clin. Med., 52 (1958) 680.
4. FREDRICKSON, D.S., ONO, K. and DAVIS, L.L., J. Lipid Res., 4 (1963) 24.
5. HOLLETT, C.R., Biochim. Biophys. Acta., 98 (1965a) 53.
6. HOLLETT, C.R., Biochim. Biophys. Acta., 98 (1965a) 61.
7. HOOD, B., BEDDING, P. and CARLANDER, B., J. Atheroscler. Res., 2 (1962) 438.
8. ISHII, M., Jap. Heart J., 12 (1971) 22.
9. KLEIN, K., LEVER, W.F. and FEKETE, L.L., J. Invest. Dermatol., 30 (1958) 41.
10. KORN, E.D., Methods Biochem. Analy., 7 (1959) 145.

11. LEVY, S.W., Rev. Can. Biol., 17 (1958) 1.
12. MITCHELL, F.R.A., Lancet, 1 (1959) 1969.
13. NIKKILA, E.A., J. Scand. Clin. Lab. Invest., Suppl. 5 (1953) 8.
14. NOVAK, M., J. Lipid Res., 6 (1965) 431.
15. ROBINSON, D.S., Compr. Biochem., 18 (1970) 51.
16. ROBINSON, D.S. and FRENCH, J.E., Quart. J. Exptl. Physiol., 42 (1957) 151.
17. VAN HANDEL, E. and ZILVERSMIT, D.B., J. Lab. Clin. Med., 50 (1957) 152.
18. WAGH, P.V. and ROBERTS, B.I., Biochemistry, 11 (1972) 4222.
19. WAGH, P.V. and ROBERTS, B.I., Absts. 9th International Congress of Biochemistry, Stockholm, 1973, Abs. 9e3.

DISCUSSION OF THE PAPERS BY OLIVECRONA ET AL AND WAGH

WAGH

I have a question for Dr. Olivecrona. He mentioned about another lipase which also adheres to the affinity column. I wonder whether it has the same pH optima and activity inhibited by 1 M NaCl?

OLIVECRONA

This enzyme is often called the liver lipase, which is not a good term because there are other lipases in the liver. In contrast to the lipoprotein lipase this "liver lipase" is not inhibited by salt, and has a higher activity against a fat emulsion than against lipoproteins. The pH optimum is in the same range as for lipoprotein lipase. When you add whole post-heparin plasma in a lipoprotein lipase assay you will measure both enzymes. Depending upon the exact conditions of your assay you will measure more of one or more of the other. It appears that with many assays, you measure more "liver lipase" than lipoprotein lipase.

FAREED

I want to ask about the conditions you used in the assay on this LPL and the Km value for the LPL, have you characterized it in terms of its Ki value.

WAGH

Well, our assay system, first of all, was simply 0.2 ml of the post-heparin plasma, Ediol, which is a coconut oil stabilized emulsion and ammonia buffer. The inhibitor was dissolved in the buffer, and the final pH of all the incubations was 8.1 or 8.2. Now regarding the Km, I do not want to speak too much because of the nature of the substrate involved in this. We add that much of the Ediol to get our proper kinetics.

FAREED

Was there any base hydrolysis without any enzyme at this pH of 8.5?

WAGH

At zero time velocity there are about .01 or .02 OD units, so it was very minimal.

FAREED

So, after extended times like 30 minutes of incubation there was no hydrolysis of the coconut triglycerides in the absence of the plasma?

WAGH

That's right.

HEPARIN AS AN INHIBITOR OF MAMMALIAN PROTEIN SYNTHESIS

Jack GOLDSTEIN, Alan A. WALDMAN and Gerard MARX

Nucleic Acids Laboratory, New York Blood Center, New York, New York 10021 and Department of Biochemistry, Cornell University Medical College, 1300 York Avenue, New York, New York 10021 (USA)

Heparin is a naturally occurring polyanion found in many tissues. The biologic properties of heparin that have been most extensively investigated are its anticoagulant activity and its interaction with lipoprotein lipase (for representative reviews, see Freeman (4) and Ehrlich and Stivala (3)). Various reports suggest that heparin and possibly other acid mucopolysaccharides also play a role in the regulation of cell growth (14). We have investigated the effect of heparin on the translation of natural mammalian messenger RNA (mRNA) in mammalian cell-free systems (6, 19-21), and we present here some of our findings concerning heparin's action as an inhibitor of the translation of exogenous rabbit globin mRNA in a Krebs ascites cell-free system.

Figure 1 presents the results of studies on the effect of heparin on protein synthesis in the Krebs ascites cell-free incubate, the assay system we use. Its properties have been elaborated by several laboratories (10, 12, 13), and it is the system currently being used extensively to study the translation of isolated mammalian mRNA's.

In the absence of added heparin, the addition of rabbit globin mRNA to the incubation mixture causes a stimulation in protein synthesis to well above the endogenous rate (Fig. 1). Increasing the concentration of heparin results in a continual decrease in the extent of stimulation by mRNA, until, at a heparin content of 40 μg/ml, the stimulation due to the presence of added mRNA disappears completely.

Throughout the range of heparin concentration tested, there is a minimal effect on the level of endogenous protein synthesis. Because endogenous protein synthesis in the Krebs ascites cell-free

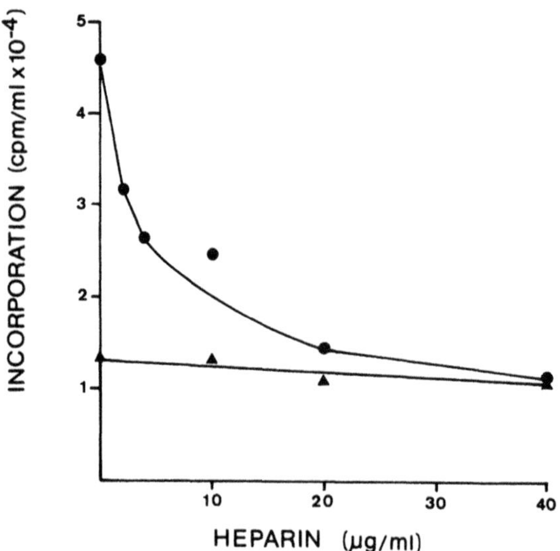

Fig. 1. Effect of heparin on protein synthesis in Krebs ascites cell-free system. Heparin was added at the concentrations noted to reaction mixtures containing L-[^{3}H]leucine, Krebs ascites cell-free extract, and either no added rabbit globin mRNA (Δ---Δ) or added (at 20 µg/ml) rabbit globin mRNA (o---o). Incorporation is defined as incorporation of L-[^{3}H]leucine into material insoluble in trichloroacetic acid. For further details, see Waldman and Goldstein (20).

system is known to represent only elongation and termination of peptide chains (13), whereas translation of added mRNA depends entirely on initiation, these data indicate that heparin acts to inhibit initiation of protein synthesis.

Further analysis of similar incubations indicated that the synthesis of α- and β-globin chains was inhibited equally and that heparin inhibition was not reversed by increasing mRNA content (19).

To confirm our conclusion that heparin was acting at initiation, we tested the effect of adding aurin tricarboxylic acid (ATA), a known inhibitor of initiation (7). The results of such additions are presented in Fig. 2.

As can be seen, increasing concentrations of ATA in reaction mixtures containing mRNA resulted in inhibition of the mRNA-dependent stimulation of protein synthesis (Fig. 2). Again, the level of endogenous protein synthesis was not affected. The similarity of

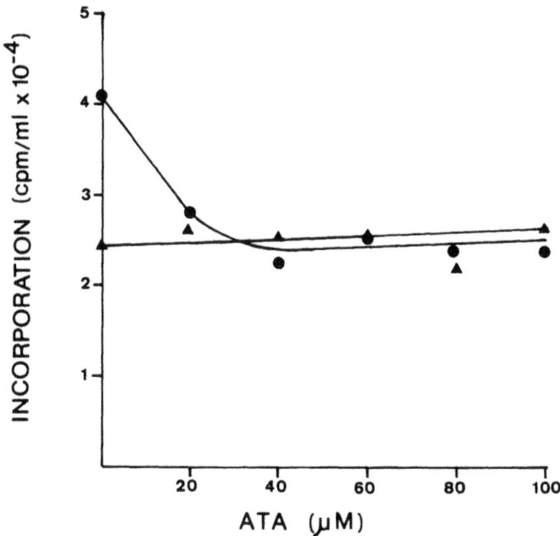

Fig. 2. Effect of aurin tricarboxylic acid (ATA) on protein synthesis in the Krebs ascites cell-free system. ATA was added at the concentrations noted to reaction mixtures containing L-[^{3}H]leucine, Krebs ascites cell-free extract, and either no added rabbit globin mRNA (Δ---Δ) or added (at 20 μg/ml) rabbit globin mRNA (o---o). Incorporation is defined at in Fig. 1. For further details, see Waldman and Goldstein (20).

the mode of inhibition exhibited by heparin and ATA indicates that heparin indeed acts as an inhibitor of initiation of protein synthesis.

We have obtained similar results in rat liver cell-free systems (6) and in mixed systems containing Krebs ascites cell supernatant and rabbit reticulocyte ribosomes (20).

To see whether the effect of heparin was specific, we examined a series of purified naturally occurring sulfated mucopolysaccharides, which were kindly provided by Dr. Cifonelli, and synthetic sulfated polyanions, namely:

natural:	synthetic:
heparan sulfate	dextran sulfate
chondroitin-4-sulfate	Sulfon
chondroitin-6-sulfate	

keratan sulfate-1
keratan sulfate-2
heparin
dermatan sulfate

Heparin sulfate (a mucopolysaccharide closely related to heparin), chondroitin-4-sulfate and chondroitin-6-sulfate, keratan sulfates -1 and -2, dextran sulfate, and Sulfon (a sulfonated polystyrene) were all inactive as inhibitors. Only heparin and dermatan sulfate were found to be active, with dermatan sulfate having only approximately 50% the potency of heparin.

Thus, the heparin effect seems specific to the molecule, and is not due to a nonspecific polyanion effect.

Currently, our research is proceeding along two lines. The first concerns the relation of heparin's structure and chemistry to its biological activities. Our initial approach here has been an attempt to remove sulfate groups selectively. Several laboratories have shown that mild acid treatment of heparin results in its progressively losing sulfate groups, especially N-sulfate (8, 11, 18). By using such a procedure, we have been able to dissociate heparin's anticoagulant activity from its ability to inhibit protein synthesis (21), as presented in Fig. 3. In agreement with data from other laboratories (18), approximately 50% of heparin's anticoagulant activity is lost during the first hour of acid treatment; this increased to 90% 3 hr after the start of treatment. Conversely, heparin's ability to inhibit protein synthesis is still unaffected at 4 hr, and 50% loss of this inhibitory ability does not occur until 8 hr after the start of acid treatment. As also reported by other workers (11, 18), only N-sulfate groups were removed during the first few hours of desulfation, whereas heparin exposed to desulfating conditions for 24 hr lost both N- and O-sulfates (21). It was found that 4.5-hr desulfation, which caused almost complete loss of heparin's anticoagulant activity but had minimal effect on its ability to inhibit protein synthesis, removed approximately five of a total of 17 N-sulfate groups per molecule of heparin. When hydrolysis was continued for 24 hr, approximately 4.5 O-sulfate and an additional four N-sulfate groups per molecule were removed. This resulted in a complete loss of heparin's ability to inhibit protein synthesis (21). Partial resulfation of amino groups of inactive desulfated heparin (adding four N-sulfates per molecule to a sample treated for 20 hr) almost completely restores its ability to inhibit protein synthesis, whereas only negligible anticoagulant activity is recovered (21). It appears, then, that differences in the number and type of sulfate groups in the heparin molecule play a significant role in determining both its anticoagulant and its protein synthesis inhibitory activities.

Our second line of work concerns the site of heparin's inhibitory

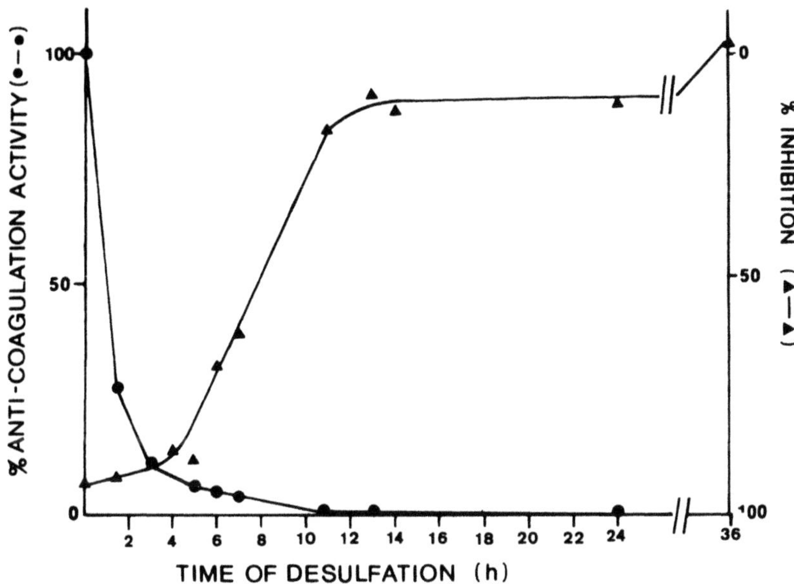

Fig. 3. Effect of desulfation on activities of heparin. Heparin was desulfated for the times noted and assayed for protein synthesis inhibitory ability (Δ---Δ) and for anticoagulant activity (o---o). Inhibition of protein synthesis was determined in the Krebs ascites cell-free system (19). Anticoagulant activity was determined by activated partial thromboplastin time, and 100% anticoagulant activity is defined as that observed using the 0 hr desulfated heparin. For further details, see Waldman et al. (21).

activity. Because our evidence indicated that initiation of protein synthesis was involved, we prepared by published procedures a crude preparation of initiation factors (16). These factors are a complex group of proteins involved in the various stages of the initiation of protein synthesis. When this crude initiation factor preparation (which also contains proteins without initiation factor activity) was added to the assay system with exogenous mRNA, it markedly stimulated protein synthesis. This stimulation was sensitive to heparin. However, considerably more heparin was necessary to inhibit the stimulation of protein synthesis in the presence of the crude initiation factor preparation than in the presence of mRNA alone. This indicated that heparin could interact with initiation factors.

Other workers have shown that the antithrombin-heparin cofactor (5, 15) and the lipoprotein lipase from milk (9), both of which are known to interact with heparin under physiologic conditions, will also bind selectively to heparin covalently attached to Sepharose 4B. We therefore decided to study the behavior of initiation factors on such heparin affinity columns. As can be seen in Table I, (which shows the distribution of initiation factor activity), when the crude initiation factor preparation is added to a column containing activated cyanogen bromide-Sepharose (line 1), all the activity recovered emerges with the starting buffer. The same result is obtained with a column consisting of partially desulfated heparin (line 2). The desulfated heparin, which was inactive as an inhibitor of protein synthesis, was prepared by a 20-hr mild hydrolysis treatment. As discussed earlier, this results in removal of approximately nine N-sulfates and five O-sulfates from the molecule. In contrast, however, when a crude initiation factor preparation is added to a heparin-Sepharose column under optimized conditions (line 3), all the activity binds to the column and can be eluted in buffer containing 0.5 M KCl. This eluted activity was also found to be heparin-sensitive.

Figure 4 presents the results of submitting a crude initiation factor preparation and the various heparin-Sepharose column fractions to sodium dodecylsulfate polyacrylamide gel electrophoretic analysis (1). It is clear that, as indicated earlier, the crude initiation factor preparation (first gel on left) contains a fair number of

TABLE I

Distribution of Initiation Factor Activity After Chromatography on Sepharose Affinity Columns[a]

Column Containing	Fraction Eluted with[b]		
	0.12 M KCl	0.5 M KCl	1.0 M KCl
(1) No heparin	+	-	-
(2) Desulfated heparin	+	-	-
(3) Heparin	-	+	-

[a] 2 mg of crude initiation factor in 0.12 M KCl applied to each column.
[b] + indicates the presence of initiation factor activity.

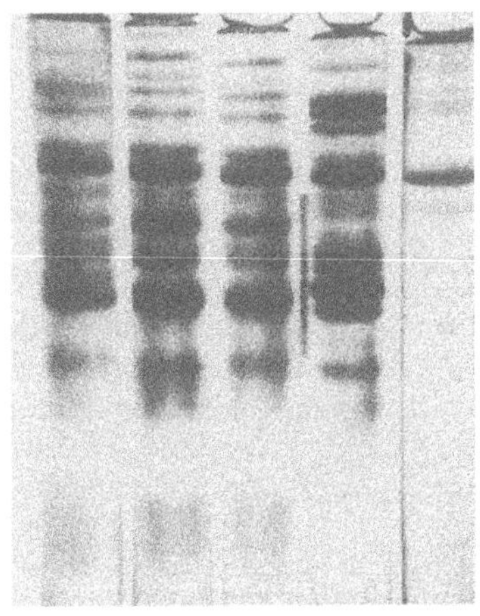

Fig. 4. Sodium dodecylsulfate (SDS) gel electrophoretic analysis of crude initiation factor fractions obtained after heparin-Sepharose affinity chromatography. Crude initiation factor was prepared as described by Schreier and Staehelin (16), and chromatographed on heparin-Sepharose affinity columns. SDS gel electrophoresis was performed (1), and the gels were stained with 0.05% coomassie blue. From left: gel 1, crude initiation factor preparation; gel 2, material passing directly through heparin-Sepharose column; gel 3, material eluted from column after washing with application buffer [buffer D containing 0.12 M KCl (16)]; gel 4, material eluted from column after washing with buffer D containing 0.5 m KCl; gel 5, material eluted from column after washing with buffer D containing 1.0 M KCl.

proteins. Although not shown, comparable electrophoretic patterns were also obtained from the material that passed through the control columns. The second and third gels represent material that did not bind to the heparin-Sepharose and contained 61% of the total protein applied to the column. The fourth represents the 0.5 M KCl eluate and contained 23% of the applied proteins. The last gel, which represents the 1 M KCl eluate, shows essentially one band and contained less than 10% of the applied protein. The 0.5 M KCl eluate (the fourth gel) contained all the recoverable initiation factor activity, as mentioned earlier. There is a striking difference in the intensity of the coomassie blue stain of the protein bands between the upper region of gel four and similar regions of the other gels. The relatively low electrophoretic mobilities exhibited by the proteins in the enriched region of the gel agree well with reports of high molecular weights for various preparations of initiation factors (2, 17).

In summary, our results indicate that heparin, a naturally occurring substance, can act on the initiation of protein synthesis to inhibit translation of a natural mammalian mRNA in a mammalian cell-free system; that this effect is related to the number of N- and O-sulfate groups in the molecule; and, that one way heparin may cause this inhibition is by interacting with one or more of the factors needed for initiation of protein synthesis. Whether heparin exerts a similar effect *in vivo* remains to be determined.

ACKNOWLEDGEMENTS

This work was supported by National Institutes of Health grant, HL-09011-11.

REFERENCES

1. BRENNESSEL, B. and GOLDSTEIN, J., Vox Sang., 26 (1974) 405.
2. CASHION, L.M. and STANLEY, W.M., Proc. Natl. Acad. Sci., 71 (1974) 436.
3. EHRLICH, J. and STIVALA, S.S., J. Pharm. Sci., 62 (1973) 517.
4. FREEMAN, L., Amer. J. Cardiol., 14 (1964) 3.
5. GENTRY, P. and ALEXANDER, B., Biochem. Biophys. Res. Commun., 50 (1973) 500.
6. GOLDSTEIN, J., WALDMAN, A.A. and MARX, G., Abst. IX Intl. Congr. Biochem., 175 (1973).
7. GROLLMAN, A.P. and HUANG, M.T., Fed. Proc., 32 (1973) 1673.
8. HELBERT, J.R. and MARINI, M.A., Biochem., 2 (1963) 1101.
9. IVERIUS, P.H., LINDAHL, U., EGEBRUD, T. and OLIVECRONA, T., J. Biol. Chem., 247 (1972) 6610.
10. JACOBS-LORENA, M. and BAGLIONI, C., Biochem., 11 (1972) 4970.
11. LEVY, I. and PETRACEK, F.J., Proc. Soc. Exp. Biol. Med., 109 (1962) 901.
12. MATHEWS, M.B., Biochim. Biophys. Acta, 272 (1972) 108.
13. MATHEWS, M.B. and KORNER, A., Eur. J. Biochem., 17 (1970) 328.
14. REGELSON, W., Adv. Cancer Res., 11 (1968) 223.
15. ROSENBERG, R.D. and DAMUS, P.S., J. Biol. Chem., 248 (1973) 6490.
16. SCHREIER, M.H. and STAEHELIN, T., J. Mol. Biol., 73 (1973) 329.
17. SCHREIER, M.H. and STAEHELIN, T., Nature New Biol., 242 (1973) 35.
18. STIVALA, S.S., YUAN, L., EHRLICH, J. and LIBERTI, P.A., Arch. Biochem. Biophys., 122 (1967) 32.
19. WALDMAN, A.A. and GOLDSTEIN, J., Biochem., 12 (1973) 2706.

20. WALDMAN, A.A. and GOLDSTEIN, J., Biochim. Biophys. Acta, 331, (1973) 243.
21. WALDMAN, A.A., MARX, G. and GOLDSTEIN, J., Biochim. Biophys. Acta, 343 (1974) 324.

DISCUSSION OF GOLDSTEIN *ET AL.* PAPER

THOMPSON

What was the effect of added magnesium to your protein synthesis effects?

GOLDSTEIN

We actually do our assay in a magnesium containing system, and have determined the magnesium optimum for this system. Furthermore, our results rule out the possibility that magnesium is artifactually responsible for heparin's protein synthetic inhibitory effect.

FAREED

What molarities of heparin inhibit your system? Did you make a concentration study? Because heparin is a naturally occurring compound, I was just wondering if it's concentration was measured.

GOLDSTEIN

Well, the only parameters we have are essentially those I indicated in the first slide; that in our assay system where we usually use about 20 gamma of exogenous messenger RNA, we can get, this depends again on how potent our ascites cell-free system is, 50% inhibition with about 5 to 10 gamma of heparin. Usually it's on the order of 5 gamma, and you saw that at 40 gamma all exogenous messenger RNA activity was completely wiped out.

THE EFFECT OF HEPARIN ON OXYGEN TRANSPORT FROM BLOOD TO TISSUES

H. ENGELBERG

465 North Roxbury Drive - Suite 1003, Beverly Hills, California (USA)

Ballistocardiographic improvement was observed after heparin administration to patients with coronary atherosclerosis in 1952. It was suggested that improved myocardial oxygenation resulted from the disappearance of chylomicrons and the larger low-density lipoproteins from the blood after the injection of heparin (1). This suggestion was verified by studies of forearm arteriovenous oxygen differences (3) and total oxygen consumption (2) in atherosclerotic patients. Tissue oxygen consumption was reduced in approximately 50% of subjects who were known to have coronary atherosclerotic disease but who were not in heart failure. After the injection of heparin, there was an average increase in oxygen consumption of 32%, whereas no increase occurred in the same subjects after injection of saline or administration of oral anticoagulants. The increase oxygen extraction occurred about 1-2 hr after intravenous injection of heparin and coincided with the time of maximal lipid clearance. These initial observations of a relation between plasma lipids, tissue oxygen consumption, and the effect of heparin have been confirmed in animals and man using various techniques (7-12, 15).

Although the mechanism underlying this relation is unclear, several pathways have been suggested (2). Lipid films upon the surface of erythrocytes or the intimal endothelium may interfere with oxygen diffusion. Lipid macromolecules may bind proteins, thus decreasing their enzymatic or other activity. Lipemia also largely prevented the increase in coronary flow and myocardial oxygen extraction normally found after exercise (13).

In 1968, Groover (6), using an ear oximeter and breathholding for at least 30 sec, measured the rate of desaturation of oxyhemoglobin *in vivo*. He showed that this was a rough index of oxygen consumption and corresponded with simultaneous arterial oxygen changes. He found that the oxyhemoglobin desaturation rate was decreased in patients after myocardial infarction, compared with normal controls. It was increased 1 hr after the injection of heparin, although there was a decrease in the first 15 min after heparin injection.

In the last few years, I have applied Groover's method. The breath is held in mid-inspiration for 30-60 sec. A stop-watch is started exactly at the end of inspiration, and the oxygen content of the earlobe blood is read every 15 sec. The rate of oxygen decrease (oxyhemoglobin desaturation) is calculated by using the 15-sec reading as the baseline. The initial zero time reading is ignored, because it takes approximately 15 sec for blood to reach the ear from the lungs. Thus, after the first 15 sec of breath-holding, the ear blood oxygen starts to decrease, reflecting only the exit of oxygen from blood to tissues, inasmuch as fresh oxygen is no longer entering the blood in the lungs. The absolute readings of oxygen saturation do not have to be exactly calibrated, because it is the rate of change that is important. In most instances, this rate of oxygen decrease per second is calculated from 15 to 45 sec after breathholding, as shown in Table I. Ten separate determinations were made in each person, and the results were averaged. A total of 43 subjects with known coronary atherosclerotic disease were studied - 12 before and shortly after a small intravenous dose of heparin, and 21 before and 24 hr after a larger dose of subcutaneous heparin. None of the subjects was in congestive heart failure.

The results are summarized in Table II. There was an increase in the rate of oxygen transfer from blood to tissues (as indicated by the rate of oxyhemoglobin desaturation), which averaged 28% after intravenous heparin and 19% 24 hr after subcutaneous heparin. Table III shows that this response was not uniform. The rate of oxyhemoglobin desaturation increased in eight of 12 subjects after intravenous heparin and in 17 of 31 patients after subcutaneous heparin. Table IV shows that the rate of oxyhemoglobin desaturation was increased by heparin only in subjects who had lower desaturation rates before heparin was given. Table V shows the findings in one patient who had a very low control value and who was followed for 3 days after subcutaneous injection of 20,000 units of heparin. It is apparent that the increase rate of oxygen transfer from blood to tissues persisted for 48 hr after heparin, but had returned to the original control value after 72 hr.

The data presented here substantiate the findings of previous studies. Over half the patients with known coronary atherosclerotic disease, but not in congestive heart failure, have decreased tissue oxygenation, which is improved after heparin. Apart from the contribution of hypoxia to the development and progression of atherosclerosis, the most serious effects probably occur in the myocardium itself.

TABLE I

Example of Experimental Design Showing Oxygen Readings from Lote of Ear During Breathholding

Length of Breathholding, Sec	% OXYGEN SATURATION		Rate of O_2 Decrease Per Second – (Calculated from Values at 15 and 45 sec)	
	Control, Before Heparin	24 Hours After 20,000 Units Heparin Subcutaneously	Before Heparin	After Heparin
0	90.2	90.0		
15	90.1	90.3		
30	89.2	88.9	.073	.097
45	87.9	87.4		
60	85.3	84.0		

TABLE II

Effect of Heparin on Rate of Oxygen Transfer from Arterial Blood to Tissues

Dose and Route of Heparin	Number of Subjects	Average Rate of O_2, Decrease Per Sec		Average Increase in Rate of O_2 Transfer from Blood to Tissues
		Before Heparin	After Heparin	
2500 Units, Intravenously	.12	.068	.087	28%
20,000 Units, Subcutaneously	.31	.064	.076	19%

TABLE III

Effect of Heparin on Rate of Decrease of Arterial Oxygen

Dose and Route of Heparin	Number of Subjects	NUMBER OF SUBJECTS		
		Rate Unchanged	Rate Decreased	Rate Increased
2500 Units Intravenously	12	3	1	8
20,000 Units Subcutaneously	31	9	5	17
TOTALS	43	12	6	25

TABLE IV

Effect of Subcutaneous Heparin on Rate of O_2 Transfer, by Control Values

Effect of Heparin on Rate of O_2 Transfer from Blood to Tissues	No. of Subjects	Average Rate of O_2 Decrease in Blood/Sec (Before Heparin)
No Change or Decrease	14	.072
Increase	17	.057

TABLE V

Effect of Heparin Over 3 Day Period on Rate of Oxygen from Arterial Blood to Tissues

20,000 Units Heparin, Subcutaneous	Rate of Decrease of Arterial Oxygen/Sec
Control - Before Heparin	.046
24 Hours After Heparin	.068
48 Hours After Heparin	.075
72 Hours After Heparin	.051

The heart normally extracts about 70% of the oxygen from its blood supply, compared with 21% for the overall systemic circulation. Thus, an increase in coronary blood flow is the only mechanism to satisfy increased myocardial oxygen demand, inasmuch as the oxygen extraction rate is already so high. However, skeletal muscle, which uses only 22% normally, increases its oxygen extraction rate to 80-90% with maximal exercise, in addition to undergoing an increase in blood flow. Consequently, if there is an additional barrier to efficient tissue oxygenation, the myocardium would be more sensitive to this.

Although a decrease in arterial oxygen saturation has been found in man after intravenous injection of fat emulsions (15), in our earlier studies (3) there was no change in arterial oxygenation after lipid clearing by heparin. The increased arteriovenous oxygen difference after heparin resulted entirely from a lower venous oxygen content. This indicated increased oxygen removal by the peripheral tissues and an amelioration of the hypoxic condition after heparin administration.

In 1958, I discussed fully the contribution of the larger triglyceride-rich lipoproteins to the formation of lipid films on endothelial and red blood cells and the barrier to the diffusion of oxygen that such films create (2). Recently, evidence (5) has been presented that an increase in plasma proteins and cholesterol will cause a significant decrease in oxygen diffusion. A plasma layer only 3 μm thick offered at least as much resistance to oxygen transfer as the red-cell membrane (14). Thus, the more recent studies corroborate the interference of lipid films with oxygen diffusion. It is precisely the larger lipoproteins that disappear from blood after heparin injection.

We have recently confirmed (4) that patients with coronary and valvular heart disease have a decreased total and "effective" or nutrient coronary blood flow. However, only those with coronary heart disease also had a decrease in the myocardial oxygen utilization rate. This was associated with a significant anaerobic trend, indicating less efficient energy-producing mechanisms. It may well be that the increased serum lipids usually present in patients with coronary disease contributed to this decreased myocardial oxygen utilization rate and that it would be improved by heparin.

The findings of our recent study show that heparin increases the rate of oxygen transfer from blood to tissues in atherosclerotic persons in whom it is below normal. Inasmuch as hypoxia promotes the atherogenic process, its correction by heparin may be one of the most important mechanisms whereby heparin therapy retards the progress of atherosclerotic disease (4). In addition, the correction of even minor degrees of hypoxia should improve myocardial function in patients with coronary heart disease.

REFERENCES

1. ENGELBERG, H., Amer. J. Med. Sc., 224 (1952) 487.
2. ENGELBERG, H., Amer. J. Med. Sc., 236 (1958) 175.
3. ENGELBERG, H. and KUHN, R., Angiology, 7 (1956) 73.
4. ENGELBERG, H., KUHN, R. and STEINMAN, M., Circulation, 13 (1956) 489.
5. GAINER, J.L. and CHISHOLM, G.M., Atherosclerosis, 17 (1974) 135.
6. GROOVER, M.E., Angiology, 19 (1968) 299.
7. JOYNER, C.R., HORWITZ, O. and WILLIAMS, P.G., Circulation, 22 (1960) 901.
8. KUO, P.T. and JOYNER, C.R., J. Amer. Med. Assoc., 158 (1955).
9. KUO, P.T. and JOYNER, C.R., J. Amer. Med. Assoc., 163 (1957) 727.
10. KUO, P.T., WHEREAT, A.F., ALTMAN, A.A. and CARSON, J.C., Circulation, 20 (1959) 722.
11. KUO, P.T., WHEREAT, A.F. and HORWITZ, O., Amer. J. Med., 26 (1959) 68.
12. REGAN, T.J., BINAK, K., GORDON, S., DeFAZIO, V. and HELLEMS, H.K., Circulation, 23 (1961) 55.
13. REGAN T.J., TIMMIC, G., GRAY, M., BINAK, K. and HELLEMS, H.K., J. Clin. Invest., 40 (1961) 624.
14. SINHA, A.K., Doctoral Dissertation, Univ. of California, San Francisco, 1969.
15. TALBOTT, G.D. and FRAYSER, R., Nature, 200 (1963) 684.

SECTION 3

CLINICAL ASPECTS

CLINICAL IMPLICATIONS OF HEPARIN

Stanford WESSLER

Department of Medicine, The Jewish Hospital of St. Louis and the Washington University School of Medicine, St. Louis, Missouri (USA)

The one clinical use for heparin (among the many suggested) that has remained unchallenged for the last quarter-century is its potential effectiveness as an antithrombotic agent.

My remarks are related to why I chose to insert the work "potential" before "effectiveness" when referring to the antithrombotic action of heparin. Although the *in vitro* efficacy of this anticoagulant, in animals and in man, is well documented, heparin has not materially reduced the prevalence of thrombosis in this country. Some of the contributing factors creating this apparent paradox are: the mechanisms involved in the deposition of arterial and venous thrombi are different, knowledge of the action of heparin on clotting proteins and platelets is incomplete, and the requirement for parenteral administration and the lack of a satisfactory method for regulating drug dosage remain handicaps. In addition, and perhaps as important as any of the preceding reasons, is the clear need for a more realistic understanding of the pathophysiology of thrombosis in the design of clinical trials to evaluate heparin as an antithrombotic agent.

Scope of Thrombosis Problem

The national impact of thrombosis can, for convenience, be divided into four discrete parts. First, arterial thromboembolism, largely because of its intimate relation to atherosclerosis, is a major contributor to the first, third, and fourth leading causes of death in the United States - acute myocardial infarction, stroke, and kidney disease, respectively.

Second, venous thromboembolism is responsible for the hospitalization of approximately 300,000 patients a year in the United States and for the deaths of more than 50,000 of them. Venous thromboembolism is apparently the most common nonsurgical cause of death in patients hospitalized for major orthopedic procedures, an important nonobstetric cause of postpartum death, and a major contributor to mortality in the large population of elderly people with chronic cardiac and pulmonary disease. Indeed, of hospitalized adult patients who die, careful autopsy examination discloses evidence of venous thromboembolism in more than 60%.

Third, thromboembolism induced by contact of the blood with foreign surfaces is the major unsolved problem in the development of artificial organs for extra corporeal circulation or for implantation within the heart and blood vessels. The problem is a major obstacle to advances in clinical care involving heart valves, extracorporeal cardiopulmonary bypass, cardiac assistance (both extracorporeal and intravascular), respiratory assistance through extracorporeal circulation, intravascular catheters, electrodes, artificial hearts, artificial blood vessels and arteriovenous shunts, vascular sutures, and all other applications in which the blood comes in to contact with a nonbiological surface. Artificial devices can be used today only in applications that involve short periods of contact with the blood, unless their surface areas are very small.

Finally, the profound importance of diseases related to or caused by microcirculatory disturbances, although it cannot be expressed in statistical terms, can best be appreciated through a partial list of the diseases in which such disturbances are significantly involved. The microcirculation is often a contributory or even primary mechanism of disease in hypertension, stroke, diabetes, cancer, infection, inflammatory disease, autoimmune disease, host-graft rejection, hemolytic anemia, drug toxicity, mismatched blood transfusion, liver disease, such diseases of the gastrointestinal tract as pancreatitis, and such diseases of the genitourinary system as glomerulonephritis. Major advances in our understanding of the function of the microcirculation, particularly microcirculatory thrombosis, may result in substantive gains in many of these subjects.

Thus, clinically, obstruction of three major components of the circulation - by arterial, venous, and microcirculatory thromboembolism - contributes materially to early death in our society. Although the three vascular components have common factors, each has its peculiarities.

If, for example, atherosclerosis could be prevented or ameliorated (and the role of heparin in this process has yet to be fully explored), arterial thromboembolism would be practically eliminated as a major contributor to arterial occlusion, except for the problem of thrombosis on prosthetic devices within the arteries and the chambers of the heart. However, although only a handful would argue that thrombosis initiates atherosclerosis, the interrelation between these two processes is not resolved. The majority might maintain that deposits of platelet debris and fibrin contribute to the growth of the atherosclerotic lesion and that the sudden gelation of fluid blood on atheromatous plaques often represents the terminal event that converts a partially narrowed lumen into a total vascular occlusion. Even here, there is some disagreement as to whether coronary arterial thrombosis precedes, rather than follows, acute myocardial infarction. The majority opinion, however, is that coronary thrombosis initiates 80-90% of transmural, as distinguished from subendocardial, myocardial infarcts. One could therefore readily anticipate that any compound, such as heparin, that could prevent the formation of arterial thrombi would profoundly diminish the clinical ravages of atherosclerosis, without affecting the atherosclerotic lesion itself. As a result, many in our society might enjoy an added decade of life. Intimal injury remains the key substrate for arterial thrombosis; and, in physiologically end-arterial circuits - such as occur in the heart, brain, and kidney - thrombosis need not be extensive to be lethal. Whether heparin can effectively prevent or limit thrombosis on artheromatous plaques, however, remains to be determined.

Venous thromboembolism is essentially unrelated to atherosclerosis. More than 95% of pulmonary emboli originate in the deep leg veins. However, rather than peripheral venous thrombosis, it is embolization to the lung that is the primary threat to life. Thus, whereas venous intimal injury may indeed by present, it is the extent of propagation of the thrombus, which usually depends on retarded blood flow and proteolytic activation of clotting factors, that leads to massive pulmonary embolism.

The microcirculation also has its unique properties. This compartment of the vasculature consists of all small blood vessels - endothelium, subendothelial structures, the extravascular supporting and surrounding matrix, the extravascular fluid environment, the arteries and the tissues that they supply, and the blood that circulates through the vessels. Within this dynamic microcirculatory environment, many vital processes are mediated. These include tissue oxygenation and energy supply; removal and transport of the intermediate products of metabolism; maintenance of vessel-wall permeability or impermeability; storage or release of initiators and modulators of local and systemic hemostatic host-defense mechanisms; and regulation of the transport of cellular products, such as hormones and proteins, for use elsewhere. Thrombus formation developing from inadequate flow or transport in the microcirculation inevitably results in functional disruption of some or all of these processes. The site, extent, and cause of the disrupted flow determine the type and degree of functional disability manifested by the organism. It is in the microcirculation, particularly, that endotoxin, the Shwartzman reaction, and many other immunologic pheonomena play critical roles in the development of intravascular coagulation.

Current Status of Antithrombotic Therapy

A number of conditions for which anticoagulants have been recommended are listed in Table I. For the conditions of venous thrombosis, pulmonary embolism, arterial embolism and cerebral embolism, the data reflect an unequivocal benefit from anticoagulant therapy. For the conditions of - acute myocardial infarction and transient ischemic attacks - the benefit from anticoagulant therapy is limited or questionable. For all the other conditions, no clinical benefit has yet been demonstrated. Table II lists several categories of patients who are believed to be at increased risk from thromboembolism. Only among patients with rheumatic heart disease and atrial fibrillation and patients with heart-valve prostheses are the benefits convincing. Among patients with fractured hips or disseminated intravascular coagulation and patients in the postoperative state, the benefits are marginal.

It seems clear that the main prophylactic value of heparin and the coumarin compounds is not against the predominantly platelet thrombi that form on atheromatous plaques, but rather in the prevention of red thromboemboli, of which erythrocytes and fibrin strands are the predominant constituents. These latter thrombi are found in the veins and on the endothelium of the heart, giving rise to pulmonary and systemic emboli, respectively. This view, moreover, provides a reasonable interpretation of the observed discrepancies in the efficacy of heparin in various cardiovascular disorders associated with thromboembolic episodes. Finally, it is evident that new ways to use heparin or new forms of heparin or other antithrombotic agents are needed to limit arterial platelet aggregation without compromising hemostasis.

Actions of Heparin

Although it was discovered during World War I and was shown to be an effective antithrombotic agent before World War II, the physiologic roles of heparin have yet to be defined. The attractive proposition that heparin acts as a natural anticoagulant remains unconfirmed, despite its presence in mast cells. The "clearing" of postprandial lipemic serum by the heparin-activated enzyme, lipoprotein lipase, has not been established as a normal function of heparin. Still to be identified is the mechanism of heparin-induced osteopenia, a toxic effect of long-term high-dosage adminstration.

TABLE I

Benefit from Anticoagulants

Definite	Slight or Marginal	None Demonstrable
Venous thrombosis Pulmonary embolism Arterial embolism Cerebral embolism	Acute myocardial infarction Transient ischemic attacks	Pulmonary thrombosis Pulmonary hypertension Arterial thrombosis Intermittent claudication Old myocardial infarction Coronary failure Angina pectoris Stroke - complete Stroke in progress

TABLE II

Categories of Patients at Special Risk from Thromboembolism

Definite	Slight or Marginal	None Demonstrable
Atrial fibrillation with rheumatic heart disease Heart-valve prosthesis	Hip fracture Disseminated intravascular coagulation Postoperative state	Postpartum state Reversion of atrial fibrillation to regular sinus rhythm Shock Congestive heart failure Administration of estrogens Sickle-cell disease Polycythemia vera Arterial bypass grafts Glomerulonephritis Host-graft rejection

The significance of other reported biological actions of heparin - including its enhancement of vascular permeability, its efforts on the Shwartzman and Arthus phenomena, and its inhibition of many well-defined enzyme systems, both within and outside clotting scheme - still await clarification into an integrated whole.

Heparin is a compound with chemical heterogeneity caused by inherent molecular instability and by polydispersion. These phenomena, coupled with the fact that the drug can be obtained from the lung and from the gut mucosa of several animals by different extraction and purification procedures and the lack of a specific assay for plasma heparin, help to explain, at least in part, the variations observed in the _in vitro_ and _in vivo_ potencies of different commercial heparin preparations. Such variations may pose difficulties in comparisons of experimental or clinical results based on different heparin preparations and different laboratories. Until these issues are resolved, uncertainties will persist regarding the therapeutic efficacy of the various commercial heparin preparations.

As might be anticipated, there have been a variety of claims and counterclaims concerning the action of heparin on the coagulation system - heparin prevents platelet aggregation and, in the presence of adenosine diphosphate, facilitates aggregation; heparin potentiates and fails to potentiate fibrinolysis; in some animal models, heparin contributes to, rather than prevents, thrombus formation; and, in man, heparin occasionally aggravates peripheral arterial insufficiency and even, paradoxically, induces peripheral emboli. We must also face the confused state of the laboratory control of heparin therapy. Dosage schedules have been based on the greatest anticoagulant effect achievable without inducing hemorrhage, rather than on an experimentally demonstrable antithrombotic effect. In fact, the so-called therapeutic range has been judged to be somewhat above (but ideally just slightly above) the hemorrhagic dosage.

Physicians who wish to use heparin are confronted with a literature that contains conflicting statements in regard to the preferable animal source of the drug, the unit dose, and the route, frequency, duration, and regulation of administration. For example, at least three approaches to dosage regulation are in use: a preselected schedule that requires no laboratory control, whole-blood clotting time, and activated partial thromboplastin time.

It is appropriate to repeat that heparin remains our most effective antithrombotic agent - certainly in venous thromboembolic disease. This may suggest that its margin of effectiveness and safety may be great enough to override, statistically, the specific problems just described. It is also reassuring that heparin, unlike its synthetic congeners, is nontoxic, at least in short-term use. And even in large doses, heparin has no substantive ill effect on peripheral or coronary circulation, respiration, renal or hepatic function, or blood chemistry.

The anticoagulant action of heparin is believed to depend on its inhibition of reactions involved in thromboplastin elaboration and thrombin formation and on its enhancement of the thrombin inhibitory action of a natural plasma antithrombin.

Minidose Heparin

In recent years, work in several centers, including our laboratory (by E. Thye Yin), has produced an increasing body of data defining the pivotal thrombogenic role of factor X, which functions at the beginning of the final common pathway (which all known coagulation activators must follow) and converts the zymogen to the activated species that participates in several reactions, leading through thrombin to platelet aggregation and fibrin formation.

Our interest in minidose heparin stemmed from work by Walter Seegers (14-16) and later in our own laboratory (21, 22, 24-26) and others (2, 4, 9, 10, 12, 13). This work identified a potent, naturally occurring inhibitor of activated factor X (antithrombin III, heparin cofactor) in human plasma. There are several indications that minidose heparin might be of prophylactic value as an antithrombotic agent. First, the anticoagulant effect of the inhibitor of factor Xa *in vitro* is profoundly augmented by trace amounts of heparin. Second, 1 μg of the inhibitor, by neutralzing 32 units of factor Xa, indirectly prevents the generation of 1,600 NIH units of thrombin. Third, to neutralize this amount of thrombin directly, 1,000 μg of inhibitor are required. Fourth, more heparin is necessay to block the thrombin-fibrinogen reaction via the inhibitor than is required to neutralize factor Xa. Finally, factor Xa is a more potent thrombogenic agent than thrombin itself (23).

The conept of low-dose heparin was initially referred to about 25 years ago by DeTakats (5), and then by Bauer (1) and Leggenhager (8); it was first used by Sharnoff (17). Our theory of the efficacy of minidose heparin prophylaxis among surgical patients is not based on anticoagulation in the classic sense, because no state of hypocoagulability is induced. Circumstantial evidence strongly suggests that during and after operation a state of hypercoagulability, not previously present, is initiated, but that it remains "nonthrombotic" so long as the rate of factor Xa neutralization exceeds its rate of generation. Because this neutralization reaction rate depends on the inhibitor concentration, the latter becomes rate-limiting. Thus, as the inhibitor becomes increasingly utilized, a point will be reached at which some factor Xa will escape its inhibitor, combine with lipid, calcium ion, and factor V, and rapidly generate large quantities of thrombin that, once formed, cannot be prevented by low doses of heparin from converting fibrinogen to fibrin. It is, in essence, the presence of small amounts of plasma heparin, which augments severalfold the rate of normal factor Xa neutralization before the development of hypercoagulability, that may prevent venous thrombosis in patients undergoing operation.

In the past, the use of antithrombotic therapy has required a balance in risk - the likelihood of thromboembolism against the hazard of drug-induced hemorrhage. The necessity of this choice has led to an interest in defining risk groups through which it might be determined that the likelihood of thromboembolism is great enough to justify the risk of inducing hemostatic failure. Whereas this approach has met with some success among selected surgical patients, it has eliminated from protection all patients who are at low risk but who might nevertheless unpredictably develop pulmonary emboli.

Minidose heparin administered to hemostatically competent adults eliminates the drug risk of hemorrhage, as well as the need to monitor heparin dosage. Published trials in postoperative patients support the conclusion that the surgeon has an antithrombotic approach tailor-made for the primary care of surgical patients that is comparable, in safety, with penicillin prophylaxis for rheumatic fever or a vaccine to prevent poliomyelitis - thousands of people can be treated to prevent a single fatality. In the long run, it is only through such primary prevention in the adult population at large that the deadly impact of thrombosis can be lessened. However, although minidose heparin trials have shown a significant reduction in postoperative calf thrombi, as measured by isotopic venous limb scans, they have not yet shown a significant reduction in lethal pulmonary emboli (20).*

Clinical Trials of Antithrombotic Therapy

At the opening of my remarks, I mentioned the design of clinical trials. The body of ideas and experience associated with clinical trials points to numerous design characteristics that appear necessary to ensure repeatability of results, although sufficiency of case material is still a poorly appreciated hurdle.

If one accepts that approximately two of every 1,000 adults subjected to a major general surgical procedure will die from pulmonary emboliism, how large a trial is required? On the basis of calculations by Mr. Jerome Cornfield of George Washington University

*Discussed in greater detail in the next presentation.

(personal communication), if the event rate (the number of fatalities attributable to pulmonary embolism) is 0.002 and if an 80% gain (or reduction) in mortality from therapy is the anticipated endpoint, then a heparin-treated population of 28,000 subjects and 28,000 controls would be required to establish an unequivocal benefit from the drug at $p < 0.01$.

Venous thrombosis is often difficult to recognize clinically. When the high frequency of venous thromboembolism surfaced as a result of the use of contrast materials (11), isotopes (20), and meticulous autopsy dissection (6), fatal pulmonary embolism was found, by comparison, to be surprisingly infrequent although still important in actual numbers of deaths, (comparable with the annual mortality from automobile accidents in the United States).

When a patient is said to have died from a pulmonary embolus, however, what is the usual evidence for such a statement? All physicians can accept the conclusion that a pulmonary embolus caused death, if the autopsy reveals massive fresh pulmonary emboli without disease in the heart and lungs. But, in patients with cardiac and pulmonary pathology, the determination of whether the pulmonary embolism was causal, contributory, or incidental to death is arbitrary at best. Few trials have defined the clinical, physiologic, and pathologic criteria for determining that a pulmonary embolus was the primary cause of death. Therefore, such statements must be interpreted cautiously.

The population size in clinical trials of therapy for thromboembolism is of such importance that I wish to refer to another problem of clinical trials - the use of anticoagulants in reducing mortality from acute myocardial infarction, a major arterial disease in which it is hoped that heparin may play an important role.

The original assumptions for the proposed beneficial effect of anticoagulants on the course of acute myocardial infarction were a reduction in systemic and pulmonary emboli, a reduction in reinfarction by prevention of new coronary thrombosis, and the prevention of infarct extension that results from continued propagation of thrombosis along the coronary arterial tree. The second and third mechanisms might be expected to have a substantial effect on mortality from myocardial infarction. Careful studies, however, have uniformly failed to show any effect of anticoagulants on either reinfarction or infarct extension. Accordingly, the only benefit to be hoped for are from prevention of embolization, which might produce at most a 2% reduction in deaths from acute myocardial infarction.

The results of a few simple calculations highlight some of the consequences of the fact that anticoagulant therapy affects only a small fraction of all deaths from acute myocardial infarction. If it were possible to identify accurately all thromboembolic mortality as a single determinant, separate from other causes of death secondary to acute myocardial infarction, in an anticoagulant trial, a substantial number of patients would be required. Thus, as shown in the upper half of Table III using standard sample-size formulas and a 20% overall mortality rate for acute myocardial infarction with a 2% mortality from thromboembolism and an 80% effective therapy, a calculated 1,580 patients (treated plus controls) would be needed to obtain a mortality reduction at $\underline{p} < 0.05$ under such hypothetically ideal circumstances. As the assumed event rate goes down, the required sample size increases. If the assumed reduction in mortality is 60%, rather than 80%, the necessary samples are more than doubled. Moreover, when a significance of 0.01, rather than 0.05, is demanded, required sample sizes increase from 1,580 to 2,910 (19).

In an actual trial, however, the figures in the upper part of Table III cannot be used, because thromboembolic deaths must be pooled with deaths from other causes, thereby masking the benefit or lack of benefit from anticoagulant therapy. Thus, using the data in the lower half of Table III, if 18 of 100 patients with acute myocardial infarction treated with anticoagulants die during hospitalization from mechanisms other than thromboembolism, and if two of the 100 die from thromboembolic phenomena, the observed decrease in deaths related to anticoagulant therapy, with an 80% reduction in thromboembolic mortality, would not be a drop from 2% to 0.4%, but rather from 20% to 18.4%. To be reasonably certain that such a reduction did not occur by chance requires markedly inflated trial populations. Thus, the number of patients (treated and controls) needed to substantiate an anticoagulant-induced 80% reduction in thromboembolic mortality from a control value of 2% would necessarily increase from 1,580 to 20,760. For a significance of 0.01, the sample size would increase from 2,910 to 38,220 (19).

That this analysis is not academic, however, is attested to by the sheer magnitude of the mortality from acute myocardial infarction, which is unrivaled by any other disease entity in this country. Even if we ignore, for the purposes of this discussion, the extent of asymptomatic coronary atherosclerosis, at least 1,000,000 persons in the United States do experience a major clinical episode of coronary heart disease and more than 600,000 die of coronary heart disease each year (18, 27). Moreover, some 500,000 patients are hospitalized each year for an acute myocardial infarction, with an in-hospital mortality ranging from 15 to 30% (27).

If we assume that only 1% of these 500,000 hospitalized patients would succumb to thromboembolism and that 80% of them could be spared from thromboembolic death by anticoagulants, then 4,000 lives would be saved each year. Similarly, if we assume that another 1% would sustain a disabling systemic embolus (which in essence would refer to embolic strokes) and that 80% of these emboli could be prevented,

TABLE III

Sample Sizes of Patient Populations (Anticoagulant-Treated plus Controls) Required to Detect Significant Reductions in Mortality from Thromboembolism among Patients with Acute Myocardial Infarction, with Overall Mortality of 20%.[a]

% Mortality among Control Patients		Numbers of Patients (Treated plus Controls) Required to Detect Reduction in Thromboembolic Deaths by Anticoagulants			
Thromboembolic Deaths Alone		*p* = 0.01; power = 0.05[b]		*p* = 0.05; power = 0.10	
Thromboembolic Mortality	Other Mortality	60% Reduction	80% Reduction	60% Reduction	80% Reduction
10%	-	1,130	550	620	300
5%	-	2,360	1,140	1,280	620
2%	-	6,040	6,910	3,280	1,580
1%	-	12,170	5,870	6,610	3,190
0.5%	-	24,440	11,780	13,270	6,400
All	Deaths				
10%	10%	2,470	1,320	1,340	720
5%	15%	10,560	5,810	5,740	3,160
2%	18%	68,480	38,220	37,210	20,760
1%	19%	277,200	155,300	150,600	84,380
0.5%	19.5%	1,115,000	626,000	605,800	340,100

[a]Derived from tables in Halperin *et al.* (7).
[b]Probability of rejecting the null hypothesis when it is false.

there would be 4,000 additional patients saved. Thus, effective antithrombotic therapy would protect 8,000 people in this country each year from death or disability. But clinical trials large enough to demonstrate these benefits have not been undertaken. For comparison, it has been estimated that, before widespread vaccination, an average of 10,000 people in this country developed severe paralytic poliomyelitis each year (3). Field trials to prove the efficacy of vaccination, however, used over 500,000 treated and control subjects.

Perhaps the principal reason for the failure to demonstrate, beyond reasonable doubt, a statistically significant decrease either in thromboembolic deaths or in residual strokes among patients with acute myocardial infarction has been the small numbers of patients that the trials have included. The acceptance of these small numbers reflects, in turn, an apparently limited understanding of the clinical implications of the pathophysiology of thromboembolism. In contrast with the high prevalence of thromboembolic events, these forms of intravascular coagulation only infrequently cause death or major disability in acute myocardial infarction. It is this disparity between the high prevalence of lesions, on the one hand, and the low incidence of associated mortality or disability, on the other, that has raised havoc with trial design.

Comments

This overview was undertaken to provide a partial bridge between the presentations and discussions of the first 2 days of this symposium and those that follow. The issue we face is not whether heparin has an antithrombotic action, but whether this action can be more effectively harnessed. The answer, I am certain, will derive initially not from applied investigations, but rather from research unfettered by targeted programs. For directed research means that our goals will be limited to conquering nature, rather than understanding it. In this regard, it should be recalled that it was only after viruses were grown in tissue culture that the seemingly incidental, but actually monumental, conquest of poliomyelitis became possible.

In thrombosis and hemostasis, as in virology, research has yielded more unexpected than anticipated therapeutic gains. Heparin was recognized by a medical student seeking a procoagulant activity in liver. The anticoagulant, Dicumarol, was discovered during the search for the basis of hemorrhagic-spoiled-sweet-clover disease in cattle. The incidental observation that the antihemophilic globulin was the last plasma protein to go into solution as frozen plasma thawed led to the development of factor VIII concentrates, which revolutionized the treatment of hemophilia. Further efforts to define the heparin molecule and to elucidate its modes of action may lead, by serendipity, to improved prophylaxis of thromboembolism, and even of atherosclerosis.

ACKNOWLEDGEMENTS

This work was supported in part by the National Heart and Lung Institute through a Specialized Center of Research in Thrombosis HL 14147 and grant 5 R01 HL 11470, and by an American Heart Association grant-in-aid.

REFERENCES

1. BAUER, G., Proceedings of the First International Conference on Thrombosis and Embolism, Basel, B. Schwabe, (1954) p. 721.
2. BIGGS, R., DENSON, K.W.E., AKMAN, N., BORRETT, R. and HADDEN, M., Brit. J. Haemat., 19 (1970) 283.
3. Communicable Disease Center, Poliomyelitis Surveillance: Report #283, Atlanta, 1964.
4. DAMUS, P.S., HICKS, M. and ROSENBERG, R.D., Nature, 246 (1973) 355.
5. DETAKATS, G., JAMA, 142 (1950) 527.
6. FREIMAN, D.G., SUYEMOTO, J. and WESSLER, S., N. E. J. Med., 272 (1965) 1278.
7. HALPERIN, M., ROGOT, E., GURIAN, J. and EDERER, F., J. Chron. Dis., 21 (1968) 13.
8. LEGGENHAGER, K., Helv. Chir. Acta, 24 (1957) 316.
9. MARCINIAK, E., Brit. J. Haem., 24 (1973) 391.
10. MARCINIAK, E. and TSUKAMURA, S., Brit. J. Haemat., 22 (1972) 341.
11. NICOLAIDES, A.N., KAKKAR, V.V., FIELD, E.S. and RENNEY, J.T.G., Brit. J. Radiol., 44 (1971) 653.
12. ROSENBERG, R.D., Circulation, 49 (1974) 603.
13. ROSENBERG, R.D. and DAMUS, P.S., J. Biol. Chem., 248 (1973) 6490.
14. SEEGERS, W.H., Ann. N.Y. Acad. Sci., 146 (1968) 593.
15. SEEGERS, W.H., COLE, E.R., HARMISON, C.R. and MONKHOUSE, F.C., Canad. J. Biochem., 42 (1964) 359.
16. SEEGERS, W.H. and MARCINIAK, E., Nature, 193 (1962) 1188.
17. SHARNOFF, J.G., Surg. Gyn. Obst., 123 (1966) 303.
18. STAMLER, J., BEARD, R.R., CONNOR, W.E., deWOLFE, V.G., STOKES, J., WILLIS, P.W., LILIENFELD, A.M., DAWBER, T.R., DOYLE, J.T., EPSTEIN, F.H., KULLER, L.H. and WINKELSTEIN, W., Circulation, 42 (1970) A 55.

19. WESSLER, S., KLEIGER, R.E., CORNFIELD, J. and TEITELBAUM, S.L., Arch. Int. Med., in press.
20. WESSLER, S. and YIN, E.T., Circulation, 47 (1973) 671.
21. YIN, E.T. and WESSLER, S., Thromb. Diath. Haemorrh., 21 (1969) 398.
22. YIN, E.T. and WESSLER, S., Biochem. Biophys. Acta, 201 (1970) 387.
23. YIN, E.T. and WESSLER, S., Thromb. Diath, Haemorrh., 20 (1968) 465.
24. YIN, E.T., WESSLER, S. and STOLL, P., J. Biol. Chem., 246 (1971) 3703.
25. YIN, E.T., WESSLER, S. and STOLL, P., J. Biol. Chem., 246 (1971) 3712.
26. YIN, E.T., WESSLER, S. and STOLL, P., J. Biol. Chem., 246 (1971) 3694.
27. YU, P.N., BIELSKI, M.T., EDWARDS, A., FRIEDBERG, C.K., GRACE, W.J., JANUARY, L.E., LIKOFF, W., SCHLERLIS, L. and WEISSLER, A.M., Circulation, 43 (1971) A 171.

LOW-DOSE HEPARIN IN THE PREVENTION OF VENOUS THROMBOEMBOLISM - RATIONALE AND RESULTS

V.V. KAKKAR

Department of Surgery, King's College Hospital Medical School, Denmark Hill, London SE5 8RX (United Kingdom)

Venous thromboembolic disease is a common and increasing cause of morbidity and mortality in hospital patients. It has been estimated that approximately 21,000 patients die each year from this cause in the United Kingdom, and 47,000-142,000 in the United States (10, 19). Apart from the immediate risk to life, one must also consider the late sequelae of this disease - swelling of the legs, varicose veins, ulceration, and other trophic changes that represent an equally distressing situation (15).

If the mortality due to pulmonary embolism and the misery due to the postphlebitic syndrome are to be significantly reduced, the disease must be detected at an early stage, when treatment may be effective; an even better approach would be to develop a simple method of prophylaxis that could eliminate this condition. Many attempts have been made to prevent thrombosis by using drugs that interfere with the coagulation mechanism. One promising approach is the use of low-dose subcutaneous heparin, which has been claimed to prevent thrombosis without increasing the risk of bleeding. The rationale for the use of low-dose heparin and the results of recent clinical trials are reviewed very briefly in this paper.

Venous thrombi are generally regarded as an expression of blood coagulation and fibrin formation in the presence of venous stasis. The crucial step in the formation of venous thrombi is considered to be the generation of thrombi in areas of stasis, in that thrombin has the dual ability to aggregate platelets and to convert fibrinogen to fibrin (22, 28). It is therefore probable that, if thrombin formation can be blocked at an early stage, the development of venous thrombi will be prevented.

During the activation of intravascular coagulation, thrombin is produced from its precursor, prothrombin, by the concentrated action of factor V and activated factor X (factor Xa) in the presence of a surface lipid and calcium ions. Thrombin thus formed initiates the development of fibrin clot by its specific action on arginine-glycine peptide bonds of fibrinogen (3), and at the same time it is inactivated by an antithrombin - a specific inhibitor that is normally present in the plasma. The anticoagulant effect of heparin has been shown to occur only in the presence of a plasma component known as heparin cofactor (4). The cofactor combines with heparin to form an active complex that accelerates, by 50 - 100 times, the blocking of the enzymatic action of thrombin. On the basis of experiments performed with plasma deficient in antithrombin III (6) and with purified material (24), it has been suggested that antithrombin III and heparin cofactor are the same substance (18). Recent work by Yin and Wessler (32) has demonstrated the presence of naturally occurring inhibitor of activated factor X (anti-Xa) in human plasma and serum. They have suggested that anti-Xa, antithrombin III, and heparin cofactor activity are probably properties of the same protein moiety (33). They have also shown that the activity of this inhibitor is enhanced by trace amounts of heparin; 1 µg of the inhibitor of activated factor X, by inhibiting 32 U of activated factor X, indirectly prevents potential generation of 1,600 NIH U of thrombin. Activated factor X occupies a key position in the intrinsic and extrinsic coagulation mechanism. The enzymatic coagulation sequence functions as a biologic amplification system; accordingly, it would be reasonable to suggest that, if small amounts of heparin are circulating in the blood, these will effectively neutralize activated factor X by enhancing the inhibitor activity. In other words, if "hypercoagulability" were treated with heparin before initiation of intravascular coagulation, less antithrombotic agent would be required than if therapy were begun after thrombin formation had occurred. The best support for this hypothesis has been provided by the failure of low doses of heparin in patients subjected to emergency operations for fracture of the femoral neck (V.V. Kakkar, unpublished observations). In these patients, the coagulation sequence has already been activated beyond the stage of thrombin generation before the small amounts of heparin administered could be effective.

Recent observations by Rosenberg and Damus (25) have suggested that all seven proteases produced within the coagulation cascade are neutralized by this inhibitor and that heparin accelerates each of these interactions. Thus, of the serine proteins directly or indirectly involved in the conversion of prothrombin to thrombin, four proteins are ultimately activated to serine proteases (factors XIIa, XIa, IXa, and Xa) and two proteins are cofactors for these proteolytic events (factors V and VIII) (25). It is therefore possible that small amounts of heparin, by accelerating inhibitor activity, may block the activation not only of factor X, but also of factors

XI and IX. These observations need confirmation.

Although there is no _in vitro_ evidence that heparin in small amounts can enhance the dissolution of a thrombus, the decrease in availability of thrombin formed under such circumstances might retard activation of the fibrin-stabilizing factor (factor XIII). This would explain the accelerated disintegration of thrombin observed in patients treated preoperatively with heparin, compared with thrombi formed in a control group (11). The loose network of fibrin formed in the presence of heparin traces may be more accessible to either fibrinolysis or mechanical breakdown through force alone.

It has also been claimed that some of the effect of low-dose heparin prophylaxis might be mediated through platelets. In the studies reported by O'Brien _et al_. (23), patients after a major operation showed a decrease in platelet aggregation response to adenosine diphosphate (ADP) and thrombin, compared with their preoperative response. It was suggested that a decreased response to ADP and thrombin is related to the cause of postoperative deep venous thrombosis. Heparin (5,000 IU subcutaneously) prevented the immediate postoperative decrease in aggregation.

In summary, the exact mode of action of low-dose heparin prophylaxis remains to be defined. Current evidence strongly suggests that heparin may act at multiple sites within the coagulation system.

Review of Recently Published Data

The suggestion that small doses of heparin should be used for preventing venous thrombosis is not new: Bauer (2) in 1954 and Leggenhager (16) a few years later recommended heparin prophylaxis for all high-risk patients undergoing major surgery. More than a decade ago, Sharnoff _et al_. (27) suggested that low-dose heparin administered before, during, and after surgery may be effective in preventing postoperative thromboembolism. The regimen they recommended consisted of the administration of 10,000 IU of heparin subcutaneously at midnight before surgery; in their view, this dose provided an anticoagulant effect lasting approximately 12 hr. If surgery extended beyond this period, the coagulation time was determined and, if it was "critically short," additional heparin (rarely more than 2,500 IU) was administered subcutaneously during operation. At the completion of the operation, the coagulation time was measured and usually 2,500 IU of heparin were administered subcutaneously every 6 hr until the patient was fully active or discharged. The findings of an uncontrolled study (26) published in 1970 suggested that this regimen was effective in preventing fatal pulmonary embolism. However, it had several drawbacks: it was not a randomly allocated, controlled trial, and the results were therefore not suitable for statistical analysis. Furthermore, with the regimen recommended by Sharnoff and co-workers, the amount of heparin to be administered every 6 hr had to be tailored to the whole-blood clotting time. Obviously, it would be impractical to adopt this form of prophylaxis on a wide scale.

Within the last 3 years, several studies have investigated the efficacy of low-dose heparin in preventing postoperative deep venous thrombosis (DVT), as assessed by the ^{125}I-labeled fibrinogen technique. Their results are summarized in Table I. These studies can be broadly divided into two main groups: controlled clinical trials in which the efficacy of a standard regimen of subcutaneous heparin in the prevention of DVT and/or pulmonary embolism was assessed, and comparative controlled trials in which heparin was compared with other drugs claimed to have beneficial effects.

Controlled clinical trials. - Kakkar *et al*. (13) assessed the effectiveness of a standard regimen of heparin prophylaxis in 53 consecutive patients over the age of 50 who were undergoing inguinal hernia repair. The regimen consisted of 5,000 IU of heparin administered subcutaneously starting 2 hr before surgery. This interval for the initial heparin injection was selected because the findings of a quantitative and sensitive assay for heparin showed that it took 30 min for heparin in this dose to become demonstrable in the plasma, and it reached a peak within 2 hr. This would allow for the heparin effect to be at its height before the operative incision. DVT was detected, by the ^{125}I-fibrinogen test, in seven (25%) of the control patients; this was significantly reduced (4%) in 26 similar patients who received heparin. There was no unusual operative or postoperative bleeding.

Using the ^{125}I-fibrinogen test, Williams (31) assessed the efficacy of the Sharnoff regimen in 56 patients over the age of 50 who were subjected to major abdominal surgery. A 41% incidence of DVT in the control group was reduced to 15% in the heparin-treated patients; all four patients in the heparin group who developed thrombi were from a group of seven patients subjected to prostatectomy.

Gordon-Smith *et al*. (8), using the ^{125}I-fibrinogen test and heparin prophylaxis in a manner similar to that of Kakkar, divided 161 patients over the age of 40 who were admitted for major elective surgery into three groups in a prospective, randomized trial. Group 1 received no heparin and had a 41% incidence of DVT; thrombosis; group 2 received only three doses (every 12 hr) of 5,000 IU of heparin subcutaneously, starting before surgery, and had a 13.5% incidence of DVT; and group 3, given 5,000 IU of heparin every 12 hr starting before operation and continuing for 5 days, had an 8.3% incidence of DVT. Groups 2 and 3, both separately and together, had a signifi-

TABLE I

Effect of Prophylactic Low-Dose Heparin on the Incidence of Postoperative Deep Venous Thrombosis as Assessed in Controlled Clinical Trials

Investigators	Control Group		Treated Group		Statistical Significance
	No. Studied	No. Developing DVT	No. Studied	No. Developing DVT	
Kakkar *et al.* (13)	27	7 (26%)	26	1 (4%)	$0.05>p>0.25$
Williams (31)	29	12 (41%)	27	4 (15%)	$0.02>p>0.01$
Gordon-Smith *et al.*[a] (8)	50	21 (42%)	52 48	7 (13.5%) 4 (8.3%)	$p<0.003$ $p<0.001$
Kakkar *et al.*[b] (12)	39	17 (42%)	39 133 50	3 (8%) 13 (9.7%) 20 (40%)	$p<0.001$
Nicolaides *et al.* (20)	122	29 (24%)	122	1 (0.8%)	$p<0.000003$
Ballard *et al.* (1)	55	16 (29%)	55	2 (4%)	$p<0.001$
Gallus et al. (7)	118	19 (16%)	108	2 (2%)	$p<0.003$

[a] A trial comparing 2 different regimens.
[b] Double-blind randomly allocated trial.

cantly lower incidence of positive scans than the control group, but the difference between the two heparin-treated groups - perhaps because of the limited number of patients - was not statistically significant.

In a second study by Kakkar and his associates (12) low-dose heparin was evaluated in 261 patients, divided into four groups. A prospective, double-blind, randomized trial was carried out in 78 patients over the age of 40 who were subjected to major elective abdominal and orthopedic surgery. The regimen of heparin prophylaxis consisted of 5,000 IU subcutaneously, begun 2 hr preoperatively and continued every 12 hr for 7 days. The frequency of DVT was 42% in the 39 control patients and 8% in the 39 patients receiving heparin. None of the 78 patients developed pulmonary emboli. The same heparin regimen was administered to another group of 133 consecutive patients over the age of 40 who had major elective surgery. The overall incidence of DVT in this group was 9.7%. The results were less encouraging in patients undergoing total hip replacement: four (27%) of 15 such patients developed thrombosis. Similarly, the results were unsatisfactory in 50 patients subjected to emergency surgery for fracture of the femoral neck. In this group, all patients received heparin, and the incidence of thrombosis was 40%. In the entire series of 226 "at-risk" patients, however, only one patient had clinically recognized emboli that proved fatal; in the five other deaths, no pulmonary emboli were found at necropsy.

Nicolaides *et al*. (20) investigated the effectiveness of Kakkar's heparin prophylaxis regimen in 251 surgical patients over the age of 40 who were undergoing major abdominal and thoracic surgery. The 24% incidence of DVT in the control group was reduced to 0.8% in the heparin-treated patients. In addition, heparin prophylaxis reduced the frequency of the dangerous extending thrombi, which are often responsible for pulmonary emboli, from 7.4% to nil.

Ballard *et al*. (1) studied the effect of prophylactic heparin in 110 patients who were undergoing major gynecologic operations and who were randomly assigned to two equal groups. The treated patients were given 5,000 IU of heparin subcutaneously 1 - 2 hr before operation and every 12 hr thereafter for 7 days. There was no significant increase in operative and postoperative bleeding in the heparin-treated patients, and only two of the 55 patients in this group developed DVT compared with 16 in the comparable control group. The difference in the incidence of isotopic DVT was highly significant ($p < 0.001$).

Finally, Gallus *et al*. (7) modified Kakkar's regimen by giving 5,000 IU 2 hr before surgery and then repeating this dose three times, rather than twice, daily, beginning 8-10 hr after the preoperative dose. In 226 patients undergoing major elective surgery, he reduced the incidence of DVT from 15% to 2%. In 46 patients with hip fractures, heparin treatment also reduced a 48% incidence of DVT to 13%. Clinically significant bleeding was not increased in heparin-treated surgical patients, although the blood requirements of patients who received transfusions were moderately increased and treated patients had a slightly lower postoperative hematocrit. In addition, the blood requirements of patients who received transfusions were moderately increased after elective abdominothoracic surgery, by a mean amount of 518 ml. Although Gallus and colleagues did not regard increased bleeding as a major problem on a regimen of 5,000 IU three times a day, it is likely that the amount of heparin used in the maximum that can be given to surgical patients without significant hemorrhagic complications.

In nearly 1,200 patients entered in seven trials - most of them randomized - the incidence of DVT in those treated with low-dose heparin was significantly lower than that in the control patients. Despite the heterogeneity of the patients and the varied nature of the surgery, the prophylactic effectiveness of low-dose heparin was such that the overall incidence of DVT was three times higher in the untreated patients. This success was achieved with a more or less standard dose of heparin (10,000 or 15,000 IU/day) and without significant hemorrhagic complications. The facts that a standard regimen is not totally effective and that it is not yet certain that preventing postoperative isotopic DVT will also prevent death from pulmonary embolism should not detract from recognition of the highly significant advance in the prophylaxis of postoperative venous thrombosis that these studies collectively represent.

The failure of low-dose heparin in the prophylaxis of DVT among patients undergoing hip surgery (particularly total hip replacement) was disappointing (Table II). However, this result could have been anticipated; in traumatic operations of this magnitude, the stimulus to intravascular coagulation, in terms of tissue damage (especially the release of intramedullary fat and the elaboration of activated clotting factors that may be absorbed into the circulation), must be considerable. It seems reasonable to suggest that failure of heparin prophylaxis in these patients may be due to the inadequacy of the dosage used.

It is interesting to note that low doses (10,000 IU daily), as used by Kakkar _et al_. (12) also failed to protect patients subjected to emergency operations for fracture of the femoral neck. However, Gallus _et al_. (7), who gave a 50% higher dose of heparin (15,000 IU/day instead of 10,000 IU), succeeded in reducing postoperative DVT in this group of patients to 15%, compared with a control incidence of 48%. Our observations (5) and those of others, such as Nicolaides _et al_. (21) and Dechavanne _et al_. (unpublished data), also indicate that a dosage of 5,000 IU of heparin three times a day is much more effective in protecting patients undergoing total hip replacement.

TABLE II

Incidence of Deep Venous Thrombosis with Different Regimens of Low-Dose Heparin in Patients Undergoing Total Hip Replacement

Investigators	Control Group		Heparin Group			Statistical Significance
	No. Studied	No. Developing DVT	Regimem	No. Studied	No. Developing DVT	
Kakkar _et_ _al_. (12)	18	7 (38.8%)	B.D.	15	4 (26.6%)	$p > 0.05$
Corrigan _et_ _al_. (5)	35	12 (34%)	T.D.S.	37	4 (10.8%)	$p < 0.01$
Nicolaides _et_ _al_. (21)	27	11 (40%)	T.D.S.	25	1 (4%)	$p = 0.00016$
Dechavanne et al. (unpublished data)	20	8 (40%)	T.D.S.	20	1 (5%)	$p < 0.025$

Excessive bleeding has not been observed with this regimen, although different findings have been reported by Hume et al. (9). These recent observations indicate that, when the coagulation sequence has already been activated beyond the stage of thrombin generation (as happens in patients with hip fractures before heparin prophylaxis is started) or when the stimulus to thrombin generation is overwhelming, 5,000 IU of heparin every 12 hr represents inadequate prophylaxis. However, when neither of these conditions is present, a regimen of low-dose heparin twice a day represents for the surgeon "an antithrombotic approach tailor-made for primary care for patients subjected to operation, comparable in safety to penicillin prophylaxis for rheumatic fever or to a vaccine to prevent poliomyelitis, whereby thousands of individuals can be treated to prevent a single patient fatality" (30).

Controlled comparative trials. - Many of the recent trials are directed toward comparing the efficacy of low-dose heparin with that of other drugs that have already been claimed to be highly effective in preventing venous thromboembolism. In a recent prospective trial reported from the Netherlands, 50 patients were given low-dose heparin and another 50 patients were given an oral anticoagulant (Sintrom) after mainly abdominal surgery (Table III) (29). The incidence of DVT as diagnosed by the ^{125}I-fibrinogen test, was 2% in the heparin group and 18% in the oral anticoagulant group. The unexpectedly high incidence of thrombosis in patients receiving oral anticoagulants may be a result of starting therapy after surgery. Our studies of the natural history of postoperative DVT (14) have shown that most thrombi detected by the radioactive fibrinogen test become manifest within the first 1-2 postoperative days; inasmuch as oral anticoagulants take longer than that to achieve an antithrombotic effect, it is not surprising that the incidence of DVT was so high in these patients. A true comparison must await the results of randomized, controlled trials in which oral anticoagulants are administered for 36-48 hr before surgery.

In another randomly allocated recent trial, the efficacy of low-dose heparin was compared with that of dextran-70 infusion (Table III), begun before surgery and continued during the first 3 postoperative days (17). Of the 382 patients included in this study, 130 acted as controls, 126 received heparin, and 126 received dextran infusion. Computer analysis showed that the three groups were well-matched, with respect to factors likely to influence the incidence of DVT. The incidence of thrombosis was 37% in the control group, 12% in those receiving 5,000 IU of heparin twice a day, and 25% in those receiving dextran-70 infusion.

The efficacy of low-dose heparin has also been compared with agents known to affect platelet function. Sixty patients over the age of 50 who were undergoing Charnley-Muller total hip replacement

TABLE III

Comparative Trials of Low-Dose Heparin Prophylaxis

Investigators	Group	No. Studied	DVT	Statistical Significance
Van Vroonhoven *et al.* (29)	Oral anti-coagulants	50	9(18%)	$p < 0.025$
	Heparin	50	1(2%)	
Macintyre *et al.* (17)	Control	130	48(37%)	
	Heparin	126	15(12%)	$p < 0.025$
	Dextran-70	126	31(25%)	
Dechanvanne *et al.* (unpublished data)	Control	20	12(60%)	
	Aspirin + dipyridamole	20	10(50%)	
	Heparin	20	1(5%)	$p < 0.025$

for osteoarthritis were randomly allocated to: a control group; a group receiving a combination of 150 mg of dipyridamole and 1.5 g of aspirin per day; or a group receiving 5,000 IU of calcium heparin (Calciparine, Laboratoire Choay) 2 hr before surgery, then every 12 hr for 48 hr, and finally every 8 hr for 8 days. In the control group, 12 (40%) developed DVT, whereas the incidence was 5% in those receiving heparin and 50% in those receiving dipyridamole and aspirin. The differences observed were statistically significant. Excessive blood loss during surgery was not observed in any of the patients.

Discussion

The goal must be the total prevention of venous thromboembolism. Patients who require prophylaxis are in three main groups: those who are confined to bed for long periods, either at home or in chronic illness medical wards; surgical patients, particularly those undergoing orthopedic procedures and "high-risk" patients undergoing abdominal surgery; and obstetric patients with complicated pregnancy. To provide effective prophylaxis for these groups of patients, oral anticoagulants are commonly used. However, the disadvantages of oral anticoagulants are the risk of hemorrhage and the need for strict laboratory control, which have undoubtedly contributed to the relatively low acceptance of this form of prophylaxis among surgeons in general - at least, in the United States and the United Kingdom.

A form of drug therapy that is both effective and without the drawbacks of oral anticoagulants would therefore meet a real need. Ideally, any agent used for the prophylaxis of deep venous thrombosis should be well tolerated by the patient, be devoid of side effects, require no special monitoring and produce no bleeding in a clinical situation in which the patient is subjected to trauma. The last requirement means that any tendency to excessive intravascular coagulation should be prevented without interfering with normal hemostasis.

Using the sensitive and accurate ^{125}I-fibrinogen technique, seven recently published clinical trials (five of them randomized) have each demonstrated, despite variabilities in design, that low-dose heparin prophylaxis begun before surgery and continued through the first 7-10 postoperative days significantly reduces the incidence of DVT. In each of these studies, heparin administration was initiated before surgery; this may well be the key to prophylactic success.

There were nearly 1,200 patients (treated and control) involved in these seven trials (Table I); all were over 40, and most were subjected to major surgery. The overall incidence of 20% isotopic DVT in the control group was reduced to less than 10% in the patients receiving heparin. Can these findings be taken as evidence that low-dose heparin prophylaxis will prevent postoperative pulmonary emboli and, therefore, death? Further analysis of these trials fails to provide an unequivocal answer to this question. In this heterogeneous group of 1,200 patients, many of whom were presumed to be at high risk, only five were diagnosed as having pulmonary emboli; three were in the heparin-treated group, and one death was attributed to pulmonary embolism. If these figures are representative of patients over the age of 40 subjected to major surgery, they indicate that the incidence of fatal and nonfatal pulmonary embolism is low. Accordingly, a patient population of many thousands would be required for a prospective, multicenter, randomized trial to establish the prophylactic value of low-dose heparin in the prevention of postoperative pulmonary embolism.

Such a trial is now in progress. Patients over the age of 40 who are undergoing major elective surgery are being randomly allocated to a control or treatment group. To date, 3,500 patients have been admitted to the trial - 1,800 in the control group and 1,700 in the treatment group. Computer analysis shows that the two groups are well matched for age, sex, presence of malignancy, type of operation performed, and other factors likely to influence the incidence of thromboembolism. Forty-three patients have died in the control group and 36 in the heparin group; autopsy examination in these showed that 11 patients in the control group had massive fatal pulmonary embolism, but only one in the heparin group. The difference between the two groups is statistically significant ($0.01 < p < 0.05$). These early results in a limited number of patients suggest that the regimen of low-dose heparin prophylaxis investigated is effective in preventing fatal pulmonary embolism in surgical patients.

Data from trials already reported or still in progress have raised important ethical problems. In the future, is it justifiable to withhold a form of prophylaxis that has been shown to be quite safe and effective? What should be the role of a practicing physician? Should he await the outcome of these trials, or has sufficient information already been collected to justify the recommendation of low-dose heparin prophylaxis for general use?

Published evidence and clinical experience now indicate that low-dose heparin prophylaxis should be recommended as primary prevention for all adults who are subjected to major abdominal, pelvic, or thoracic - but not, as yet, orthopedic - surgery. A standard regimen fulfills most of the criteria demanded of an ideal prophylactic agent: it is well tolerated by the patient, is free of side effects and requires no monitoring other than that the patient receives the drug appropriately, and does not produce excessive bleeding when the patient is subjected to major tissue trauma.

REFERENCES

1. BALLARD, R.M., J. Obstet. Gynaec. Brit. Cwlth., 80 (1973) 469.
2. BAUER, G., Proc. First International Conference on Thrombosis and Embolism, B. Schwabe, Basle, 1954, p. 721.
3. BLOMBACK, B., BLOMBACK, M., HESSEL, B. and IWANAGE, S., Nature, 215 (1967) 1445.
4. BRINKHOUS, K., SMITH, H.P., WARNER, E.D. and SEEGERS, W.H., Amer. J. Physiol., 125 (1939) 683.
5. CORRIGAN, T.P., KAKKAR, V.V. and FOSSARD, D.P., Brit. J. Surg., 61 (1974) 320.
6. EGEBERG, O., Thromb. Diath. Haemorrhag., 14 (1865) 473.
7. GALLUS, A.S., HIRSH, J., TUTTLE, R.J., TREBILCOCK, R. and O'BRIEN, S.E., New Engl. J. Med., 288 (1973) 545.
8. GORDON-SMITH, I.C., GRUNDY, D.J., LE QUENSNE, L.P., NEWCOMBE, J.F. and BRAMBLE, F.S., Lancet, 1 (1972) 1133.
9. HUME, M., KURIAKOSE, T.X., ZUCH, L. and TURNER, R.H., Arch. Surg., 107 (1973) 803.

10. HUME, M., SEVITT, S. and THOMAS, D.P., Venous Thrombosis and Pulmonary Embolism, Harvard University Press, Cambridge, Mass., (1970) p. 3.
11. KAKKAR, V.V., Bull. Swiss Acad. Med. Sci., 29 (1973) 235.
12. KAKKAR, V.V., CORRIGAN, T.P., SPINDLER, J., FOSSARD, D.P., FLUTE, P.T., CRELLIN, R.Q., WESSLER, S. and YIN, E.T., Lancet, 2 (1972) 101.
13. KAKKAR, V.V., FIELD, E.S., NICOLAIDES, A.N., FLUTE, P.T., WESSLER, S. and YIN, E.T., Lancet, 2 (1971) 699.
14. KAKKAR, V.V., HOWE, C.T., FLANC, C. and CLARKE, M.B., Lancet, 2 (1969) 230.
15. KAKKAR, V.V., HOWE, C.T., LAWS, J.W. and FLANC, C., Brit. Med. J., 1 (1969) 810.
16. LEGGENHAGER, K., Helv. Chir. Acta, 24 (1957) 316.
17. MACINTYRE, I.M.C., VASILESCU, C., JONES, D.R.B., RUCKLEY, C.V., HARPER, D.R., MACPHERSON, A.I.S., DOWNIE, J., BRADFORD, W.P., RAINE, P.M., McNICOL, M.F., GRANT, J., CRISPIN, J., CLARK, W.B., WALLS, A.D.F., WALLWORK, J., TOTHELL, P., SIMPSON, J.D., PAREKH, M., FERRINGTON, C.M., LAUDER, N. and PRESCOTT, R.J., Lancet, 2 (1974) 118.
18. MONKHOUSE, F.C., FRANCE, E.S. and SEEGERS, W.H., Circ. Res., 3 (1955) 397.
19. MORRELL, M.T., TRUELOVE, S.C. and BARR, A., Brit. Med. J., 2 (1963) 850.
20. NICOLAIDES, A.N., DUPONT, P.A., DESAI, S., LEWIS, J.D., DOUGLAS, J.N., DODSWORTH, H., FOURIDES, G., LUCK, R.J. and JAMIESON, C.W., Lancet, 2 (1972) 890.
21. NICOLAIDES, A.N., DUPONT, P., PARSONS, D., APPLEBURG, M., HORAN, F.T., ESAH, K.M. and WALKER, C.J., Brit. J. Surg., 61 (1974) 320.
22. NIEWIAROWSKI, S. and THOMAS, D.P., Nature (London), 212 (1966) 1544.
23. O'BRIEN, J.R., JAMIESON, S., ETHERINGTON, M. and KLABER, M.R., Lancet, 2 (1971) 1301.
24. PORTER, P., PORTER, M.C. and SHANBERGE, J.N., Biochemistry, 6 (1967) 1854.
25. ROSENBERG, R.D. and DAMUS, D.S., J. Biol. Chem., 248 (1973) 6490.
26. SHARNOFF, J.G. and DE BLASIO, G., Lancet, 2 (1970) 1006.
27. SHARNOFF, J.G., KASS, H.H. and MISTICA, B.A., Surg. Gynec. Obstet., 115 (1962) 75.
28. THOMAS, D.P., Nature (London), 215 (1967) 298.
29. VAN VROONHOVEN, T.J.M.V., VAN ZIJL, J. and MULLER, H., Lancet, 1 (1974) 375.
30. WESSLER, S., Circulation, 47 (1973) 661.
31. WILLIAMS, H.T., Lancet, 2 (1971) 950.
32. YIN, E.T. and WESSLER, S., J. Biol. Chem., 246 (1971) 3694.
33. YIN, E.T., WESSLER, S. and STOLL, P., J. Biol. Chem., 246 (1971) 3712.

DISCUSSION OF KAKKAR PAPER

HORNER

I think most of us have been very impressed by the data we heard today and therefore I have a concerned question to address, not to the speakers, but to representatives of the heparin manufacturing industry in the audience. If mini-dose heparin should become standard practice, can they handle the demand?

SILVERGLADE

I do not believe that if this became standard procedure that there would be a sufficient supply of heparin available at this time based on the shortage of raw materials.

BAUE

There would be a heparin crisis then?

SILVERGLADE

Yes, at that time. There is an apparent shortage, but not of crisis proportions.

ENGLEBERG

I want to ask Dr. Kakkar if there is any follow-up on what happens to the leg vein thrombi that do not produce pulmonary embolism. Have there been any studies a month later with radioactive fibrinogen? Do those thrombi in the leg veins lyse? Do they disappear? Because maybe there is some morbidity and ill health based upon impairment of circulation even though pulmonary emboli did not occur. I wonder if there have been follow-up studies in that regard?

KAKKAR

I think such studies are now in progress. Whether they undergo spontaneous lysis, how often and how quickly, and those that do not lyse, what are the ill effects, is something we have been studing by looking at valve functions, measuring their pressure, studying their tissue profusion by using either xenon clearance or radioactive 24 clearance. The study is now in progress to follow these patients for the next five to ten years and see what happens. Most of these patients are now entering their trial and at least we are convinced and so, are using low doses of heparin. We should be able to answer your question in another two years.

LEVIN

What is the basis for the selection of patients who are over 40 years old?

KAKKAR

The basis for that really has been the work of the pathologists and especially, Dr. Simon Savage from Birmingham. He showed in his studies that morbidity is much higher over the age of 40 than below 40. Once you are subjecting a group of a population where you want to show the differences quite clearly, you pick-up the high risk group and there are people who are now studying over the age of 60 where the incidence is even higher.

LEVIN

Your comment about excessive blood loss and an increase in hematomas in patients receiving low dose heparin therapy reminded me of Salzman's article in the New England Journal of Medicine two or three years ago, in which he also reported a high incidence of post-operative bleeding, following the use of presumably benign medications. You did not define excessive blood loss. Was this reflected in an increased requirement for transfusion? If so, what are the implications of an increased need for blood in patients receiving low dose heparin?

KAKKAR

First, I'm really not sure what Dr. Salzman's views are. Secondly, we are doing a lot of things in this big trial. There are centers using radiolabeled red cells to measure the loss accurately. There are a group of centers who are recording very accurately the blood loss which is collected through drains and a group of patients which are now being followed where the blood requirement is being estimated. When I said the subjective impression of surgeons, and being a surgeon I know what the surgeons attitude is, it is that when a patient bleeds "it must be that damn heparin", whether the patient had it or not. It is very difficult when you are talking about subjective impressions. If you take objective impressions and objective evidence, that is being collected, and that is why the trial is in progress, then we will define it very clearly, whether it is safe or not. I think these premature trials that are done with 10 patients or 12 patients and then an article is presented in some journal are confusing.

BROZOVIC

I would like to comment very briefly on the heparin levels in patients. We have done patients over 60 who were waiting for hip replacements and we measured the heparin levels hourly after the first injection. In one out of 6, a very tiny woman, heparin levels in blood after 5,000 subcutaneously (as far as we can ascertain, it was not a mistake in the dose) were .6 of a unit which are anticoagulant levels and which could have caused bleeding complications. The other comment that I'd like to make is that over half of the patients which we have studied at 2 and 8 hours after heparin injection have satisfactory blood heparin levels, if you define satisfactory levels as between .05 and .2 of a unit at eight hours. According to this they should not require tedious regime. Less than half require tedious regime. This may reduce the heparin shortage a little bit.

KAKKAR

I studied about 400 patients now with blood heparin levels and there are all sorts of things one has to take into account. If you study the activity, partial thromboplastin time, before the patient goes into surgery, you find very interesting things occasionally. One of the things may be differences in heparin utilization in different individuals, as it has been brought out in this meeting. It maybe different at different times in the same individual. So what we are doing is using a standard regime making certain that it covers a wide range of operative procedures.

LOWENBERG

I was very impressed Dr. Kakkar with the correlation between the phlebography and the radioactive uptake studies. I was wondering if you'd mention briefly what technique you find the most helpful in the phlebography. It is a ascending, do you use the pumping action of the calf or just what, to show-off the clots best?

KAKKAR

The technique which we use consists of having the patient in a horizontal position and using two pneumatic tourniquets, one tourniquet is applied at the level of the ankle, and it is inflated to a diastaltic pressure. The idea of this is that when you inject the contrast media through the vein on the dorsum of the foot, the contrast medium is directed into the deep venous system and you only fill the superficial veins. The other tourniquet is applied in the mid-thigh region. This is inflated to well above systolic pressure. The idea of using this tourniquet is that you create enough pressure so that the veins distend and they fill nicely. Then you inject contrast medium slowly and watch the progress of the contrast medium on a image intensifier on a television screen, rotating the leg to make sure all the veins are filled. If there is any difficulty, the patient can plantiflex and dorsiflex the foot three to four times so that there is good mixing of a contrast medium with the blood and you take the pictures, release the mid-thigh tube, follow the contrast medium into the illiac veins and you can take pictures at that level. We find this technique very useful.

CHAPA

I would like to comment on the use of subcutaneous heparin. Dr. Sawyer has treated 300 patients with chronic thrombophlebitis with the injection of 20,000 units of heparin daily and the length of treatment has been from one month to 60 years. We have found only 4 patients who had pulmonary emboli while on treatment and the other complication that we had was minor bleeding episodes.

GLUECK

I have two questions. I was not quite clear Dr. Kakkar from your presentation whether you think mini heparin is effective in hip fractures. I am constantly asked this by my surgical colleagues. And point two, if a patient has previously had a really bad episode of thrombophlebitis, do you use mini dose heparin or "therapeutic" doses of heparin?

KAKKAR

I am sorry for not making that point clear. We have excluded from these studies, which I have reported today, those with fractured neck of the femur. Only in one study that we have investigated in 50 plus patients, did we find that the regimen that we had recommended was not effective in patients undergoing emergency surgery for fractured neck of the femur. Now in patients with a previous history of thrombophlebitis, I will still use low dose heparin in tedious regime.

MESSMORE

I have a question regarding studies that may have been done on medical patients who have a predisposition for venous thrombosis and pulmonary embolism analogous to the surgical patients: for instance, the acute myocardial infarction patient, or the patient with congestive heart failure, can anything now be recommended with respect to mini dose heparin *versus* no therapy *versus* larger dose heparin, followed by coumadin?

KAKKAR

Sorry, my topic was mini dose heparin and surgery, though we have not limited ourselves to that. We have studied nearly 300 cases admitted to the myocardial coronary care unit and it really is very effective in preventing leg vein thrombosis in these patients. Also, the study, that was reported by Professor Douglas from Aberdeen, showed that an incidence of about 22% was reduced to about 5%. Again, very effective. Now another joint study has been organized on a fairly large basis to study the reduction in the incidence of fatal pulmonary embolism in patients after myocardial infarct using a tedious regime of low doses of heparin. Early results were very effective in that group. The only group where it seems to be less effective are those who are in shock, those who have had cardiac arrest. They really are in shock for a long period and they need more heparin than can be delivered by 5,000 units of heparin tedious.

BAUE

I have one quick question, Mr. Kakkar, about the incidence of positive leg scans. You found a 30 to 40% incidence of positive leg scans in postoperative patients over 40 and your first slide showed an increasing incidence of problems of fatal pulmonary emboli in your patient population in England. The studies in Canada have indicated 20% positive scans. Studies in the United States have been in small numbers of patients, limited by the controls that we need for patient approval for this type of study, but the incidence has been about 8 to 10% positive leg scans. Do you have any thoughts about these differences in incidence in the different patient populations?

KAKKAR

I think that the studies that are reported from Canada, which look at surgical patients as a whole group, found their incidence to be 22%. But if you read the paper carefully, they are including patients over the age of 27 but not over the age of 40. Then you start looking at what happened in the hip fracture patients the incidence is 48%, while I am saying it is 40%. When you go into detail, you find of 40 having major abdominal surgery, the incidence is slightly higher. So Canada is not a safe place, I can tell you. When you look at results from the States, I think that this is not true. In the studies reported by Dr. Hume, from Boston, the incidence is about 25 to 27%. If you look at the studies reported by Mr. Sapogos, from Albany, his incidence was running at about, if my recollection is correct, 25%. This is the usual thing. It is to some extent a reflection of what the patient group is being subjected to. I have heard from various people, that the incidence is the same in the States, too. When what you have is a group of patients where some sort of prophylactic measures are being used very effectively, then the incidence drops a little. But on the whole I have looked through about 400 centers where radioactive fibrinogen tests are being used, I know of only a few centers like in Kenya, Uganda, Khartom, and few other places where the incidence is remarkedly low and I will accept that. Otherwise in the Western Hemisphere the incidence is remarkedly similar.

LONG-TERM USE OF MINIDOSE HEPARIN IN POST-MYOCARDIAL INFARCTION

Menard M. GERTLER, Hillar E. LEETMA, Russell J. KOUTROUBY, and Elyse D. JOHNSON
New York University Medical Center, Institute of Rehabilitation Medicine, New York, New York 10016 (USA)

The value of heparin has been generally accepted in the prevention and treatment of venous thromboembolic phenomena. However, the use of heparin in the prevention of recurrent myocardial infarction (MI) in patients who have already experienced an MI is still being debated.

The results of studies designed to determine the efficacy of oral anticoagulant versus heparin therapy after MI have generally fallen into three main categories: a decrease in mortality with coumadin or Dicumarol therapy (3, 4, 12), a decrease in mortality with heparin therapy (6, 9) and no decrease in mortality with either oral anticoagulants or heparin and thus routine use of anticoagulants in post-MI patients is opposed (5).

The differences in the results in the literature are due to variations in methodology - e.g., in time of initiation of anticoagulant therapy, in dosage, in duration of therapy, and in the endpoints selected. The lack of uniformity in dosage and duration of therapy can account for the differences in morbidity and mortality rates reported. Without a common methodology, it is difficult to assess the effectiveness of oral anticoagulant versus heparin therapy.

The controversy among investigators will probably continue until a well-controlled long-term trial is instituted on a multicenter basis to compare heparin with various regimens of oral anticoagulant therapy after acute MI.

The present study was designed to evaluate the efficacy of daily subcutaneous minidoses of heparin as a therapeutic agent in secondary prevention in patients who have recovered from at least one MI. A group of patients was started on heparin therapy almost immediately after their acute event. Another group was placed on heparin because they were doing poorly on oral anticoagulant agents.

The impetus for the study and for the use of heparin prophylactically in coronary heart disease patients was based on published data that evaluated the effect of oral anticoagulants on oxidative phosphorylation and myocardial contractility. It has been reported that coumadin and Dicumarol are uncouplers of oxidative phosphorylation (8), whereas heparin is not. The uncoupling action of Dicumarol causes histotoxic anoxia in the myocardium and thereby decreases myocardial contractility, whereas heparin increases myocardial contractility (7). The latter observation is not surprising, in that heparin increases both myocardial oxygenation and the oxygen-carrying capacity of hemoglobin. Due to the uncoupling action of oral anticoagulants, the administration of coumadin would only worsen the already existing anoxic state of an infarcted myocardium. In addition, it was observed that small doses of heparin increase contractility while large doses of heparin decrease it, thus supporting our theory that minidose heparin is far superior to the oral agents as an anticoagulant in the anoxic state.

It is our contention that minidoses of heparin, given over an extended period following acute MI, will greatly decrease the risk and the severity of recurrent MI and that the benefits of heparin therapy outweigh those of oral anticoagulants.

Our study showed that the administration of subcutaneous minidoses of heparin as secondary prevention of coronary heart disease resulted in a statistically significant decrease in the morbidity and mortality of patients after MI.

Trial Design

The study included 55 male and six female patients with documented MI. The diagnosis of acute MI was made according to the stringent criteria set forth by the World Health Organization, considering history and electrocardiographic and laboratory changes (16). Patients with uncontrolled hypertension or diabetes mellitus requiring insulin therapy, with a history of active peptic ulcer, with bleeding tendencies or with sensitivity to heparin were excluded from the study.

To evaluate the effect of minidoses of heparin and oral anticoagulants (coumadin and Dicumarol) on morbidity and survival in the secondary prevention of MI, we divided the study population into three groups:

The heparin group consisted of 26 males and four females with a mean age of 54 years at the time of MI. In 22 patients, the subcutaneous sodium heparin therapy was initiated at the time of the acute event; in eight patients, therapy was started at least 2 years after the acute event because of either late referral or severe attacks of angina pectoris that did not respond to nitroglycerine The patients in the heparin group have been treated for an average of 3.3 years. In all cases, heparin has been administered by the patients themselves. The initial dosage was 5,000 units twice a day for an average of 30 days and then was decreased to an average dose of 40 units/lb of body weight once a day. The clinical progress of these patients has been followed, with periodic determination of the kaolin-activated partial thromboplastin time, liver and lipid profiles, calcium and phosphorus levels, resting electrocardiograms, exercise-stress testing on a motor-driven treadmill, and the consumption of nitroglycerine tablets. Skeletal surveys were accomplished yearly.

The coumadin group consisted of 17 males and one female with a mean age of 56 years at the time of acute MI. In 16 patients, coumadin was started at the time of the acute event; in two patients, it was started 6 years after the first MI. The average duration of coumadin therapy was 5.7 years. The dosage of coumadin had been titrated to maintain a prothrombin time approximately twice normal. The clinical progress of the patients has been followed in a fashion similar to that of the heparin group.

The coumadin/heparin group consisted of 12 males and one female with a mean age of 52 years at the time of acute MI. They were initially on coumadin for an average of 3.3 years. Because of numerous reinfarctions, symptoms of angina pectoris, or the worsening of resting electrocardiograms or exercise-stress tests, the coumadin therapy was discontinued and the patients were placed on minidoses of subcutaneous heparin - 5,000 units twice a day for 30 days, and then decreased to an average dose of 40 units/lb of body weight once a day. The average duration of heparin therapy in this group has been 2.5 years so far. The patients' clinical progress has been monitored as outlined above.

Trial Results

Table I summarizes the findings on recurrent MI in three groups on anticoagulant therapy. There was no statistically significant difference in age at onset of acute MI among the three groups.

The ratios of incidence years among the three groups were compared with the following results: the coumadin group had an incidence of recurrent MI 4.65 times greater than the heparin group; the incidence in the coumadin/heparin group was 3.11 times greater

TABLE I

Incidence of Recurrent Myocardial Infarction During Anticoagulant Therapy

Group	No. Patients	Age at Onset	Duration of Therapy, years	Recurrent MI's per Person	Recurrent MI's per Person per Year [a]
Coumadin	18	56	5.7	1.06	0.186
Heparin	30	57	3.3	0.133	0.040
Coumadin/heparin	13	52	3.3/2.5	0.86/0.21	0.261/0.084

[a] When multiplied by 100, these values yield recurrence rate per year.

during the years on coumadin than on heparin. The incidence in the coumadin/heparin group was 1.40 times greater during the years on coumadin than in the coumadin group. Thus, the patients in the coumadin/heparin group were more symptomatic and at greater risk while on coumadin, which necessitated the institution of minidose heparin therapy (Table II).

The incidence rate is based on a linear scale between recurrence of MI and the time elapsed from the first acute episode. However, assuming that the probablity of recurrence of MI increases in nonlinear proportion to the time after the initial event, a projected incidence rate would be a more valid criterion for comparison among the groups. It was found that a projection of the number of recurrent MI's per person per year in the heparin group still does not approximate the recurrence of MI in the coumadin group - projected incidences of 0.069 and 0.186, respectively, which give a projected incidence rate of 2.70. This value shows a recurrence rate of MI twice as great in the coumadin group than in the heparin group.

When the incidence rate of recurrent MI is projected for the coumadin portion of the coumadin/heparin group to enable comparison

TABLE II

Standard and Projected Ratios of Recurrent Myocardial Infarction

Compared Groups	Standard Ratios	Projected Ratios	Number of Years Projected
Coumadin: heparin	4.65:1	2.70:1	5.7
Coumadin/heparin (Coumadin period: heparin period)	3.11:1	4.10:1	3.3
Coumadin/heparin (Coumadin period: Coumadin)	1.40:1	2.42:1	5.7

TABLE III

Survival and Mortality During Anticoagulant Therapy

Group	No. in Group	Survivors		Deaths		Deaths from MI		Deaths from other causes	
		No.	%	No.	%	No.	% of Deaths	No.	% of Deaths
Coumadin	18	5	28	13	72	10	77	3	23
Heparin	30	29	97	1	3	1	100	0	0
Coumadin/heparin	13	8	62	5	38	2	40	3	60

with the incidence rate of the coumadin group, the former is found to have a rate 2.42 times greater than the latter. When the incidence rates within the coumadin/heparin group are projected so that the duration of therapy for each drug is equivalent, the incidence rate during coumadin therapy is 4.10 times greater than for heparin.

The greatest advantage of minidose heparin over coumadin is evident from Table III which summarizes the data on survival and mortality. Of the patients in the coumadin, heparin and coumadin/heparin groups, 72%, 3%, and 38%, respectively, died. The chi square analysis of death and survival in both the heparin and the coumadin/heparin groups compared with the coumadin group, proved to be statistically significant ($X^2 = 98.6$, $X^2 = 22.0$, respectively) at $p < .001$.

In Table IV the number of recurrent MI's in each group and the resulting mortality are expressed for as many as four subsequent episodes. It was found that 39% of the coumadin group was free from recurrent MI. This result is mirrored in the coumadin portion of the coumadin/heparin group, in which 38% of the patients were also free from subsequent episodes. However, a large majority of the heparin group and the heparin portion of the coumadin/heparin group, were free from recurrent MI's - 93% and 85%, respectively. It is interesting to note that six of the seven patients in the coumadin group succumbed to their second infarct. This high mortality rate leads one to believe that prevention of subsequent MI is of paramount important in post-MI patients.

TABLE IV

Incidence of Recurrent Myocardial Infarction and Resulting Mortality During Anticoagulant Therapy

Group	Number of Recurrent MI's [a]				
	0	1	2	3	4
Coumadin (N=18)	7	7 (6)	2 (2)	1 (1)	1 (1)
Heparin (N=30)	28	1 (0)	0	1 (1)	0
Coumadin/heparin (N=13)	5/11	7/2 (2)	0/0	1/0	0/0

[a] Numbers in parentheses are deaths resulting from a recurrent MI.

Discussion

The data presented reflect only those patients who were taking their prescribed medication - either coumadin or heparin - consistently. The stringent criteria for selecting patients for inclusion in this study reduced the population totals by half.

The statistical analyses of the data on the patients in the heparin group do not include some interesting facts that add to the importance of the use of heparin in the secondary prevention of MI. Nine of the post-MI patients in this study were initially offered coronary bypass surgery. They have been treated instead with a medical regimen of subcutaneous minidose heparin. More than 5 years later, eight of the nine patients are still surviving. Only one patient in the heparin group has died from a recurrent MI, whereas in the coumadin group, six deaths occurred at the second MI, one at the third, one at the fourth and one at the fifth MI. The patient who died while on heparin had been classified as high risk in 1947. He later developed inferior vena caval thrombosis, and the vein was subsequently ligated. He had experienced at least four documented episodes of pulmonary embolism, which left him with right heart failure. The patient was placed on heparin after his first acute MI and then experienced two more episodes within a 2.5 year period. He died after his fourth MI.

Heparin therapy has many advantages over oral anticoagulants. One cannot assume that the degree of efficacy of heparin and oral anticoagulants are comparable since their modes of action differ. Oral anticoagulants depress the hepatic synthesis of nearly all enzymes concerned with clotting - such as factors II, VII, IX, and X, which are vitamin K-dependent - and thus decrease the net amount of these plasma proteins in circulation (10). Heparin, however, is a naturally occurring anticoagulant and interferes with the coagulation process in the following manner:

1. It prolongs the one-phase prothrombin time (13).

2. It complexes with fibrinogen. In a strict sense, this complex is not a natural substrate of thrombin; as a result, a much smaller amount of fibrin monomer can be formed with the same amount of thrombin (1, 2).

3. It complexes with antithrombin III enabling antithrombin III to inactivate the clotting enzymes - e.g., thrombin or factor Xa - much faster (15, 17, 18).

4. It complexes with factor XIII and thus will not give the final fibrin clot the proper cross linkages (11).

5. It reduces platelet adhesiveness (14).

Therefore, heparin has an antithrombotic effect, in addition to its anticoagulant effect; coumadin has solely an anticoagulant effect.

Another major advantage of heparin is its proven clearing effect on serum triglycerides through the mediation of the enzyme lipoprotein lipase, which in turn affects the procoagulation scheme.

Other advantages to heparin therapy merit attention. The therapeutic range and the toxic range of heparin are so far apart that, with normally effective therapeutic doses, the chances of bleeding complications are remote. Furthermore, heparin is readily eliminated by the kidney and liver. This property of heparin enables one to administer simultaneously with caution, such drugs as aspirin, Butazolidin (phenylbutazone), Tandearil (oxyphenbutazone), etc., which ordinarily would be considered synergistic in their action with the oral anticoagulants. In addition, since the therapeutic and toxic ranges are so far apart, frequent monitoring is not as critical as with oral anticoagulants. Thus, physicians who have inadequate facilities for laboratory control of oral anticoagulants can readily take advantage of heparin.

Therefore, the two main disadvantages of heparin - that it must be administered parenterally and that it is appreciably more expensive to use than oral anticoagulants - do not justify the use of oral anticoagulants, which have been shown to be less effective in the secondary prevention of myocardial infarction.

REFERENCES

1. ABILDGAARD, U., Scand. J. Haemat., 5 (1968) 440.
2. ABILDGAARD, U., Scand. J. Haemat., 5 (1968) 432.
3. BERRYMAN, G.H., et al., JAMA, 189 (1964) 555.
4. BJERKELUND, C.J., Acta Med. Scand. Suppl., 330 (1957) 1.
5. CHARLETON, R.A., SANDERS, C.A. and BURACK, W.R., New Eng. J. Med., 263 (1960) 1002.
6. ENGELBERG, H., KUHN, R. and STEINMAN, M., Circulation, 13 (1956) 489.
7. GREEN, J.P. and NAHUM, L.H., Circulation Research, 5 (1957) 634.
8. GREEN, J.P., SONDERGAARD, E. and DAN, H., J. Pharmacol. and Exper. Therap., 119 (1957) 12.
9. GRIFFITH, G.C., ZINN, W.J. and ENGELBERG, H., et al., JAMA, 174 (1960) 1157.
10. KAZMIER, R.J., SPITTELL, J.A., THOMPSON, J.J. and OWEN, C.A., Arch. Int. Med., 115 (1965) 667.
11. KUDRYASHOV, B.A. and LYAPINA, L.A., Probl. Germatol. Pereliv. Drovi., 17(4) (1972) 28.

12. LOVELL, R.R.H., DENBOROUGH, M.A., NESTEL, P.J. and GOBLE, A.J., Med. J. Austral., 2 (1967) 97.
13. MARDER, V.J., Thromb. Diath. Haemorrh., 24 (1970) 230.
14. McDONALD, L. and EDGILL, M., Lancet, 1 (1961) 884.
15. PORTER, P., PORTER, M.C. and SHANBERGE, J.N., Clin. Chim. Acta. 17 (1967) 189.
16. World Health Organization, World Health Organization Technical Report Series, (1959) 168.
17. YIN, E.T., WESSLER, S. and STOLL, P.J., J. Biol. Chem., 246 (11) (1971) 3703.
18. YIN, E.T., WESSLER, S. and STOLL, P.J., J. Biol. Chem., 246 (11) 3712.

WALL-BONDED HEPARIN -- HISTORICAL BACKGROUND AND CURRENT CLINICAL APPLICATIONS

Vincent L. GOTT

Department of Surgery, The Johns Hopkins University, Baltimore, Maryland 21205 (USA)

Several medical devices in clinical use depend on a surface of wall-bonded heparin for their proper function. This report will summarize the development of these heparinized surfaces a portion of which has been previously related to a biomaterials symposium (3), and review the current status of their clinical application. The next presentation, by Dr. Richard Falb, will deal with the chemistry and dynamics of the surfaces.

The development of heparinized surfaces resulted from some rather serendipitous events that occurred in the early 1960's. With the advent of open-heart surgery in the mid-1950's, there was a major effort in many centers to develop suitable prosthetic valves to replace diseased valves. In spite of the development during this period of many different types of artificial valves, even by 1960 there were hardly any published reports of long-term canine survival after valve replacement. Many of these valves seemed to work satisfactorily in a valve tester; but, when they were placed in a canine heart, they usually ceased to function within a few hours because of the development of massive thrombi on the prosthesis. Our own experience with new types of prosthetic valves confirmed this serious problem with thrombosis. Therefore, we felt that it was imperative to screen a number of different polymers, metals, and coatings before proceeding with furthur valve development.

We designed a small prosthetic ring whose dimensions permitted it to be placed easily in the thoracic aorta of a dog weighing approximately 20 lb (5). Rings of a number of different polymers and metals that had been used in our prosthetic valves were constructed, placed in the thoracic aorta, and then examined after a month. All these rings were found to be patent, with only a thin layer of fibrin on the surface. These results did not simulate the findings previously noted with prosthetic valves placed in the canine mitral area. It was felt that a more severe test would be the placement of these rings in the inferior vena cava. When this procedure was carried out, nearly all the plain polymer and plain silicone rings were completely thrombosed within 2 hr. It was decided that this simple in vivo test would allow the screening of a number of materials, some of which might be suitable for valve construction. After implanting many different polymers and metals in the canine vena cava, we became discouraged because of the consistent observation of thrombus.

We then became aware of the studies of Sawyer and Page, who demonstrated that the intima of a blood vessel had a negative charge with respect to adventitia (14). This seemed to be significant, in that Abramson (1) had discovered in 1925 that all blood cells were negatively charged at the pH of blood and that the cells moved toward the positive pole when placed in an electrophoretic chamber. The importance of Abramson's findings was not fully appreciated until the demonstration of the negative charge on the intima (14). The importance of charge was emphasized when Sawyer and co-workers showed that thrombi could be initiated under a positive electrode placed on a blood vessel and that the development of thrombus could be inhibited by an electronegative charge (15).

On the basis of these findings, we thought it worthwhile to evaluate the effect of a battery charge on prosthetic surfaces. Because all the polymers for valve construction are nonconductive, it was elected to paint the plastic surface with an industrially used conductive lacquer. We had wanted to coat a prosthetic mitral valve with the conductive lacquer, and then apply a negative battery charge; but it appeared simpler to use the vena cava rings for this study. Therefore, polycarbonate rings were coated with a conductive lacquer containing colloidal graphite, and an insulated wire was attached to the outer surfaces of the rings and then to the negative pole of a 9.6-Volt battery. The positive pole of the battery was connected by means of a wire to the latissimus dorsi muscle. It was gratifying to note the absence of thrombi when the first of these rings with a negative charge was removed from the canine vena cava after 2 hr. We had not previously observed a thrombus-free caval ring in our laboratory.

Several additional rings were implanted in a similar way with a negative battery charge; the results were identical with those of the first ring. One day, however, we noted, before removing a ring with a supposed negative charge, that the wire to the battery had broken; the ring was free of thrombi . We repeated this several times and verified that a battery charge was not required for the prevention of thrombi formation.

Initially, the resistance to thrombus formation of the graphite coating was attributed to physical properties of the graphite itself. Months later, we learned that it was not simply the graphite surface, but rather that all the vena cava rings, wired and unwired, had been sterilized in a common hospital disinfectant - benzalkonium chloride - and had always been rinsed in heparin before placement in the canine cava.

It was the study represented in Fig. 1 that fairly conclusively demonstrated the importance of the cationic surfactant, benzalkonium chloride, in bonding heparin to a graphite surface. Polycarbonate rings coated with graphite alone were almost totally obstructed with thrombi after 2 hr in the canine vena cava. The results were similar with graphite-coated rings treated with benzalkonium alone and with heparin alone. Only after sequential immersion in benzalkonium and heparin did the rings remain free of thrombi for 2 hr - actually for several months (6).

Figure 2 depicts in a schematic way the graphite-benzalkonium-heparin (GBH) surface as we envision it. The graphite, with its significant adsorptive properties, appears to bond firmly the cationic detergent, benzalkonium chloride, with its potent surface-active properties. This bonding between graphite and benzalkonium chloride appears to be simple physical adsorption of a surfactant onto an adsorptive surface. The quaternary ammonium radicals of the benzalkonium, in turn, electrostatically bond the heparin, with its negative sulfate radicals.

To obtain further information on the stability and durability of the GBH surface, graphite-coated rings were treated with ^{14}C

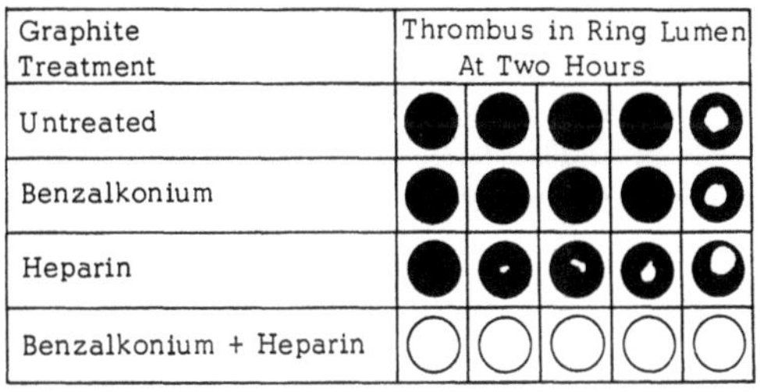

Graphite Treatment	Thrombus in Ring Lumen At Two Hours				
Untreated					
Benzalkonium					
Heparin					
Benzalkonium + Heparin					

Fig. 1. Thrombus formation in graphite-coated plastic vena cava rings after various treatments of the graphite surface. Only the rings treated with graphite-benzalkonium-heparin (GBH) were free of thrombi after 2 hr in the vena cava. Reprinted with permission from Whiffen et al. (19).

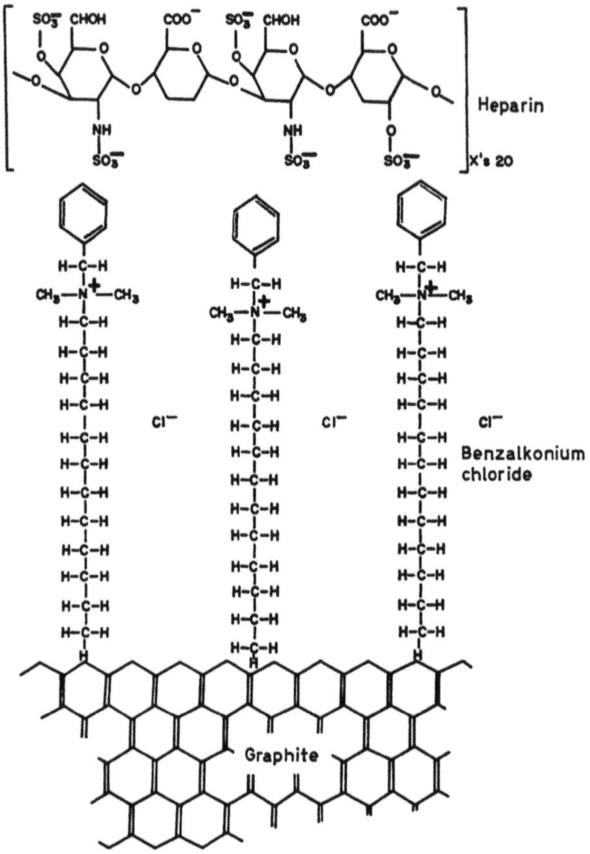

Fig. 2. Schematic illustration of the proposed relation of graphite, benzalkonium chloride, and heparin on a prosthetic surface.

-labeled benzalkonium chloride and then heparin (18). The rings were placed in the inferior vena cava and removed after exposure to the venous blood for periods varying from 6 min to 6 months. The rate of elution of the benzalkonium cation appeared to be exponentially related to the amount of cation present, at least during the first 2 months of blood exposure. However, there was a significant amount of benzalkonium cation on the graphite surface after 2 months placement in the canine cava.

Studies were also carried out (17) with radioactive heparin incorporated in the GBH surface (Fig. 3.) When this type of surface was placed in the canine atrium, a significant quantity of radioactive heparin remained on the prosthetic device even after 3 months.

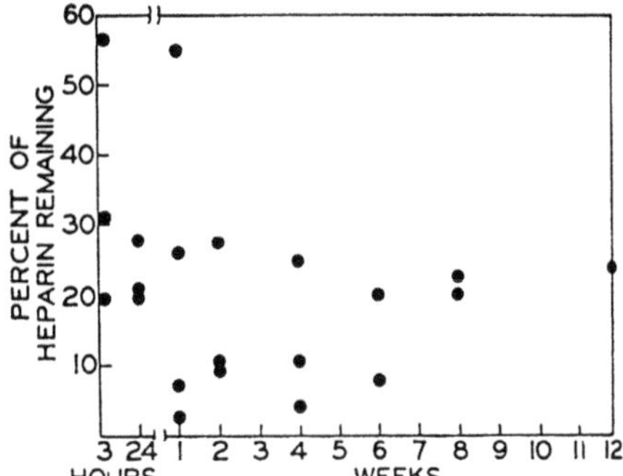

Fig. 3. Proportion of initial surface heparin found to remain on 21 GBH-coated plastic surfaces after exposure for various periods to blood flowing through the right atria of dogs. Reprinted with permission from Whiffen and Beeckler (17).

There appeared to be a high rate of elution of poorly bound heparin during the first 3 hr, but then the rate of elution quickly diminished; and a mean of 23% of the heparin remained on the surface after 24 hr. The mean remaining after 1 - 12 weeks of blood exposure ranged between 14 and 25%.

The GBH surface was thus shown to have good long-term thromboresistant properties, and it has been interesting to postulate why this surface works so well. Figure 4 illustrates a possible mechanism for thrombus repulsion by the GBH surface. The benzalkonium chloride radicals are shown as golf tees on the graphite surface. The heparin radicals are shown as "scrunched-up" cigarettes, with bonding between the quaternary ammonium radicals of the benzalkonium chloride and the sulfate radicals of the heparin. This illustration suggests that, after immersion of the GBH surface in the bloodstream, there is adsorption of plasma proteins in such a way that the proteins are not denatured. The surface, in turn, becomes pacified by these relatively unaltered protein molecules, so that the blood, in a sense, sees itself and thrombus formation is not initiated. It is known that plasma proteins are adsorbed on every surface, heparinized or not. At least four research groups have shown recently that heparinized surfaces preferentially adsorb heparin cofactor. Figure 4 (prepared a few years ago) should depict a significant amount of heparin cofactor adsorbed to the surface. No doubt, the leaching of significant amounts of heparin in the first few hours is important in the resistance to thrombus formation of these surfaces. This long-term resistance must be related, however, to a pacifying effect of adsorbed protein.

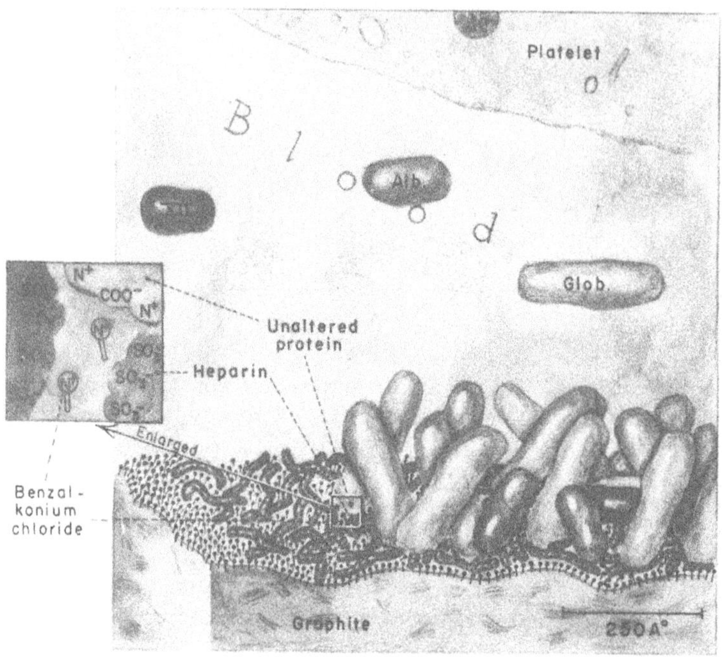

Fig. 4. Schematic diagram showing how relatively unaltered plasma protein molecules might be adsorbed on a GBH surface. There may be adsorption of the plasma proteins in such a way that the proteins are not denatured, and the surface would then become pacified by these relatively unaltered protein molecules.

By the mid 1960's, it had been demonstrated that a GBH surface had significant resistant properties to thrombus formation and that it should be adapted for clinical medical devices. This surface was applied to the rigid polycarbonate housing of a hinged-leaflet prosthetic valve (4) that was used in more than 100 patients at the University of Wisconsin Hospital between 1963 and 1965. The prosthesis functioned very well, and the GBH surface appeared to impart to the valve housing good long-term qualities of resistance to thrombus formation. However, the flexible Silastic leaflet could not be coated with a resistant surface, and thrombus often formed on the undersurface of the leaflet in the area of blood stagnation. A valve of this design, although it has excellent hemodynamic properties, appears to be much more prone to thrombus formation than a ball or disk valve, because of the turbulent and stagnant flow on the downstream side of the leaflet. Because approximately 10% of these valves developed thrombi, usually on the Silastic leaflet, their use was discontinued in 1965.

At about the same time, it appeared that the GBH coating might have significant value if applied to a polyvinyl shunt used to bypass aneurysms of the descending thoracic aorta temporarily during surgical resection. Before 1965, the principal method for bypassing such aneurysms involved a pump system to shunt blood from the left atrium to the femoral artery, this allowing the aorta to be clamped above and below the aneurysm with continued perfusion of the critical abdominal organs. This system, however, also required systemic heparinization, which contributed greatly to oozing and bleeding from the large operative field and from the prothesis. This, of course, increased the morbidity and mortality associated with the operative procedure.

I believed that the pump and the systemic heparinization could be eliminated if a temporary shunt with wall-bonded heparin were used. Therefore, we constructed a shunt system with polyvinyl tubing coated with GBH for use in resection of aneurysms of the descending thoracic aorta (16). Figure 5 shows the first such use of this shunt in 1965. The heparinized shunt between the subclavian artery and the femoral artery allowed satisfactory perfusion of the critical abdominal organs and the distal spinal cord and greatly simplified this operative procedure.

The GBH coating had some disadvantages, however, when used on this temporary shunt. The coating was opaque, so the blood flowing in the shunt could not be seen, and tubing coated this way could not be occluded with a metal clamp, for fear that the graphite layer might be damaged. Fortunately, at the same time, an improved heparinized surface became available for this type of temporary vascular shunt. This newer surface, developed by Falb and his associates at Battelle Columbus Laboratories (7), represented a significant improvement over the GBH surface. It is called the TDMAC-heparin coating and can be applied to silicone rubber and most polymers (TDMAC = tridodecylmethylammonium chloride, a powerful cationic surfactant somewhat similar to benzalkonium chloride). The unique property of the TDMAC-heparin surface is that the graphite layer can be eliminated. The TDMAC is combined with a solvent; when a polymer is dipped into this solution, the solvent swells the polymer slightly and carries the cationic surfactant into the material. There is then physical bonding of the surfactant to the surface of the polymer. Immersion in heparin leads to a firm electrostatic bonding of heparin to the TDMAC molecules.

This new TDMAC-heparin surface has distinct advantages over the GBH surface: It is transparent; it is not easily traumatized with tube clamping; and it allows a firmer bonding of greater quantities of heparin than the GBH surface. Obviously, the TDMAC-heparin coating was a great step forward in the development of a heparinized

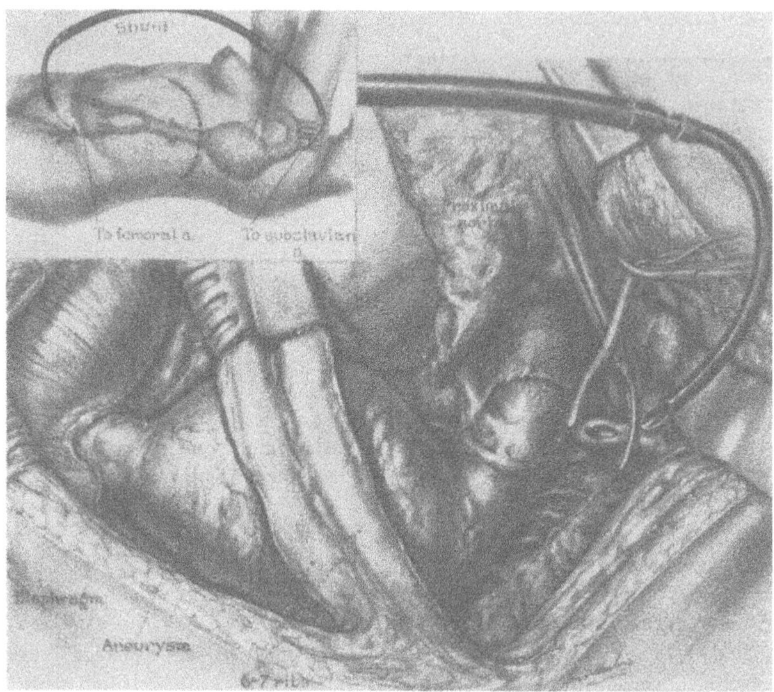

Fig. 5. A. Exposure of a very large aneurysm of the descending thoracic aorta through the bed of the left fifth rib and seventh interspace; the insert shows the site of proximal cannulation in the left subclavian artery and the site of distal cannulation in the left common femoral artery.

surface, and it was quickly applied to a new type of polyvinyl shunt for the temporary bypass of thoracic aneurysms (12). This shunt is tapered at each end to allow easy cannulation of the subclavian artery proximally and the femoral artery distally, and it has a large diameter in the central portion to minimize the resistance to blood flow. It has a unitized design with no connectors, is transparent, and is excellent for this type of surgery.

This shunt with the transparent TDMAC-heparin coating has been used by many surgeons for vascular procedures on the thoracic aorta and the innominate artery. The largest experience with the heparinized shunt has been obtained at Barnes Hospital of St. Louis by Ferguson and colleagues (T.B. Ferguson, personal communication). Before this shunt was available, they used left atrial-femoral

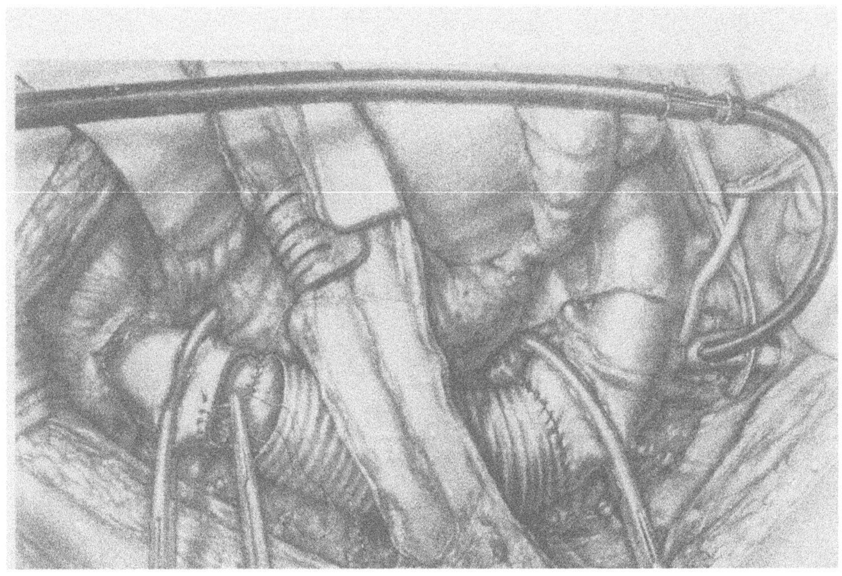

Fig. 5. B. With the GBH-coated polyvinyl shunt in place, a Dacron graft is being anastomosed to the residual descending thoracic aorta. Reprinted with permission from Valiathan *et al.* (16).

arterial bypass with systemic heparinization for resection of aneurysms of the descending thoracic aorta and of eight consecutive patients between 1960 and 1970, only two survived the operation. Most of the deaths seemed to be related to excessive bleeding and oozing secondary to systemic heparinization (8). They have now used this TDMAC-heparin coating on shunts in 15 patients who underwent elective resection of an aneurysm of the descending thoracic aorta: only one died (Ferguson, personal communication).

This type of heparinized shunt appears to be particularly valuable in patients who sustain trauma to the thoracic aorta or to the great vessels arising from the aortic arch. An excellent example of its value is illustrated in Fig. 6 (12). The patient had sustained massive injuries to the abdomen, head, and chest, with a resulting avulsion of the innominate artery from the arch of the aorta. Because of the significant head, ocular, and abdominal injuries, the patient could not be heparinized for a pump bypass. A heparinized shunt from the ascending aorta to the carotid artery was used while the injury site of the innominate artery was isolated, resected, and grafted.

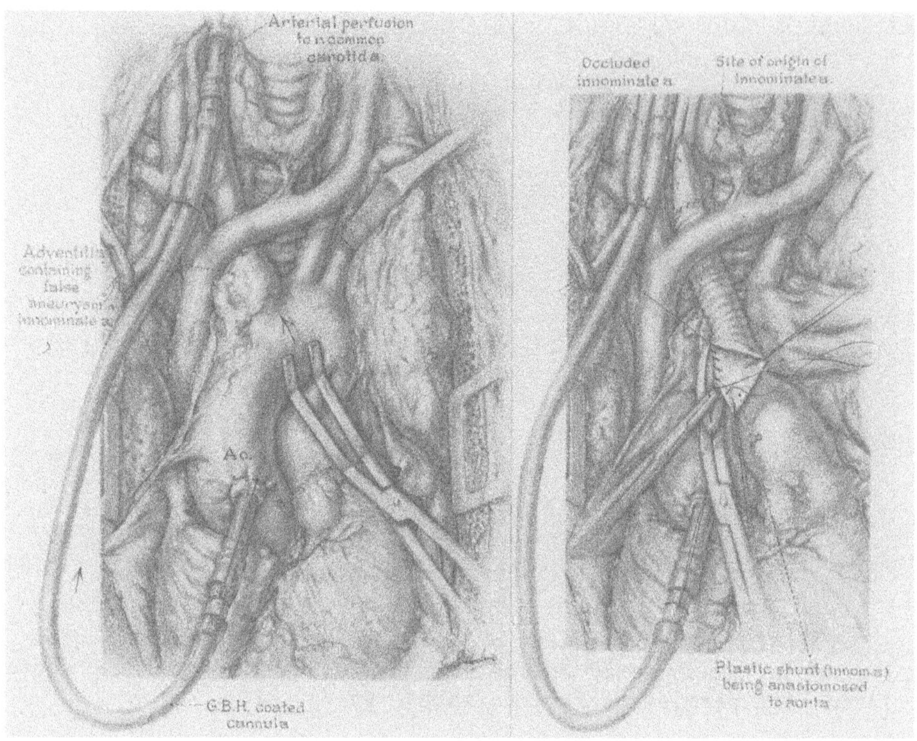

Fig. 6. Left, a GBH-coated shunt was used to bypass a traumatic aneurysm of the innominate artery. Right, entry into the false aneurysm revealed complete avulsion of the innominate artery from the arch; the defect in the aorta was oversewn, and the innominate artery was reconstructed with a Dacron prosthesis. Reprinted with permission from Murray et al. (12).

The major clinical application of heparinized surfaces today is combination with the temporary vascular shunts just discussed. Over 700 of these shunts have been coated by the Battelle scientists for approximately 100 vascular surgeons around the world. More than 100 patients have undergone vascular repair with a heparinized-shunt bypass, and no major problems have been reported as resulting from their use.

Unfortunately, in spite of the demonstrated value of this type of heparinized shunt and the apparently lower risk of surgery, such a shunt with wall-bonded heparin is not commercially available. The Food and Drug Administration (FDA) has ruled that a shunt with wall-bonded heparin is similar to a new drug; therefore, before it can be distributed commercially for clinical use, it must be subjected to the expensive premarket testing accorded drugs. Surgeons may purchase only uncoated shunts from the medical device manufacturer and then either coat them themselves or send them to Battelle Laboratories for coating. Data from Falb's laboratory show the total lack of toxicity of this type of surface; it is hoped that the FDA will approve this coating for wide commercial use.

Falb and his associates at Battelle have now coated dozens of types of devices for physicians and surgeons around the world (9), including prosthetic valves (11), vena caval emboli filters (10), membrane oxygenators (13), vascular grafts, artificial hearts, artificial kidneys, arteriovenous dialysis cannulae, and many kinds of blood catheters. A number of these devices have been used clinically, with significant benefit derived from their heparinized surfaces.

I think the next major clinical application of heparinized surfaces will be on catheters used for heart and vascular catheterization. Each week, more than 5,000 patients in this country undergo vascular and cardiac catheterization, and Amplatz and co-workers of the University of Minnesota School of Medicine have shown that approximately 50% of the catheters develop a fibrin or thrombus coating during catheterization (2). This problem was not appreciated earlier, because, when a catheter is pulled out of an artery or vein, the thrombus is wiped off the catheter surface and it can embolize to some distal site in the vascular system. As a result of Amplatz's studies, this problem is now more fully understood, and many centers use total-body heparinization for these catheterization procedures. Because there is potential morbidity and even mortality related to systemic heparinization in these patients, there should be an advantage to using a catheter with wall-bonded heparin. Amplatz currently immerses his catheters in a solution of benzalkonium and heparin (no graphite) before the procedure; and excellent resistance to thrombus formations have been obtained with his surface, compared with catheters with plain polymer surfaces. Falb and his associates have also been working with Amplatz on the TDMAC-heparin coating of cardiac catheters. Other cardiovascular radiologists are also evaluating these catheters with heparinized surfaces, and it is not inconceivable that, within a year or two, this type of catheter may be used in thousands of patients each week.

In summary, it appears that significant progress has been made during the last 14 years, since heparinized surfaces were first developed. The original GBH surface was a fairly crude method of bonding heparin to a prosthetic device. Fortunately, it was quickly supplanted by the outstanding TDMAC-heparin surface developed by the Battelle scientists. This latter surface provides an excellent method for producing a nontoxic, long-lasting thromboresistant coating for many different polymers. In looking back, though, over the last 14 years, it appears to me that the adaptation of these surfaces for clinical devices has been extremely slow. Undoubtedly, there will be greater clinical use of these coatings in the future, and I hope that there will be further improvement in the stability and durability of heparinized surfaces and that they will be applied to an even greater number of medical devices.

ACKNOWLEDGEMENTS

This work was supported by USPHS grant HL 09997, Department of Surgery, The Johns Hopkins University School of Medicine, Baltimore, Maryland.

REFERENCES

1. ABRAMSON, H.A., J. Exp. Med., 41 (1925) 445.
2. CRAMER, R., MOORE, R. and AMPLATZ, K., Radiology, 109 (1973) 585.
3. GOTT, V.L., Bull. NY Acad. Med., 48 (1972 216.
4. GOTT, V.L., DAGGETT, R.L., WHIFFEN, J.D., KOEPKE, D.E., ROWE, G.G. and YOUNG, W.P., J. Thorac. Cardiovasc. Surg., 48 (1964) 713.
5. GOTT, V.L., KOEPKE, D.E., DAGGETT, R.L., ZARNSTORFF, W. and YOUNG, W.P., Surgery, 50 (1961) 382.
6. GOTT, V.L., WHIFFEN, J.D. and DUTTON, R.C., Science, 142 (1963) 1297.
7. GRODE, G.A., ANDERSON, S.J., GROTTA, H.M. and FALB, R.D., Trans. Amer. Soc. Artif. Int. Organs, 15 (1969) 1.
8. KRAUSE, A.H., FERGUSON, T.B. and WELDON, C.S., Ann. Thorac. Surg., 14 (1972) 123.
9. LEININGER, R.I., CROWLEY, J.P., FALB, R.D. and GRODE, G.A., Trans. Amer. Soc. Artif. Int. Organs, 18 (1972) 312.
10. MOBIN-UDDIN, K., TRINKLE, K.J. and BRYANT, L.R., Surgery, 70 (1971) 914.
11. MOBIN-UDDIN, K., UTLEY, J.R., BRYANT, L.R., DILLON, M. and WEISS, D.L., Ann. Thorac. Surg., 17 (1974) 351.
12. MURRAY, G.F., BRAWLEY, R.K. and GOTT, V.L., J. Thorac. Cardiovasc. Surg., 62 (1971) 34.
13. REO, W.J., WHITLEY, D. and EBERLE, J.W., Trans. Amer. Soc. Artif. Int. Organs, 18 (1972) 316.
14. SAWYER, P.N. and PAGE, J.W., Amer. J. Physiol., 175 (1953) 103.
15. SAWYER, P.N., SUCKLING, E.E. and WESOLOWSKI, S.A., Amer. J. Physiol., 198 (1960) 1006.
16. VALIATHAN, M.S., WELDON, C.S., BENDER, H.W., TOPAZ, S.R. and GOTT, V.L., J. Surg. Res., 8 (1968) 197.

17. WHIFFEN, J.D. and BEECKLER, D.C., J. Thorac. Cardiovasc. Surg., 52 (1966) 121.
18. WHIFFEN, J.D. and GOTT, V.L., J. Surg. Res., 5 (1965) 51.
19. WHIFFEN, J.D., YOUNG, W.P. and GOTT, V.L., J. Thorac. Cardiovasc. Surg., 48 (1964) 317.

SURFACE-BONDED HEPARIN

R.D. FALB, R.I. LEININGER, G. GRODE and J. CROWLEY

Battelle Memorial Laboratories, Columbus, Ohio (USA)

The first report of surface-attached heparin was by Gott and coworkers in 1965 (8). In this work, heparin was attached through formation of a complex with a quaternary ammonium salt that was adsorbed on graphite. The discovery of the thromboresistant properties of heparinized surfaces was inadvertent: the original intent of the work was to evaluate graphite. The graphite surface was treated with a quaternary ammonium salt to sterilize it, and the surface was rinsed in heparin as a simple precautionary measure. Gott _et al_. showed that the marked thromboresistance of the graphite-benzalkonium-heparin (GBH) surface was due to the presence of heparin. This held promise of enabling the use of foreign sufaces in contact with blood without systemic anticoagulation. However, in practice, the GBH coating could not be applied to flexible materials; thus, its use was limited.

Synthesis of Heparinized Surfaces

Shortly after the work of Gott _et al_., a group at Battelle Laboratories in Columbis, Ohio began a research program to develop methods of attaching heparin to the surfaces of a large number of synthetic and natural polymers. The primary goal of the research was to bond heparin simply and rapidly to materials - such as silicone rubber, polyvinylchloride, Teflon, and polyethylene - that were in current use as blood contacting materials. A second objective of the work was to determine the mechanism through which surface-bonded heparin inhibited clotting. The initial experimental approach to this work was the covalent attachment of a variety of quaternary ammonium salts to surfaces. Heparin bonding could then be effected on these surfaces by taking advantage of the ability of heparin to form highly nondissociable complexes with quaternary ammonium salts. The synthetic route taken in the initial stages of the work for the attachement of heparin was as follows (14):

$$\text{Polymer}-C_6H_5 \xrightarrow[\text{Al Cl}_3]{CH_3OCH_2\ Cl} -C_6H_4-CH_2\ Cl \xrightarrow[(CH_3)_2]{C_6H_5-N-} -C_6H_4-CH_2-\overset{C_6H_5}{\underset{CH_3}{N^{+}}}-CH_3\ \ Cl^-$$

Polystyrene was selected as the first material for heparin attachment because of the ease with which it underwent electrophilic substitution at the surface. Its properties were not suitable for implantation, so procedures were then developed for heparinization of silicone rubber, polypropylene, polyethylene, and natural rubber by the radiation grafting of styrene to the surfaces of these materials. After grafting, the sequence of reactions shown was followed to bring about heparin binding.

A second, more direct route to attach heparin to polymer surfaces was achieved by the radiation grafting of 4-vinylpyridine or dimethylaminoethylmethacrylate to the surfaces, as shown in the following reaction scheme:

$$\text{Surface}-H \xrightarrow{h\nu} \text{Surface}-CH_2CH_2-C_5H_4N \xrightarrow{CH_3I} \text{Surface}-CH_2-CH_2-C_5H_4\overset{\oplus}{N}-CH_3$$

$$\text{Surface}-CH_2-CH_2-\overset{O}{\overset{\|}{C}}-O-CH_2CH_2-N(CH_3)_2 \xrightarrow{CH_3I} \text{Surface}-CH_2-CH_2-\overset{O}{\overset{\|}{C}}-O-CH_2-CH_2-\overset{\oplus}{N}(CH_3)_3\ \ I^-$$

The surface-bound quaternary ammonium salts complexed large amounts of heparin and had properties similar to those of the styrene-based surfaces.

A further method of attaching heparin involved the use of a plasticizer that contained amino groups. For example, polyvinylchloride was plasticized by the addition of Hydrin rubber (a copolymer of epichlorohydrin and ethylene oxide that is reacted with various monofunctional or bifunctional amines).

An alternate route to heparinization of silicone rubber was developed, in which the heparin binding site was attached to the silica filler. In this process, cross-linked silicone rubber was reacted with aminopropyltriethoxysilane to yield a covalently attached amino group. This amine surface could be even further reacted with methyliodide to effect quaternization or treated directly with heparin at a pH less than 7.0.

The methods discussed above resulted in varying amounts of bound heparin, as determined by radioisotopic labeling techniques, (4) and shown in Table I.

The effect of the attached heparin on thrombogenicity was evaluated in several ways. In an *in vitro* coagulation assay, recalcified blood in contact with a heparinized surface did not clot during a test period of several hours. Later, when the blood was placed in a glass container, clotting occurred normally. The nonthrombogenicity of the surface as judged by this assay could not be related to the amount of heparin attached, except that surfaces containing less than 1.0 μg/cm^2 were thrombogenic. In some instances, surfaces with amounts of heparin as high as 5 μg/cm^2 were thrombogenic, presumably

TABLE I

Bonding of Heparin to Polymers by Quaternary Ammonium Salts

Polymer	Treatment Method	Amount of Heparin Bond, μg/cm^2
Silicone rubber	Chloromethylation	15.7
Silicone rubber	Styrene grafting, chloromethylation	5.6
Polypropylene	Styrene grafting, chloromethylation	3.9
Polyethylene	Styrene grafting, chloromethylation	2.8
Hydrin rubber	--	3.9
Graphite-Benzalkonium-Heparin	--	1.7

because of nonhomogeneous coverage of the surface.

The covalent attachment of heparin to polymer surfaces was also investigated, because the ionically bonded heparin could dissociate from the quaternary ammonium groups at the surface. Indeed, as will be discussed later, ionically bonded heparin slowly dissociates from the surface in the presence of blood. The heparin molecule has carboxylic acid, hydroxyl, and amino groups available for covalent attachment. A number of reagents and heparin derivatives were used, such as the acid hydrazide of heparin, silylated heparin, ethylenimine-heparin derivative, a carbodiimide-heparin deivative, and a heparin - cyanuric chloride adduct (3). Of these, the best results, in terms of the amounts of heparin bound, were achieved with the cyanuric chloride adduct. Heparin was first reacted with cyanuric chloride to form a derivative that had approximately seven residues per heparin molecule (on the basis of a molecular weight of 15,000) and retained 85% of its anticoagulant activity. Attachment to silicone rubber was effected through first bonding the amino groups to the silicone rubber by means of aminopropyltriethoxysilane and then reacting with the heparin derivative. This method resulted in large amounts of bound heparin (50-100 μg/cm^2), which could be displaced by 4 N sodium chloride solutions. That the heparin linked with cyanuric chloride could not be displaced with salt indicated covalent bonding, in that ionically bonded heparin is almost completely displaced under similar conditions. Surprisingly, surfaces containing covalently bonded heparin did not perform well when implanted in animals. Possible reasons for this will be discussed later.

All the methods listed here for surface attachment of heparin required chemical modifications of the surface. Thus, heparinization by many of these processes could often be time-consuming and, in some cases, could result in damage to the surface, such as crazing, roughness, or opacification. In addition, most of the processes were specific for a given material. To heparinize a composite device, a separate process for each material in the device would be required. For these reasons, after the general efficacy of heparinized surfaces had been established, further work on heparinization was directed toward the development of rapid and broadly applicable methods. Because covalent attachment reactions are specific for given polymers, heparinization via adsorption of the quaternary ammonium compound was explored. Quarternary ammonium salts containing single long-chain alkyl groups are easily displaced from polymer surfaces and thus would not be acceptable. However, quaternary ammonium salts with three long alkyl groups are water-insoluble and are not easily displaced from a surface in contact with an aqueous phase.

The use of tridodecylmethylammonium chloride (TDMAC) was investigated as a means of attaching heparin to a large number of polymers by a broadly applicable and simple technique. In the original work, a solution of TDMAC in an organic solvent was used as a first step. The material was simply dipped in this solution, dried, and then exposed to 0.25% solution of heparin in water. The method proved to be generally applicable for many polymers (9) and resulted in the attachment of large amounts of heparin, as shown in Table II. The heparin attached to the surface was stable; very little was removed after 4 hr of contact with blood plasma. This technique had the advantages of simplicity and speed over previous methods of heparinization, and it enabled the treatment of composite devices containing several different polymers. Thus, an artificial heart device constructed from a silicone rubber, polypropylene, and polycarbonate could be heparinized in one simple procedure.

Further improvement has been made on the TDMAC process by the development of the one-step treatment (15). This technique takes advantage of the ability of heparin to form an organic soluble complex with TDMAC at appropriate ratios of heparin to quaternary

TABLE II

Attachment of Heparin to Polymers with Tridodecylmethylammonium Chloride

Polymer	Initial amount μg/cm^2	Fraction after 100 hr in Saline, %	Fraction after 4 hr in Plasma
Polyvinylchloride	25	100	95
Polycarbonate	4	93	90
Mylar	18	95	90
Silicone rubber	22	91	-
Polyurethane	186	100	98
Polyethylene	25	100	94
Polypropylene	8.2	92	91
Teflon	1.1	99	87

[a]As measured with ^{35}S-labeled heparin.

ammonium salt. To make the complex, an aqueous solution of heparin is shaken briefly with a toluene solution of TDMAC; during this time, the complex forms and dissolves in the toluene. This complex can then be used to heparinize polymers in a simple one-step process that consists of dipping the material for a short time into a 1% solution of the complex and then drying it in air. The duration of exposure varies according to the polymer, but for most materials it is only a few seconds. The resulting surfaces are stable and can be sterilized either by autoclaving or by using ethylene oxide.

Characteristics of Heparinized Surfaces

The interactions of heparinized surfaces with blood have been evaluated by a number of *in vivo*, *ex vivo*, and *in vitro* systems. When whole blood is placed in a test tube containing a heparinized surface, it does not clot; however, if the blood is then poured into a glass tube, clotting occurs within 5 min. This shows that the heparinized surface does not work simply by anticoagulating the blood. *Ex vivo* experiments by other investigators (16) have shown the TDMAC-heparinized surfaces to be the most thromboresistant of all materials investigated thus far. In these experiments, blood was passed directly from a dog into a chamber that was coated with test materials and in which a vortex was generated. The vortex stimulated clot formation, and the amount of clot was influenced by the type of surface.

The thrombogenicity of these surfaces has been evaluated in numerous animal studies. In one, TDMAC-heparinized rings were implanted in the venae cavae of dogs and were found free of clots (7), in marked contrast with the control surfaces, which were completely occluded. In other studies, these surfaces were used successfully in membrane oxygenators in the absence of systemic heparinization (20) and in indwelling vascular cannulae (11).

Several other investigators (2, 10, 13, 17, 18, 19, 21) have developed heparinized surfaces in which heparin is attached by either covalent or ionic bonding. Several substrate materials have been used including Cellophane, silicone rubber, and polyethylene. In general, these heparinized materials have greatly improved thromboresistance. A detailed discussion of each of these materials is beyond the scope of this paper.

The TDMAC-heparinized surfaces are now undergoing clinical studies by many investigators. Our laboratory is currently preparing heparinized devices of different configurations. The device with the largest number of clinical trials to date is a shunt used in the repair of thoracic aneurysms (1, 12). In this technique, a treated polyvinylchloride shunt is used without systemic heparinization to divert blood from the aorta to the femoral artery while the aorta is repaired. We have prepared over 700 of these shunts and sent them to 130 clinical investigators. The TDMAC-heparin process has also been successfully used with heart valves in humans (20). In addition, we have heparinized over 100 shunts for repair of carotid aneurysms and numerous types of cannulae used for access to the cardiovascular system. The results of these studies indicate that the heparinized surfaces have very low thrombogenicity when in contrast with blood.

The mechanism by which heparinized surfaces are thromboresistant is not well understood. When heparinized surfaces are placed in contact with blood, adsorption of proteins occurs rapidly. The pattern of adsorption of the major plasma proteins does not seem to vary significantly from that of the underlying polymers (5). The adsorption and interaction of these surfaces with the components of the coagulation system has not been studied in detail. Our work, as well as work by Merrill *et al.* (19), indicates that heparinized surfaces do not activate Hageman factor, which is involved in the initiation of surface-induced coagulation.

Another characteristic of most surfaces containing ionically bound heparin is the elution of small amounts of heparin when the surfaces are in contact with blood. As mentioned previously, blood placed in a heparinized test tube clots normally; thus, normal anticoagulation does not occur. In studies with radiolabeled heparin (9), workers in our laboratory showed that heparin was released in small amounts from particulated heparinized materials placed in a column in contact with plasma. The rate of release was initially high and then diminished to a very low point within 2 hr after contact. Of the heparin in heparinized carotid-jugular shunts implanted in rabbits, 25% remained after a week (9). The shunts were patent, in spite of the loss of heparin.

Another characteristic of heparinized surfaces that we (9) and others (22) have observed is the adhesion of platelets. In *in vitro* experiments using a column packed with heparin attached to small (200 μm) particles of silicone rubber, we found that most of the platelets were removed from the initial fractions of platelet-rich plasma eluted through the column. Equilibrium was quickly established, however, and later fractions did not have platelets removed. Similarly, the surfaces of heparinized materials implanted in animals were covered with platelets, although no thrombus formed. Results of several studies, both in our laboratory and by Salzman *et al.* (22), have indicated that although platelet adhesion does occur on heparinized surfaces, later platelet metamorphosis and release of intracellular constituents do not occur. Evidently, a relatively benign layer of platelets is deposited on the surface, and this layer is not involved in further thrombogenesis.

Recent work has shown that heparinized surfaces bind heparin cofactor (6, 23). This binding is very strong and occurs in the presence of other plasma proteins. The adsorbed heparin cofactor is capable of neutralizing thrombin. This may explain in part the nonthrombogenic properties of heparinized surfaces.

In summary, the mechanism by which heparinized surfaces act is not well understood. They do not act by releasing amounts of heparin sufficient for anticoagulation. Yet, that covalently bound heparin surfaces do not perform nearly as well as those containing ionically bound heparin implies that heparin must have some degree of mobility at the surface. The most attractive hypothesis to explain the mechanism is that small amounts of heparin are released in the microenvironment of the surface and that this released heparin can effectively block thrombogenesis at the surface. The presence of heparin at the surface also determines the nature of the adsorbed protein layer and causes the deposition of a benign layer. Thus, the surface remains free of clots long after the initially bound heparin is gone.

Heparin can be attached to many different polymers by means of a TDMAC complex. The resulting surfaces, which have a marked reduction in thrombogenicity, are now finding practical use in clinical applications.

REFERENCES

1. BRENNER, W.I., ENGELMAN, R.M., WILLIAMS, C.D., BOYD, A.D. and REED, G.E., Amer. J. of Surg., May, 1974.
2. ERIKSSON, J.D., GILLBERG, G. and LAGERGREN, H., J. Biomed. Mate. Res., 1 (1967) 301.
3. FALB, R.D., GRODE, G.A., GROTTA, H.M., WRIGHT, R.A., POIRIER, R.H., TAKAHASHI, M.T. and LEININGER, R.I., U.S. Government Reports PB 175 668 (March 30, 1967), PB 183 317 (October, 1967), PB 188 108 (October, 1968), and PB 188 111 (October, 1969).
4. FALB, R.D., GRODE, G.A., LUTTINGER, M., EPSTEIN, M.M., DRAKE, B. and LEININGER, R.I., Government Report PB 173 053, (1966).
5. FALB, R.D., TAKAHASHI, M.T., GRODE, G.A. and LEININGER, R.I., J. Biomed. Mate. Res., 1 (1967) 239.
6. GENTRY, P.W. and ALEXANDER, B., Biochem. Biophys. Res. Comm., 50 (1973) 500.
7. GOTT, V.L., RAMOS, M.D., NAJJAR, F.B., ALLEN, J.L. and BECKER, K.E., Proceedings of the Artificial Heart Program Conference, National Heart Institute, (1969).
8. GOTT, V.L., WHIFFEN, J.D. and DUTTON, R.C., Science, 142 (1963) 1297.
9. GRODE, G.A., et al., PB 188 108, PB 195 727, PB 205 475.
10. HALPERN, B.D. and SHIBAKAWA, R., Heparin Covalently Bonded to Polymer Surface, Advances in Chemistry Series, Interactions of Liquids at Solid Substrates, (Ed. Gould, R.F.) Amer. Chem. Soc. Publ., (1968).

11. KOWARSKI, A., THOMPSON, R.G., MIGEON, C.J. and BLIZZARD, R.M., J. Clin. Endocrinology, 30 (1971) 356.
12. DRAUSE, A.H., FERGUSON, T.B. and WELDON, C.S., Annals of Thoracic Surg., 14 (1972) 123.
13. LAGERGREN, H. and ERIKSSON, J.C., Trans. ASAIO, 17 (1971) 10.
14. LEININGER, R.I., COOPER, C.W., EPSTEIN, M.M., FALB, R.D. and GRODE, G.A., Science, 152 (1966) 1625.
15. LEININGER, R.I., CROWLEY, J.P., FALB, R.D. and GRODE, G.A., Trans. ASAIO, 17 (1972) 312.
16. LEONARD, E.F., in Proceedings 5th Annual Contractors Conference of Artificial Kidney Program of the NIAMDD, (Ed. Dreuger, K.K.), (1972).
17. MERKER, R.L., ELYASH, L.J., MAYHEW, S.H. and WANG, J.Y.C., Artificial Heart Program Conference Proceedings, (Ed. Hegyeli, R.J.), National Heart Institute Artificial Heart Program, Washington, (1969) 29.
18. MERRILL, E.W., SALZMAN, E.W., LIPPS, B.J., GILLILAND, E.R., AUSTEN, W.G. and JOISON, J., Trans. ASAIO, 12 (1966) 139.
19. MERRILL, E.W., SALZMAN, E.W., WONG, P.S.L. and ASHFORD, T.P., J. Appl. Physiol., 29 (1970) 723.
20. REA, W.J., WHITLEY, D. and EBERLE, J.W., Trans. ASAIO, 18 (1972) 316.
21. SALZMAN, E.W., AUSTEN, W.G., LIPPS, B.J., MERRILL, E.W. and JOISON, J., Surgery, 61 (1967) 1.
22. SALZMAN, E.W., MERRILL, E.W., BINDER, A., WOLF, C.F.W., ASHFORD, T.P. and AUSTEN, W.G., J. Biomed. Mate. Res., 3 (1969) 69.
23. THALER, E. and SCHMER, G., Abstracts, 20th Annual Mtg. Amer. Soc. Art. Int. Organs, (1974) 73.
24. UDDIN, K.M., UTLEY, J.R., BRYANT, L.R., DILLON, M. and WEISS, D.L., Annals of Thoracic Surgery, 17 (1974) 351.

DISCUSSION OF FALB *ET AL.* PAPER

LEVIN

Dr. Falb, a technical question. I was interested in your data concerning the reaction between platelets and the surface. Were you using platelet rich plasma?

FALB

Yes, it was citrated, platelet rich plasma.

LEVIN

Is the type of anticoagulant critical in order to show the phenomenon of platelet adsorption?

FALB

We tried platelet adsorption in the presence of several anticoagulants: EDTA, citrate and heparin. In each case, platelets adhered onto the surface of the heparinized materials.

LEVIN

Finally, you made the point that the release reaction did not occur. Were you therefore able to elute intact platelets from the surface?

FALB

No, with the TDMAC heparin-coated materials, the platelets were absent from the initial fractions going over the column, and then they gradually built-up in subsequent fractions. If you examine the column material with a microscope, you can see that it's almost completely covered with platelets. We have also used radio-labeled platelets and gotten results that were comparable with surface counting. It stays on, it doesn't come back off.

HEPARIN-INDUCED OSTEOPENIA: AN APPRAISAL

Louis V. AVIOLI

Washington University School of Medicine and The Jewish Hospital of St. Louis, St. Louis, Missouri 63110 (USA)

Heparin has been used since 1936 for the treatment of vascular thrombotic episodes, but the association between heparin administration and osteopenia in man was not reported until 1964, when a causual notation was made regarding a spontaneous vertebral fracture in a 50-year-old man on heparin with no history of trauma (10). Jaffe and Willis (24) later reported the development of multiple symmetrical rib fractures, compression fractures of the thoracic vertebrae, and generalized skeletal demineralization in a 41-year-old man subjected to heparin at 20,000 units/day for 1 year. Earlier in the same year, Griffith _et al_. (18), reporting on 117 patients subjected to long-term heparin therapy, noted that 10 patients on subcutaneous heparin at 15,000 - 30,000 units/day for 6 months or longer developed spontaneous vertebral and/or rib fractures. These complications were not observed, however, in 107 patients receiving 10,000 units/day or less for 1 - 15 years. Bone biopsies obtained from two patients with osteopenia revealed that "the bony matrix was very soft, offering little or no resistance to the pathologist's knife" (18). A cause-and-effect relation between heparin and osteopenia was strengthened by the additional observation that five patients improved remarkably when coumarin drugs were substituted for heparin. The skeletal abnormalities were associated with normal concentrations of circulating calcium, inorganic phosphate, and alkaline phosphatase and normal urinary total hydroxyproline, although calcium excretion was low (30-80 mg/day). In 1966, Miller and DeWolfe (33) reported the development of generalized demineralization with thoracic vertebral compression fractures in a 52-year-old man treated with heparin at 40,000 units/day for approximately 4 months. Serum calcium, inorganic phosphate, and alkaline phosphatase were also normal in this case. Additional compression fractures occurred, despite cessation of therapy and the addition of calcium supplements and "vigorous androgen therapy." Three years later, Schuster and co-workers (45) described an osteopenic syndrome in a 5-1/2-year old child after 9-1/2 weeks of heparin therapy at approximately 10,000 unit/day. These authors considered the relation a direct one, although the child had also been subjected to supplemental adrenocorticoid therapy before the osteopenia was discovered.

These isolated clinical reports of bone fracture in patients on heparin therapy have been considered by many to be rather insignificant, in comparison with the large population of patients subjected to heparin therapy since 1936. And, if radiographically detectable demineralization and bone fractures are used as the only indicators of disordered skeletal metabolism in patients on long-term heparin therapy, the true incidence of heparin-induced osteopenia may never be fairly appreciated. Inasmuch as approximately 30-50% of bone mineral must be lost before demineralization becomes radiographically evident (30), the so-called osteoporosis observed by the radiologist actually represents rather marked diminution in the total mineral content of bone. The amount of skeletal decalcification necessary for the diagnosis of osteopenia varies considerably in different bones and in different parts of the same bone, depending on structural composition. Thus, it may take years for appreciable radiographic abnormalities to be noted in patients on long-term heparin therapy in dosages of 10,000 units/day or less.

By the same token, it may also require years for therapeutically induced improvement in the mineral content of the skeleton to be appreciated by routine radiographs. In this regard, it may be worth noting that, in 1963, a year before a vertebral collapse was first associated with long-term heparin therapy (10), Hughes _et al._ (23) noted that, of 53 patients receiving heparin at 25,000 - 30,000 units/day, "four patients complained of chronic muscular skeletal pain of moderate severity, which involved various levels of the spine without demonstrating any clinical, radiographic or laboratory evidence of possible etiology. No similar pain had been present previously, and the symptoms subsided shortly after the heparin was terminated." Were these authors actually reporting on symptoms resulting from progressive skeletal demineralization and vertebral microfractures undetectable by routine radiographic analysis because of its established limited sensitivity? If, in fact, heparin therapy does result in progressive bone demineralization (and the accumulated data, although still somewhat anecdotal, seem consistent in that direction), it seems appropriate to establish the mechanism(s) whereby heparin may interfere with skeletal turnover in man.

Bone functions not only as a supporting organ for the body, but also as a well-organized dynamic system consisting of mineral, a matrix of collagen fibers, and cells. The cells include osteocytes, osteoblasts, and osteoclasts and occupy 3-4% of the total volume. The osteoblasts, which originate from precursor mesenchymal cells, function primarily to produce the collagenous portion of the bone matrix. As bone matrix is deposited on a bone-forming surface, an "osteoid seam" is formed, which, after a complex sequence of poorly understood biochemical changes, is normally mineralized. During this bone maturation and mineralization process, the enzyme alkaline phosphatase can be detected, by appropriate staining techniques, within the osteoblast and in the surrounding layers of newly deposited matrix. Although an increase in plasma "bone" alkaline phosphatase (characteristically heat-labile, compared with other isoenzymes, such as placental or hepatic alkaline phosphatase) is often seen in actively growing children and in patients with disorders characterized by rapid bone turnover, the role of this enzyme in active bone formation and remodeling is still entirely conjectural.

Like the osteoblasts, the osteoclasts in bone arise from mesenchymal cells. Compared with the osteoblastic controlled rate of bone formation, osteoclastic resorption of bone is an extremely fast and efficient process. Consequently, stimulation of osteoclastic activity may result in the immediate release of large quantities of mineral from bone. Although osteoclasts are known to contain the enzyme acid phosphatase and to be stimulated by parathyroid hormone, the exact mechanisms that condition the osteoclastic resorption of bone are still unknown.

Osteocytes, localized within the substance of densely mineralized compact bone, release alkaline phosphatase, proteases, and lysosomal acid hydrolases. Histologic evidence of osteocytic resorption of compact bone in pregnancy, hyperparathyroidism, osteomalacia, and thyrotoxicosis suggests that osteocyte-mediated resorption may play a significant role in the rapid mobilization of bone mineral. In contrast, the role of the osteoclast may be more important in regulating the normal remodeling process of skeletal tissue and in the chronic resorptive response to hormones, e.g., parathyroid hormone.

The collagen fibers in bone are spatially oriented, highly organized in interlacing bundles and layers, and embedded in a gelatinous ground substance. This mucopolysaccharide gel is composed primarily of chrondroitin-4-sulfate, chrondroitin-6-sulfate, and sialic acid and accounts for 4-5% of the organic bone matrix. In fully mineralized mature bone, the collagenous protein matrix accounts for approximately 30-35% of the volume of the intercellular material, with mineral occupying most of the remainder in the form of hydroxyapatite. Although calcium and phosphate are the principal ions in hydroxyapatite, the crystal contains significant amounts of Na^{+}, Mg^{+2}, CO_3^{-2}, and citrate^{-3} ions. It has been well established that collagen fibers can act as nucleation centers for the calcification processes that attend new bone formation, and the mucopolysaccharides appear to serve as fundamental cofactors during the calcification process. The integrity of the collagen-mucopolysaccharide complex appears intrinsic to the maintenance of the events that normally condition the formation and remodeling of osseous tissue.

The biosynthesis of collagen fibrils in bone is initiated in the osteoblast by the assembly of a glycine- and proline-rich polypeptide precursor of collagen (procollagen) as ribosomal complexes containing RNA (3). Later intracellular alterations include hydroxylation of appropriate proline and lysine residues of procollagen to hydroxyproline and hydroxylysine by specific hydroxylases. Current evidence also indicates that hydroxylation of prolyl and lysyl residues occurs with translation. Cofactors essential for this hydroxylation include oxygen, Fe^{+2}, ascorbic acid, and α-ketoglutarate. The _in situ_ hydroxylation of proline appears to be a prerequisite for secretion, in that procollagen molecules accumulate within the cell when hydroxylation is prevented. The procollagen-hydroxylating enzymes are also inhibited by sulfhydryl reagents, such as _p_-mercuribenzoate (20), hydralazine (6), and Dilantin (32). Inherited defects in the procollagen-hydroxylating enzymes have been reported in man and characterized by severe scoliosis, joint laxity, hyperextensible skin, and a marked deficiency in the hydroxylysine content of the collagen molecule (27).

Before the immature procollagen molecules are secreted by the osteoblasts, galactose is enzymatically attached (by UDP-galactose transferase) to hydroxylysine residues by an O-glycosidic linkage, and glucose is attached to the galactose residue via a second enzymatic process involving UDP-glucose transferase.

After the microtubular transcellular transport of the glycosylated procollagen molecule, it is secreted from the osteoblast. Later extracellular alterations of the collagen molecule include enzymatic cleavage (by procollagen peptidase) of a portion of the molecule at the NH_2-terminal end and a series of reactions that result in cross-linkages between individual microfibrils requiring the enzyme lysine oxidase. The essential role played by the intermolecular and intramolecular cross-links is reflected by documented alterations in collagen metabolism by lathyrogenic agents, penicillamine, and thiosemicarbazides. Lathyrogenic agents inhibit lysine oxidase activity (46), penicillamine acts by reacting or chelating with the aldehydes generated by the lysine oxidase reactions (9), and thiosemicarbazides inhibit aldehyde formation, as well as binding to preformed aldehydes (42, 50). Thus, beginning with the assembly of amino acids into polypeptide chains on cytoplasmic polyribosomes, the intracellular (i.e., osteoblastic) development of the collagen molecule, its secretion, its extracellular aggregation, and its final maturation are conditioned by a number of specific enzymatic reactions.

The relation among bone mucopolysaccharides (glycosaminoglycans), high-molecular-weight protein-mucopolysaccharide complexes (proteoglycans), and bone mineralization are largely speculative. Because both the sulfate and carboxyl groups of the proteoglycans have considerable affinity for calcium, it has been postulated that the rapid depolymerization of such high-molecular-weight components in bone could increase the concentration of calcium ions locally and therefore promote calcification, inasmuch as the ion binding capacity of long-chain polymers decreases with decreasing chain length (17). This hypothesis is consistent with the analytical evidence that the proteoglycans of calcified bone contain less protein than the protein-polysaccharides of uncalcified cartilage (21). It is currently assumed that an interaction between constituents of the organic matrix of bone and either the calcium or phosphate ions in plasma is fundamental to the deposition of mineral in bone. Collagen probably plays a significant role in the mineralization process, either as a passive support for mineral or as an active nucleator (or both). A variety of "local factors", including proteoglycans, probably function as inhibitors of calcification in bone, although the role of these complexes is not fully elucidated.

With this brief review of the essential ingredients of bone and the potential interplay of mineral, collagen, cells, glycosaminoglycans, and proteoglycans, it appears appropriate to examine the potential and documented effect of heparin on maturation and mineralization of bone. In 1927, it was demonstrated that heparin derived from hepatic tissue inhibited the growth of tumor cells _in vitro_ (13). Three years later, Goerner (15) reported that heparin pretreatment interfered with the successful transplantation of Flexner-Jobling rat carcinoma tissue. Heparin and related sulfomucopolysaccharides have since been shown to inhibit cell division in a variety of biologic systems, and their propensity to inhibit cell mitosis has been established (40).

Although detailed studies on the effect of heparin on osteoblasts and on cellular proliferation in bone are meager, a number of investigations with the synthetic heparin substitute, dextran sulfate, on the cellular component of bone may be significant in this regard, although skeletal demineralization has not been reported in patients during short-term (1 - 3 weeks) dextran sulfate anticoagulation therapy (11). The anticoagulant effect of dextran sulfate, like that of heparin, is due to the strong electronegative charges carried by the acidic groups and is therefore related to the number of sulfate groups in the molecule. Heparin, although it contains fewer sulfate groups than dextran sulfate, has greater anticoagulant activity. Whereas preparations of dextran sulfate used clinically have a molecular weight of approximately 7,500, they are not homogeneous; heparin preparations are heterogeneous, with an average molecular weight of 16,000-17,000, and offer seven times the anticoagulant effect (weight for weight) of dextran sulfate. Rabbits treated with dextran sulfate at approximately 20 mg/kg for 4-5 weeks reportedly develop generalized osteoporosis with spontaneous fractures and growth retardation (22). In the cited study, histopathologic examination of decalcified slices of costochondral tissue revealed generalized bone resorption with a significant decrease in the number of osteoblasts, the latter being replaced by an abundance of osteoclasts. Although heparin, at 5-20 mg/kg for periods identical with those used for the dextran-treated animals, resulted in a decreased rate of growth in the rabbits, no comparable effects on bone cells were recorded (22). In 1965, Tourtellotte and Dziewiatkowski (51) demonstrated that the repeated administration of dextran sulfate to young rats led to a disorder of endochondral ossification characterized by a reduced amount of bone matrix in the metaphyses of long bones with sparing of the epiphyseal-cartilage plate and articular cartilage. In their study, osteoblasts were seen in adequate numbers and were for the most part structurally normal, although in some instances they appeared "somewhat shrunken with pyknotic nuclei." As had been noted in the earlier study of Hint and Richter (22), there was a proliferation of osteoclasts at sites of spontaneous fracture in the animals treated with dextran sulfate.

In 1970, Ellis and Peart (12), while evaluating the effects of heparin and dextran sulfate on the linear growth rates of cultured mouse limb bones, observed a decrease in bone growth rates and osteoblast counts of bone tissue cultured with dextran sulfate at 0.1-5.0 mg/ml. Whereas heparin in a concentration of 0.1 mg/ml failed to inhibit new bone formation, concentrations of 5.0 mg/ml also resulted in an impairment of linear growth and a decrease in osteoblastic activity. There authors concluded that "osteoporosis" induced by dextran sulfate results from a combination of diminished osteoblastic and excessive osteoclastic activity - independently of parathyroid hormone intervention.

The accumulated data, although less than adequate, suggest that both substances related to heparin,and heparin (*in vitro*, with doses higher than those achieved in the blood and tissue fluids of experimental animals on chronic heparin therapy) may interfere with osteoblastic activity in bone. Although it would be consistent with the aforementioned reports of heparin-induced alterations in cellular division, it seems premature to ascribe the heparin-induced osteopenic syndrome entirely to ostoblastic insufficiency.

The effects of long-term administration of dextran sulfate and other sulfated polysaccharides (e.g., laminarin sulfate) on the skeletons of a variety of animals have been studied extensively. The abnormality common to all reported observations is the development of osteopenia due to impaired endochondral ossification. Cortical bone mass decreases and fractures characteristically occur immediately adjacent to sites of endochondral ossification. Chemical analyses of bones of rabbits, rats, and guinea pigs treated with dextran sulfate reveal normal concentrations of calcium, phosphate, nitrogen, manganese, and copper (11). Serum calcium and phosphate concentrations are also unaffected by treatment, but the circulating alkaline phosphatase tends to decrease; the latter has been attributed to the decrease in osteoblastic activity in bone cited earlier. It has been noted that skeletal lesions induced by dextran sulfate resemble those of ascorbic acid deficiency, although experimentally induced wounds heal normally in these animals and collagen formation at fracture sites appears unimpaired (11).

These observations are difficult to interpret, because not only does ascorbic acid or heparin deficiency itself interfere with wound healing (1, 39), but heparin potentiates the effects of ascorbic acid deficiency in this regard (39). Moreover, although the bone lesions of animals treated with dextran sulfate are reportedly similar to the lesions attending experimental copper or manganese deficiency, the tissue concentrations of these elements are normal (11). One may conclude, therefore, that dextran sulfate might interfere directly or indirectly with such enzyme systems as UDP-glycosyl transferase and lysine oxidase, which are essential to the biochemical maturation of the collagen molecule, and require Mn^{+2} and Cu^{+2}, respectively, as cofactors (3).

In view of reported observations that mucopolysaccharides inhibit bone crystal nucleation and that endochondral ossification depends on chrondroitin sulfate concentration (21), and others demonstrating a decrease in hydroxyapatite and an increase in citrate in bones of heparinized rats (35), it has been postulated that dextran sulfate and related mucopolysaccharides, such as heparin, interfere with the utilization or modification of the normal matrix mucopolysaccharides of bone and that this results in defective endochondral ossification. The observations of Stinchfield and coworkers (49) on rabbits and dogs with induced skeletal defects are of interest in this regard. They showed that anticoagulant therapy (e.g., heparin and/or Dicumarol) resulted in defective bone healing and reported that the changes were associated with a decrease in the concentration of mucopolysaccharide that could be stained with toluidine blue at nonunion sites. Although this theory is provocative and consistent with preliminary tinctorial observations, it must await direct experimental confirmation.

The accumulated data relating either abnormal bone cellular activity or alterations in mucopolysaccharide metabolism to osteopenia induced by heparin and dextran sulfate are still scanty and controversial. What alternatives might there be to explain the increased fracture incidence cited for patients on long-term heparin therapy? It has been reported that heparin, as a sulfated polysaccharide, has an affinity for calcium ions (19, 36). The simple addition of 0.05 ml or 0.2 ml of heparin to normal human serum results in a 13% or 45% decrease, respectively, in the ionized calcium concentration (36). Because a decrease in circulating ionized calcium is the only known determinant of the release of parathyroid hormone in man, it has been suggested that heparin-induced decreases in ionized calcium result in parathyroid gland overactivity, stimulated osteoclastic activity, and progressive demineralization of bone (19). Moreover, a decrease in circulating ionized calcium and a stimulated release of parathyroid hormone should also attend the hyperphosphatemia observed in patients on long-term heparin therapy by Bijvoet *et al*. (7). These authors attribute the tendency to hyperphosphatemia (i.e., serum inorganic phosphate up to 6.4 mg/100 ml) to an increase in the ratio of the renal reabsorptive maximal tubular capacity for inorganic phosphate to the glomerular filtration rate. Although a decrease in circulating ionized calcium and an acquired state of secondary hyperparathyroidism should naturally result from an accumulation of heparin-Ca^{+2} complexes and inorganic phosphate in blood, this hypothesis has not been widely accepted, because calcium salts of heparin or dextran sulfate are as effective in inducing osteoporosis as the corresponding sodium salts (12). The absence of parathyroid hyperplasia and morphometric evidence of parathyroid hormone effect on the skeletons of animals or humans subjected to prolonged heparin administration also makes this theory difficult to substantiate. More definitive measurements of circulating parathyroid hormone in patients on heparin therapy should be obtained, however, before this issue is settled.

Data have also been presented to show that heparin exerts its effect on the skeleton by acting as a cofactor or permissive agent for the action of parathyroid hormone on bone. In defense of this hypothesis are *in vivo* observations that heparin administration to rats enhances the stimulation by parathyroid hormone of acid phosphatase activity of bone (34) and *in vitro* tissue-culture studies demonstrating that heparin (but not chondroitin sulfate or hyaluronic acid) enhances the amount of bone resorption obtained with suboptimal concentrations of parathyroid hormone (16). In addition to these isolated reports, which suggest that heparin and parathyroid hormone act synergistically in promoting skeletal resorption, there are those which demonstrate heparin-induced alterations in mineral metabolism in the parathyroprivic state. Heparin administration to actively growing kittens rendered hypoparathyroid surgically leads to a progressive *increase* in serum calcium, a decrease in bone formation, and an increase in bone resorption, although heparin failed to produce similar changes in adult animals under similar circumstances (26). The report suggest that the calcium-mobilizing effect of heparin on bone can be direct (e.g., in the absence of parathyroid hormone) and that young bone, with its characteristic rapid turnover, is much more susceptible. In this regard, it should be added that protamine, well recognized as an antagonist of the anticoagulant effects of heparin, also produces hypocalcemia and hypophosphatemia by direct inhibition of bone resorption (2, 25). This bone-suppressive effect of protamine is also masked *in vivo* by heparin (25). Although these data are consistent with a skeletal antagonism between heparin and protamine, they may simply reflect peripheral combinations between the strongly basic protamine and acidic heparin molecules.

Finally, in an attempt to explain the osteopenia that results from heparin administration, attention has been focused on some clinical disorders characterized either by an increase in mast cells or by inherited defects in mucopolysaccharide metabolism. These association seem to add only confusion to an ever-increasing body of conflicting data. It has been reasoned that, inasmuch as systemic mast-cell disease is associated with skeletal demineralization, there is a relation between the osteopenia and the heparin-containing mast cell (18). Although potentially contributing to the skeletal lesions of mastocytosis, heparin itself should not be considered as the sole circulating or local bone toxin, because mast cells also contain 5-hydroxytryptamine (5), proteolytic enzymes (29), and histamine (41), and at least one of these substances (histamine) can also inhibit bone remodeling and maturation (37).

A comparison between heparin-induced osteopenia and the bone lesions attending some heritable disorders of connective tissue, such as Hunter's and Hurler's syndrome, may also be misleading. Both Hunter's and Hurler's syndromes stem from inherited enzyme deficiencies (14) which result in a decreased rate of lysosomal degradation of both dermatan and heparan (<u>not</u> heparin) sulfates and subsequent elevations in the plasma concentrations and urinary excretion of these glycosoaminolycans. Although the similarity between heparin and heparan is often stressed, it should be emphasized that heparin is a distinct polymer and does not belong to the heparans, although it may contain a significant number of N-acetyl groups (8). The heparan isolated from the urine of patients with mucopolysaccharidoses may also differ markedly from that normally present in lung and aorta. Linker and Hovingh (31) have recently stressed this difference in a report describing the isolation of a heparan with a molecular weight of less than 3,000 from the urine of a patient with the Sanfillippo syndrome, a disorder of mucopolysaccharide metabolism characterized by increased urinary heparan sulfate (and not dermatan sulfate) and a deficiency of heparan sulfate sulfatase and N-acetylglucosaminidase (28, 38). It may prove inappropriate to compare these inherited disorders of mucopolysaccharide metabolism, which condition changes in endochondral calcification of the growing child (47) with the osteopenic effects of heparin seen in an adult population, in whom endochondral activity is minimal.

This review was undertaken to evaluate the present state of the art regarding the experimental documentation of heparin-induced osteopenia and the evidence of biochemical derangements that result in progressive demineralization. The accumulated data are still less than satisfactory. When bone biopsies are obtained from hypocalciuric patients on prolonged heparin therapy, they are soft and offer "little or no resistance to the pathologist's knife" (18). This observation suggests osteomalacia (poorly mineralized bone) as the underlying histologic lesion, although that is not the case when undecalcified histologic specimens from animals treated with dextran sulfate are analyzed (11). Reports of increased collagenolytic activity of rat bone-cell homogenates and decreased stability of bone-cell "lysosome-like bodies" are associated with greater than normal rates of bone matrix or collagen synthesis in man (18). We are also confronted with data that describe a decrease in bone formation in heparinized cats (26) but no change in bone matrix biosynthesis in heparinized rats (18). Theories relating in vitro formation of heparin-Ca^{+2} complexes to hypocalcemia (19) and stimulated parathyroid hormone release are to be compared with reports of heparin-induced hypercalcemia in intact animals (26). Patients with heparin-induced fractures have responded favorably to heparin withdrawal and coumarin therapy (18), although coumarin is as effective as heparin in preventing bone healing (49). Associations with inborn errors of connective-tissue metabolism have been made that grossly underestimate the molecular differences between the heparan sulfates and the heparin used therapeutically, and references to disorders characterized by an abundance of the heparin-containing most cell ignore other mast-cell constituents that may also induce alterations in the orderly sequence of bone mineralization and maturation.

A reviewer is thus confronted with a variety of in vitro and in vivo observations of the tissues of rats, mice, dogs, guinea pigs, kittens, cats, and rabbits of various ages, with different rates of bone turnover, which are subjected either to doses of heparin that could prove fatal if administered to man, or to related mucopolysaccharides, such as dextran sulfate, the response to which may bear little relation to the response to the structurally dissimilar mocule, heparin.

Future studies in this area should include practical systematic appraisals of mineral metabolism in a significant number of humans on heparin, with measurements of circulating immunoassayable parathyroid hormone, ionized calcium, alkaline phosphatase isoenzymes, phosphate clearance, and the intestinal absorption of calcium, phosphate, and vitamin D. Because heparin has been identified as a potent inhibitor of a variety of enzymatic reactions, measurements of circulating biologically active vitamin D metabolites - such as 25-hydroxycholecalciferol and 1,25-dihydroxycholecalciferol, which require intact hepatic and renal hydroxylating systems for their production - might also be useful (4). More specific evaluation of the demineralizing effect of heparin on bone could also be attempted in man, by using quantitative noninvasive methodology with detection sensitivities much greater than that of the ordinary x-ray, such as osteodensitometric analysis (48). If the bone lesions do develop on heparin dosages of less than 10,000 units/day, this approach may prove valuable in detecting patients at risk before a symptomatic, radiographically detectable fracture occurs. Using appropriate animal models, one could anticipate more definitive and better defined studies of bone remodeling and mineralization with double tetracycline labeling techniques, x-ray diffraction analysis, and well-established methods of measuring the degree of collagen maturation and bone crystal integrity (43, 44). Although studies with substances related to heparin, such as dextran sulfate, may prove fundamental to our ultimate understanding of the relation of acidic mucopolysaccharides to bone metabolism, well-designed experiments in heparinized animals - with emphasis on the characteristics of bone and mineral metabolism that are under established enzymatic control (e.g., collagen synthesis and maturation, biologic activation of vitamin D, and parathyroid-induced increases in skeletal cyclic adenosine monophosphate) - may add considerable insight into the fundamental heparin-induced abnormality. These studies, if performed on animals with concentrations of circulating heparin similar to or approximately the same as those of patients on long-term heparin therapy, could broaden existing concepts of heparin effect on bone and facilitate a more rational approach both the prevention and the therapy of a most painful fractured vertebral process or femoral neck.

ACKNOWLEDGEMENT

I wish to express my appreciation to Miss Linda Graf for her excellent secretarial and editorial assistance.

REFERENCES

1. ALDRICH, E.M. and LEHMAN, E.P., Surg. Gynec. & Obstet., 87 (1948) 26.
2. ANDERSON, J.R., TOMILINSON, W.S. and WRIGHT, J.E.C., Brit. J. Cancer, 21 (1967) 48.
3. AVIOLI, L., Kidney International, 4 (1973) 105.
4. AVIOLI, L. and HADDAD, J.G., Metabolism, 22 (1973) 507.
5. BENDITT, E.P., WONG, R.L., ARASE, M. and ROEPER, E., Proc. Soc. Exp. Biol. Med., 90 (1955) 303.
6. BHATNAGAR, R.S., RAPAKA, S.S.R., LIU, T.Z. and WOLFE, S.M., Biochim. Biophys. Acta, 271 (1972) 125.
7. BIJVOET, O.L.M., JANSEN, A.P., PRENEN, H. and MAJOOR, C.L.H., in Water and Electrolyte Metabolism" (Eds. DE GRAEFF, J. and LEYNSE, B.) Amsterdam, 1964, vol. 2, p. 151.
8. CIFONELLI, J.A. and KING, J., Carbohyd. Res., 12 (1970) 391.
9. DESHMUKH, K. and NIMNI, M.E., J. Biol. Chem., 244 (1969) 1787.
10. DIMOND, E.G., Amer. J. Cardiol., 14 (1964) 53.
11. ELLIS, H.A., J. Pathol. Bacteriol., 89 (1965) 437.
12. ELLIS, H.A. and PEART, H.M., Brit. J. Exp. Pathol., 51 (1970) 43.
13. FISCHER, A., Gewebe Züchtung, p. 396 (1927).
14. FRANTANTONI, J.C., HALL, C.W., and NEUFELD, E.F., Science, 162 (1968) 570.

15. GOERNER, A., J. Lab. Clin. Med., 16 (1930) 369.
16. GOLDHABER, P., Science, 147 (1965) 407.
17. GREGOR, H.P., Ann. N.Y. Acad. Sci., 18 (1956) 667.
18. GRIFFITH, G.C., NICHOLS, G., ASHER, J.D. and FLANNAGAN, B., Jama, 193 (1965) 85.
19. HAHNEMANN, S., Lancet, 2 (1965) 855.
20. HALNE, H., KIVIRIKKO, K.I. and SIMONS, K., Biochim. Biophys. Acta, 198 (1970) 460.
21. HERRING, G.M. and KENT, P.W., Biochem. J., 89 (1963) 405.
22. HINT, H.C. and RICHTER, A.W., Brit. J. Pharmacol., 13 (1958) 109.
23. HUGHES, M.L., MORTENSEN, F. and SHOURIE, L., Amer. Heart J., 65 (1963) 615.
24. JAFFE, M.D. and WILLIS, P.W., Jama, 193 (1965) 152.
25. JOHNSTON, C.C., GRINNAN, E.L., WILSON, H.C. and BODER, G.B., Endocrinology, 87 (1970) 1211.
26. JOWSEY, J., ADAMS, P. and SCHLEIN, A.P., Calc. Tiss. Res., 6 (1970) 249.
27. KRANE, S.M., PINNELL, S.R. and ERBE, R.W., Proc. Nat. Acad. Sci. US, 69 (1972) 2899.
28. KRESSE, H. and NEUFELD, E.F., J. Biol. Chem., 247 (1972) 2164.
29. LAGUNOFF, D. and BENDITT, E.P., Ann. N.Y. Acad. Sci., 103 (1963) 185.
30. LEROUX, G.F., Rev. Med. Liege., 28 (1973) 497.
31. LINKER, A. and HOVINGH, P., Carbohyd. Res., 29 (1973) 41.
32. LIU, T.Z. and BHATNAGAR, R.S., Proc. Soc. Exp. Biol. Med., 142 (1973) 253.
33. MILLER, W.E. and DeWOLFE, V.G., Cleveland Clinic Quart., 33 (1966) 31.
34. MILLS, B.G., MALLET, M. and VAVETTA, L.A., Proc. Soc. Exp. Biol. Med., 121 (1966) 1052.
35. MITTERLUNG, I., HÄHNEL, H., LINDENHAYOR, K., MÜHLBACH, R. and SCHMIDT, U.J., Z. Alternsforsch., 27 (1973) 71.
36. MOORE, E.W., in Ion-Selective Publication, 314 (1969) 215.
37. NORTON, L.A., PROFFIT, W.R. and MOORE, R.R., Nature, 221 (1969) 469.
38. O'BRIEN, J.S., Proc. Nat. Acad. Sci., 69 (1972) 1720.
39. OHLWILER, D.A., JURKIEWICZ, M.J., BUTCHER, H. and BROWN, J.B., Surg. Forum, 10 (1959) 301.
40. REGELSON, W. and HOLLAND, J.F., Nature, 181 (1958) 46.
41. RILEY, J.F. and WEST, G.B., J. Physiol. (London), 120 (1953) 528.
42. ROJKIND, M. and GUTIERREZ, A.M., Arch. Biochem. Biophys., 131 (1969) 116.
43. RUSSELL, J. and AVIOLI, L.V., J. Clin. Invest., 51 (1972) 3072.
44. RUSSELL, J.E., TERMINE, J. and AVIOLI, L.V., J. Clin. Invest., 52 (1973) 2848.
45. SCHUSTER, J., MEIER-RUGE, W. and EGLI, F., Deutsche Med. Wochen., 94 (1969) 2334.
46. SIEGEL, R.C. and MARTIN, G.R., J. Biol. Chem., 245 (1970) 1653.

47. SILBERBERG, R., RIMOIN, D.L., ROSENTHAL, R.E. and HASLER, M.B., Arch. Pathol., 94 (1972) 500.
48. SORENSON, J.A. and CAMERON, J.B., J. Bone Joint Surg., 49A (1967) 481.
49. STINCHFIELD, F.E., SANKARAN, B. and SAMILSON, R., J. Bone Joint Surg., 38A (1956) 270.
50. TANZER, M.L., MONROE, D. and GROSS, J., Biochem., 5 (1966) 1919.
51. TOURTELLOTTE, C.D. and DZIEWIATKOWSKI, D.D., J. Bone Joint Surg., 47A (1965) 1185.

DISCUSSION OF AVIOLI PAPER

ENGELBERG

I just want to comment on the excellent discussion on heparin and bone, that in many years of treating patients with heparin twice a week intermittently or three times a week, with doses of heparin 15 or 20 thousand units, we have not observed spontaneous fractures. Apparently it is the high daily doses of heparin over long periods of time, that has lead to this complication, and when you give intermittent doses somehow you don't have the effect, however it works, on bone resorption, at least not to the extent that it has clinically caused fractures or bone pain.

THE USE OF HEPARIN AS AN ANTI-THROMBOTIC AGENT:

A PANEL DISCUSSION

Moderator: Irving S. WRIGHT

Participants: Richard FALB
Louis B. JAQUES
V. V. KAKKAR
Robert D. ROSENBERG
Lawrence SHERMAN

WRIGHT

It is appropriate to preface this panel discussion with a few informal comments about some of the early history of heparin and the characters involved. I would like at this time to pay homage to the men who played leading roles in the early history of heparin. Jay McLean took the first major step in the production of this substance, but few know of the struggles of Jay McLean. When our young people feel discouraged about the situation in reference to research as it is now, they should bear in mind his experience. He first attended the University of California and then decided that he wanted to go east to Johns Hopkins for his medical training. His dean at the University of California wrote a somewhat lukewarm letter to the Dean of Johns Hopkins stating that he "was not the kind of man Johns Hopkins sought". As a result of this he was refused admission at Johns Hopkins. Not once but twice. He had very little money, so he worked drilling oil wells for some 15 months. Then despite his rejection notices he migrated to Baltimore where he got a job in Dr. Howell's Laboratory, doing the most undesirable work in the laboratory. He presented himself to the Registrar and the Dean who expressed surprise at seeing him after his rejection but probably because of his persistence he was accepted in the medical school. When he was a sophomore in medical school, while working in Professor Howell's laboratory, he discovered heparin. This was in 1916. The original product was a very crude substance. We were quite familiar with this discovery when I first began to work in this field in the early 1930's, but it was not feasible to use it in man. We waited patiently and sometimes impatiently for further developments in purification. From 1934 through 1936 Charles Best started his notable work in the purification of heparin with Drs. Scott and Charles. Dr. Murray acted as the surgeon who first experimented with its use in man, together with Dr. Jaques who is with us today. This was monumental work, but it should be pointed out that Dr. Eric Jorpes of the Karolinska Institute visited Dr. Best about 1935 and they collaborated and worked out some general methods of approach. Dr. Jorpes went back to Stockholm and he also purified heparin to the point where it could be used in man. In 1938 I had a young male patient who developed a migrating thrombophlebitis. He had bought a new pair of shoes, tied the shoe laces very tightly, walked several miles and developed a local thrombus in the vein over the dorsum of one foot. From this site migrating thrombophlebitis developed, continuing over a period of nine months, during which time he ran a fever almost daily. The phlebitis involved many of the most important superficial veins of the limbs and trunk. Symptomatically it appeared that the splenic veins and mesenteric veins were also involved, and after nine months of progression of his disease it seemed that he was dying. At this time I learned that Dr. Best and Dr. Murray had used heparin very cautiously in a small number of patients in Toronto. We persuaded Dr. Best to come down to New York with a large portion of the world's supply of heparin, largely on the basis that this patient presented a very difficult and challenging problem. Together, with some apprehension, we began to administer the heparin intravenously; he then became the first patient to be treated with heparin in the United States. Within a few days his temperature fell to normal, the evidence of migratory thrombophlebitis rapidly disappeared, and he was on the way to recovery. We continued this intravenously for 16 days. By that time the supply of heparin was exhausted and so were Dr. Best and I, from trying to maintain some kind of control. No one had carried a patient for this long a time. We used a modified Lee White test. He made a temporary recovery. Dr. Best returned to Toronto and the patient did well for two or three months when he started to have a recurrence of his thrombophlebitis. Once more Dr. Best generously sent a new supply. There were several more relapses when heparin was discontinued. When dicumarol was crystallized and became available in early 1940 I obtained a supply from Dr. Karl Paul Link, and we began our work with dicumarol. This patient was one of the first patients to receive dicumarol on a long-term basis, and has been on anticoagulant therapy almost continuously since that time. Every time he has discontinued dicumarol, he has, within a period of two to three months, developed migratory thrombophlebitis. We have done all manner of coagulation studies but have been unable to determine any specific abnormalities in his coagulation mechanism. I believe he has now been on anticoagulant therapy longer than any other person. In 1938, he was a normal, well-built young man. He was quite athletic. Thirty four years later in 1972, he shot the largest elephant that was shot that year in East Africa. He has just returned from his fifth African safari. There are several points of special interest. First, he has never developed thrombophlebitis at any time while he has been on anticoagulant therapy. He did have a mild subendocardial infarction with no evidence of thrombotic complications about three years ago. Now he is 66 years old, so this was not surprising. However, he made an uneventful recovery and has been on two shooting safaris since then. We have repeatedly subjected him to rather complete surveys in terms of liver and kidney profiles. At no time have there been any adverse changes in his liver, kidney or any other metabolic function, in spite of the fact that he has been on dicumerol continuously for 25 years. There have been no significant hemorrhagic complications. His prothrombin time has been held at the standard 1-1/2 to 2 times the control. This regimen has in no way interfered with his active life. He has now entered the annals of medical history*.

Dr. Kakkar answered quite a few key questions in his excellent paper but some additional questions have been raised. What placebo did you use in your control study?

KAKKAR

In the double blind trial where we included 78 patients, the placebo which was used was normal saline. They were identical ampules sealed and randomly numbered, so no one knew what they contained.

WRIGHT

Dr. Kakkar, when people have used subcutaneous heparin in this country, there have been tell-tale marks in the skin produced by multiple minor hemorrhages. Did this interfere in any way with the maintainence of a double blind investigation?

KAKKAR

The method of administration is really very important and we followed what was recommended by Dr. George Griffith. The finest possible needle should be used and heparin should be injected in the lateral abdominal wall. It is best to raise or pinch a fold of skin and by using a fine needle, give a subcutaneous injection. By using this technique, you minimize the occurrence of hematomas.

WRIGHT

Did you encounter some cases in which it became quite clear that the patients were on heparin?

---

*WRIGHT, I. S., Editor, The Discovery and Early Development of Anticoagulants: A Historical Symposium, Circulation, 19 (1959) 73.

KAKKAR

Yes, certainly. When you use a therapy over such a long period and on a large scale, of course, you can see in their abdominal wall minor hematomas and it becomes quite evident that they are on heparin. In all subsequent studies which I conducted, the control group has received no medication, so that the placebo study was a limited trial.

WRIGHT

So that was indeed the only double blind test?

KAKKAR

Yes, that's the only double blind trial that has ever been done where the radioacitve fibrinogen test has been used to assess the effectiveness of therapy.

WRIGHT

Dr. Kakkar, in this country, you know the thing that has held up the use of ^{125}I-fibrinogen has been the risk of hepatitis. What has been the incidence of hepatitis in the various series with which you are familiar in England?

KAKKAR

I can only report on the more than 2,000 patients which we have personally studied. If we consider clinical hepatitis, this has not been observed in any of the patients who have been carefully followed for one year. If you are referring to subclinical hepatitis, I only wish someone would define that for me. Another factor is, that all of these patients have had major surgery. They are receiving blood transfusions. Which are you going to incriminate, fibrinogen or blood transfusions?

WRIGHT

Do you feel then that the incidence of hepatitis is so much lower in Great Britain than it is in this country that the fibrinogen you use is not as great a risk as ours would be?

KAKKAR

I do not believe I can answer that question.

SHERMAN

We have been using labelled fibrinogen for several years now, and the approach we have employed is similar to that that the FDA is requiring, namely a very careful pedigree selection of donors, with follow-up studies for antigen and antibody on a number of the recipients. The most desirable group in whom to study potential hepatitis from a given donor might be to select people who already have antibody suggesting that a rise in antibody level would be an even better sign that the fibrinogen contains hepatitis antigen. In addition, this group of patients, who already have antibody, might be thought to be somewhat protected from developing it from the fibrinogen, if it indeed contained hepatitis virus.

WRIGHT

We have a question here which has to do with the different molecular weights of heparin. The question really has two sides or directions. First, the possibility of increasing the molecular weight producing a slow release on one end of the spectrum versus the possibility of reducing the molecular weight so that it might be possible to use it for either oral or rectal administration, which has not been achieved. Dr. Falb, will you open this discussion?

FALB

I have been intrigued by the fact that for over 10 years now, heparin attached to a solid surface, has biological activity. This work has been confirmed again and again. This phenomenon of course is useful when you are putting a foreign surface in contact with the blood. I would like to speculate that the fact that heparin attached to a surface has activity, might be used to provide a long-term effective parenteral heparin. It may be valuable to attach heparin to a high molecular weight-water soluble material, such as dextran. In this case then you would have a dextran with a molecular weight of 70,000, which would contain a large number of heparin molecules and this moiety would perhaps remain longer in circulation than heparin of molecular weight 15,000. Someone speculated yesterday that heparin perhaps compartmentalizes and is removed from the blood by going into cell walls. However, heparin has unique features in that it interacts with clotting components in circulating blood. So if you had high molecular weight heparin, perhaps it could interact within the vascular system itself. This kind of phenomenon is being explored with drugs such as insulin. Porath has made a high molecular weight dextran-insulin complex which has good insulin activity. I would speculate that this might be the way to go with parenteral heparin.

WRIGHT

Would you care to comment on the possible value of low molecular weight heparin?

FALB

I am not highly qualified to do that, except that having worked with enzymes that are effected allosterically, it strikes me as odd that heparin is an allosteric effector of an enzyme in its high molecular weight form. I can think of dozens of allosteric effectors that effect enzymes that are of very small molecular weight, for example, a lithium ion. Why does heparin attach to one small site on the molecule and yet it has to have its entire molecular weight? The direction in the future, which has already been followed in some degree is to determine the specifics, the site of heparin that attacks the molecule. If you could obtain low molecular weight material that has these characteristics, it might be absorbed through an oral route or from a rectal suppository.

JAQUES

Commenting on the possibility of higher molecular weight material, such as that attached to dextran, the one piece of information we have on this is that the various heparinoids such as dextran sulfate, and certainly a number of other heparinoids that have been prepared, are taken up by the cells of the reticulo-endothelial system like heparin. The significant difference, in terms of the long-term toxicity that has been observed with heparinoids and the symptoms described by Krondheim seems to lie in the fact that heparin is degraded when in reticulo-endothelial cells, whereas the heparinoids remain there and interfere seriously with body defense mechanisms. With regard to the smaller molecular weight compounds, the first step is to determine whether they have the necessary activities in terms of the coagulation system, the prevention of thrombosis, etc. This information should be ascertained first. The other part of this question is the issue of absorption by the intestinal tract of an oral form of heparin. These experiments are much more complex than most workers appreciate. What is required is sufficiently high concentration of heparin in the blood to increase the clotting time. When a drug is taken orally, there are the questions of possible changes in the stomach, of absorption across the mucosa, the transport, and that heparin is absorbed in the portal system, goes into the liver and is deactivated there. It will, therefore, require a much larger dose of heparin, than is the case when injected intravenously.

WRIGHT

This is very complicated and we cannot resolve it today. However, the idea is challenging and I hope will be explored further.

Dr. Kakkar, is it true that the heparin you have used is calcium heparin?

KAKKAR

This is true for all of the studies which I personally have performed and all the studies that are now being carried on in the international trial. Calcium heparin is being used as prepared from one source only, is standardized, and we have been very careful in its administration. Some physicians have reported that there has been excessive bleeding at the site of injection. If you ask them, "what are you doing", they are using vials containing 125,000 units of heparin out of which they try to withdraw 5,000 units in a .2 ml syringe. This cannot be done accurately and leads to hemorrhagic complications. But if the ampule only contains 5,000 units, and a very fine needle is used, administration is much simplier and bleeding is not a major problem.

WRIGHT

Where do you obtain calcium heparin?

KAKKAR

We get it from France.

WRIGHT

It is obtainable in France, but not obtainable in the United States.

KAKKAR

I would like to point out, however, that the other studies which I have reported have been with sodium heparin.

SEEGERS

Why calcium heparin?

JAQUES

There is some information that heparin amongst its other properties is a calcium chelator and at the local site of injection, the removal of calcium from the vessel wall would undoubtedly be a precipitating factor in purpuric hemorrhage. This is the rational basis for it, although I think it was probably empirically derived first.

WRIGHT

Perhaps we should prevail upon some of our pharmaceutical friends to become interested in providing us with calcium heparin so that its value can be studied further. Dr. Rosenberg, would you discuss the preferred measurement of the effectiveness of heparin?

ROSENBERG

I have rather extreme views about the current methods that we have for monitoring heparin in patients. All these methods for monitoring heparin miss the point, I think, because of an inaccuracy in the way we think about how heparin works. The following analogy has some historical role since Best was involved both with insulin and heparin. Look at the diabetic and ask the question, what would happen if we treated diabetics by giving them insulin, and then measured insulin levels? This would undoubtedly lead to different dosages of insulin. What we do is measure glucose, which is a measure of the effect of insulin on the patient's glucose metabolism. We have different patient populations, which might be quite heterogeneous, and if we only considered insulin levels we would probably end up throwing a significant percentage of these patients into insulin shock. In essence, we face a similar situation with heparin. We give it, we measure activated antithrombin, by whatever technique, and ask, what is the level of heparin or heparin activated antithrombin in that plasma. What we want to do is to measure the biological effect of the anticoagulant in the patient's circulation, namely the level of serine proteases and how they are effected by heparin. Until we are able to do this, we will always be caught in the trap of never being able to know how much is enough, and what is the base line level, etc. We will have real problems in extending some of the excellent surgical studies that Dr. Kakkar has reported to more general populations.

WRIGHT

Dr. Sherman, how about changes that take place in the requirement of heparin from time to time?

SHERMAN

This bears to some extent to what Dr. Rosenberg was just saying. Most clinicians are aware, and watch for, the phenomenon of a decreasing heparin requirement, by whatever test they are using for measuring heparin effect, or noticing a decreasing requirement over the first several days of something like pulmonary embolus or thrombophlebitis. Most physicians have the idea that the level of serine proteases is going down, but while this probably does occur, we still have the question as to which serine proteases we should be looking at. This is another example in which there is a great deal of individual variation. It remains an argument in favor of varying patient dosage for an individual patient.

ROSENBERG

I cannot agree with that last remark because we are ultimately going to have to measure serine proteases all along the cascade, and it may turn out that we find that we want to suppress early events in the cascade. Variations in heparin response are really a measure of the binding of heparin to other proteins and, because the partitioning is really rather slightly in favor of co-factor, we only incorporate a small amount of heparin in co-factor and changes in acute phase-reactants can easily disturb that. It still does not answer the question: what is the effect of activated antithrombin on the coagulation system?

WRIGHT

What is the potential future of heparin particularly as it applies in the field of obstetrics?

KAKKAR

I trust you will understand that I am quoting someone else's work which I consider to be very important. Mr. Manair, a surgeon who is working in Oxford and uses low doses of heparin given to women with stillbirths,theorizes that there is deposition of large amounts of fibrin at the transplacental barrier, and this produces vascular insufficiency in these patients which is responsible for stillbirths. He has followed fairly large numbers of patients very methodically and has found that when these women were on low doses of heparin (calcium heparin), they had more successful deliveries. I was very impressed with this study, so I have taken the liberty of bringing it to your attention.

WRIGHT

It would be an interesting development if someone was able to produce a satisfactory low molecular weight heparin, because one of the arguments in favor of using heparin versus the coumarin compounds in the treatment of thrombophlebitis during pregnancy, rightly or wrongly, has been its high molecular weight so it does not cross the placental barrier. If the heparin molecular weight is decreased, this will require further evaluation. I am not passing judgement on the validity of this position; I have used warfarin and Dicumarol with pregnant patients over a period of years with minimal ill effects by keeping the prothrombin time at 1-1/2 to 2 times the control level.

ENGELBERG

Since minidose heparin may prevent thrombosis in leg veins, and since thromboses in leg veins and subsequent pulmonary emboli are an important cause of death complicating many medical conditions including congestive heart failure and highly toxic infectious states, one can reason that it could be used in these areas. The minidoses used do stimulate a considerable amount of lipoprotein lipase production. They may be anticoagulant doses, but they do form lipoprotein lipase in large amounts, causing the removal of a considerable amount of triglyceride from the blood stream. If my observations are correct, there is correction of low grade anoxia by this action and the minidoses will effectively do that.

WRIGHT

Dr. Rosenberg, would you care to discuss hereditary angioneurotic edema?

ROSENBERG

We have proposed a couple of weeks ago, that we might start thinking about using heparin in acute attacks of hereditary angioneurotic edema. It has been thought, at least by the people who have treated these patients, that the primary stimuli for the edema and for the sudden death that sometimes occurs results from the liberation of plasmin, kallikrein, and activation of the complement system. It is now turning out that heparin, either through antithrombin, or by itself, can inhibit all three of these enzymes. Although there are practical drugs now for inhibiting these attacks in long-term patients, there is really no good way of dealing with the diseases in acute attack which often results in asphyxia and death. I think it may well turn out that boluses of heparin during these crises may prevent such episodes and turn them off acutely. I think the other possibility which is an interesting one, refers to α_1-antitrypsin. We have with heparin, an ability to increase the activity of antithrombin. Alpha-1-antitrypsins are another of the serine protease inhibitors, and since their specificity is not terribly narrow, an interesting possibility would be, since antithrombin does inhibit trypsin, to see whether we could stimulate the low dose heparin therapy to increase the trypsin inhibitor levels in such patients. One wonders whether we will be able to deal with emphysema or liver disease that occurs in these disorders. At the present these are speculations.

WRIGHT

We are trying to extend our horizons a bit, but the more we extend our horizons the more we expose our ignorance. This is something we must accept as one of the penalties for looking into the future. Dr. Sherman, would you like to discuss disseminated intravascular coagulation?

SHERMAN

The only comment would be to note the pendulum swing which has occurred in the past couple of years. Six or seven years ago there was great enthusiasm for the use of heparin in disseminated intravascular coagulation and it was regarded to a certain degree as a panacea. The extent to which the pendulum has swung is illustrated by a statement from a recent monography by Harker, who said that heparin is seldom, if ever, indicated in therapy in intravascular coagulation. He lumped it with EACA in that regard. Perhaps other workers would feel that that is going too far. The thrust of the point may be that while we use heparin a great deal in intravascular coagulation, we obscure the fact that we are not treating the primary disease processes involved. This has led to a certain degree of discouragement because the physician has ignored the primary process. In processes of intravascular coagulation there is a need for individual modification of dosage, route of administration, change as indicated from bolus to continuous infusion, etc., depending upon the primary disease process that the patient has. Some side effects of bleeding that people have described in, for example, leukemic patients, were caused by the fact they were using high dose intermittent heparin therapy in patients who were thrombosidopenic for other reasons than intravascular coagulation. When the patient does have adequate and functioning platelets, he will tend to hemorrhage easily. In such cases one must be cautious as with any active pharmaceutical agent.

WRIGHT

One of the problems in really classifying this as a disease syndrome is the wide range of grays and borderline diagnoses that constantly confuse the picture.

SHERMAN

Unquestionably.

WRIGHT

It is difficult for most physicians to evaluate this literature. They may get the impression that heparin is a valid treatment for these cases. Then they cannot find out exactly what the authors mean by intravascular coagulation defects. Soon they begin to use it in a wide-range of patients, often inappropriately and with poor results. This is going to have to be straightened out in order to evaluate its use in proper situations. Dr. Rosenberg, what about the antifibrinolytic effect of heparin?

ROSENBERG

I am particularly intrigued by this problem. I was rather surprised to see how rapidly antithrombin interacts with plasmin and how rapidly heparin stimulates this inhibition. We now are extending observations into whole plasma. It is interesting to look at isolated inhibitor-serine protease interactions, but one does not know what they mean until studies include the other inhibitors present in plasma. One wonders in some situations whether we may not be inhibiting the resolution of clots after long periods of giving heparin by inhibiting the fibrinolytic mechanism. Perhaps one should consider this in applying the fibrinolytic therapy, and I think if one does, then one may have to tailor the fibrinolytic therapy carefully so that heparin does not obviate this treatment. The competition between thrombin and plasmin for antithrombin in the presence of heparin may also turn out to be an important phenomenon in DIC.

SHANBERG

These past two days have been facinating in that we have gone through mental gymnastics about the structure and degradation of heparin and so forth. The most important subject is in fact therapy and the use of heparin in the treatment of various clinical conditions. We have, thus far, not contributed to the education of the common physician as to how to use heparin or what he should be trying to do with it. Most of his experiences are with coumarin drugs and he know that they are given once-a-day in a certain dosage and this is controlled by once-a-day prothrombin time determinations. He therefore tends to extrapolate this to his use of heparin. As a pathologist, and a teacher of pathologists, I find that many doctors order heparin and ask for a clotting time daily, not really understanding what they are doing. Now this may have been acceptable back in the days when we were using constant intravenous drip or infusion therapy where one clotting time was sufficient. But with the intermittent therapy this does not hold true. We do not have anything in the literature to indicate to physicians why the clotting time of any type is important, what type of clotting times are important, when they should be done, and what are they looking for. Many doctors want to know if there is a residual effect of heparin and do not care whether there was any effect to begin with. So this is one aspect that has to be understood if we are going to give heparin therapy and not just in research or controlled studies.

It is important to recognize that the use of heparin threatens normal hemostasis. We know that the hemorrhage is not occurring because the patient is heparinized, but because the patient who is well heparinized is no different from a hemophiliac and the way heparinized patients are treated in hospitals would be abhorred if such treatment were given to hemophiliacs. If low dosage and minidose heparin therapy is good for prevention, is it not also possibly good for treatment. Why is it necessary to give high doses of heparin in the treatment of thrombophlebitis? Actually, the thrombus has already occurred and we are only looking for prevention of the propagation of the thrombus. I wish to ask anybody on the panel if they feel, as we did fifteen years ago, that heparin is more important in the prevention process for its action against initiation of clotting rather than the actual later stages of formation of a clot.

KAKKAR

This is purely a matter of clinical impression and a clinical approach. If you give heparin under certain circumstances in low dosage and test for the blood heparin level in the presence of an established thrombus, the test will be negative. The stimulus to thrombosis is so strong, you must block this first before you really are achieving the degree of therapy which will prevent further extension. The dose which you initially need, a heavy dose, is to block the process so that you have some of the inhibitors available. Then low dose heparin can activate them and prevent further extension. I may be wrong, but this is my impression.

FLETCHER

Dr. Wright, I would like to make a comment on something you said, Dr. Wessler said, and Dr. Ratnoff said. It concerns the monitoring of heparin and other related therapies. At the present moment, as we all know, we are using methods that tell us what effects the drugs have on the coagulation system. It is clear that we need methods that tell us also what effect the drug has on the disease process itself. There is such a method that is being quite widely used, plasma fibrinogen chromatography. This will tell you what effect the drug is having on the disease. Dr. Wessler showed us a slide which indicates that the glomerulonephritis could sometimes be regarded as a thrombotic microcirculatory disease. We have been fascinated by this, particularly acute progressive glomerulonephritis and our results of controlling therapy by plasma fibrinogen chromatography were reported two weeks ago in the Pediatric Research Society. In fact we have had considerable success.

ENGELBERG

I have a question for you Dr. Wright. It seems to me the logic for the use of heparin in the therapy for acute myocardial infarction is a fairlv overwhelming one. There are many good reasons for using this potent anticoagulant in acute myocardial infarction. I would then like to ask, what is your answer to the statements that now appear in the literature based upon large multicenter studies, that anticoagulants in acute myocardial infarctions are worthless.

WRIGHT

In the original study by the American Heart Association Committee on Anticoagulants we examined this very carefully and those of you who will take the time to look at this volume will be edified by the pathological studies. It was found that the effect of anticoagulant therapy was on the thromboembolic complications. That was where the improvement in statistics could be obtained. The changes in the number of thrombi within the coronary arteries was not statistically significant. It was in the thrombi in the veins, the pulmonary emboli, and the arterial emboli found in many areas of the vascular system. This included a number of strokes produced by emboli from the muralthrombi of the heart. Physicians neglected to read the report carefully or forgot about that important point and they misquoted us saying that we were concerned entirely with the coronary artery thrombi. If you will read the report you will see that this was not the case and it was not the conclusion drawn. Now after some years, new workers have rediscovered that our original statement was true although they neglect to refer to it. The question now is whether, in view of the much improved general care of these patients in coronary units, can anticoagulation be expected to further reduce the mortality? As Dr. Wessler pointed out in his beautiful address, the figures are suggestive that you can reduce it, but it is by a small amount. Since most of the young physicians running coronary units are fascinated by arythmeas, the anticoagulant aspect has been deemphasized-one might even say neglected. However, pathologists are sill finding a large number of thrombi scattered all through the system of the patients that are dying. They still find muralthrombi and thromboemboli scattered throughout the arterial system. Pathologists have also found that in patients with transmural infarcts the majority who do,have thrombi in the appropriate coronary arteries. The question is, how much are you willing to pay for prevention of additional disability even if you elect not to consider the 1 or 2 percent of lives involved. Many of these untreated patients are left with considerable disability. These include strokes with permanent damage. Many of them have intermittent claudication. Appropriate anticoagulant therapy is of some value in minimizing disability. Unfortunately, figures of many of these studies have dealt almost entirely with death. It is final- it is easy for the statisticians who do not have to care for the patient as he struggles to regain his health, but must not be considered as the complete answer to the living patient's problems. When we started this study in 1946, there was no treatment for myocardial infarction. Patients remained immobile in bed, were given morphine, there was even a debate as to whether they should ever be given digitalis. We were able to show impressive gain in both mortality and disability. Part of this may very well have been due to the fact that more attention was being paid to the patients as well as the effects of the anticoagulants. It was difficult to attribute the reduction in muralthrombi and cerebral and pulmonary infarction to nursing care. But today, our patients are receiving very intensive care, and therefore, the potential improvement attributed to anticoagulants are less significant.

WESSLER

I was impressed with Dr. Shanberg's question. His question on the selection of the appropriate dose of heparin is one of the important ones that we are unable to answer at this Symposium. Recognizing that there is not any definitive response, it might be worthwhile to learn how Dr. Rosenberg and Dr. Sherman handle this problem, because doctors do ask this question and experts such as Dr. Shanberg have difficulty responding to it.

ROSENBERG

If you give a heparin drip to patients at some reasonable rate, the incidence of hemorrhage is significantly decreased. I hope that Ed Salzman does not mind my quoting this; it appears that this is probably going to turn out to be true. Statistically there is a decreased incidence of hemorrhage if you drip rather than give bolus heparin and it makes reasonable sense. I am not sure once you start dripping that it makes much difference how often you measure the levels of heparin by whatever test you are using as long as you know it is prolonging the clotting time or the PTT initially. I think there are big changes, big swings, in patients who are on drip and it is not clear if those changes are therapeutically meaningful.

SHERMAN

I would simply agree with the matter of drips and with the one added comment that I think it is important on communicating to physicians in practice, that whatever technique is employed in their hospital to measure heparin effect, that they understand the therapeutic ranges in that hospital laboratory and do not go necessarily by what they see in the literature for the simple reason that as with prothrombin times, once you start to use tests such as the activated PTT, there may be large variations in the test values based on the reagents used in the test.

WRIGHT

I have to comment on this matter because of the position I took yesterday about the continuous intravenous drips. I have no objection to this technique. From a theoretical viewpoint it is physiologically the soundest way to do it. But my experiences during visits to community hospitals with a wide range in control capacity and staffing, this becomes not only difficult but dangerous. Therefore, we have the question of being scientifically ideal, but practically dangerous. I must disagree with Dr. Rosenberg if he really means that the control of heparin dosage is unimportant. I have seen too many serious hemorrhages and even death from neglect of careful controls to accept this statement.

SHERMAN

I would make one added comment. I think that a lot of the problems that people have with continuous drip with continuous infusion could be obviated by one or two things. Either use a pump, or if you can only use a drip, use a pediatric drip.

WRIGHT

Many small hospitals do not have the staff and the technique to do this properly. This is a problem of education and staffing.

BAUE

Dr. Wright, I would like to ask you and each member of the panel, if you feel you are now ready to recommend low dose heparin as a routine preventive measure for all patients over 40 having elective surgery?

WRIGHT

We know Dr. Kakkar's position. Dr. Rosenberg?

ROSENBERG

Dr. Kakkar's figures are so impressive that one would have to accept them.

WRIGHT

Dr. Rosenberg, you think you should, but are you going to do this?

ROSENBERG

Fortunately, Ed Salzman is going to have to make that decision.

SHERMAN

I am not sure personally, for one or two particular reasons. One is the fact that there has been some evidence, the best example is a paper from Oklahoma in one of the surgical journals of last year, of great variation in the incidence of positive scans depending on the type of major surgery involved. This may be significant. We are still looking at a heterogeneous group. The other point is that there are several groups of people, Dr. Salzman included, and workers in Canada, who have been quite concerned with the particularly high rise in hematomas in orthopedic cases about the wound. They have seen, not only with heparin, but with dextran, this sort of situation where a wound hematoma and infection would really cancel the favorable results of the surgery. It may make it more risky because of, in essence, the limited effectiveness.

WRIGHT

Dr. Kakkar should have the opportunity to respond to that.

KAKKAR

I would like to ask Dr. Sherman, has he operated on one of these patients who have been given heparin first thing? I have got to be provocative at this stage. Secondly, there are a group of patients where one has to be cautious about giving low doses of heparin. These are patients having neurosurgical procedures, those who are having lemnectomies and those who are having total hip replacements with epidural anesthesia.

WRIGHT

What is your position regarding prostatectomy?

KAKKAR

Prostatectomies? This is a confusing picture at present. We are looking at this operation in the trial at King's College Hospital. We have not run into any trouble. Some other surgeons have reported that there has been increased bleeding.

WRIGHT

I have recommended this approach for a long time for selective patients, but I would not like to see it recommended as a routine for every surgical case. Dr. Kakkar has pointed out some instances in which it would be ill advised or highly questionable. However, this should be a matter for consultation. What I recommend and what our surgeons are willing to do at Cornell is usually, but not always, the same.

JOIST

I have a question regarding a practical point. Dr. Rosenberg contends that what we are measuring with various clotting tests is really the heparin level. He has likened this to measuring insulin levels in diabetes. I do not really think that is true. Is it not a fact that on a given level of heparin, regardless of what clotting time you use, the effect may vary considerably. Does that not suggest that what we are really measuring is the balance of anticoagulant and procoagulant forces, rather than heparin?

ROSENBERG

My analogy was not absolutely accurate, but it is still pretty close. If you put in an activation step so that you had to activate the insulin, then the analogy would be quite a good one. You are still measuring the activation of antithrombin outside the body by adding extraneous serine proteases, whether you do the whole blood clotting time or thrombin addition. In essence, you are measuring the biological effect of heparin on antithrombin without knowing the procoagulant balance in the patient.

MCGEHEE

When I am asked to talk to physicians about the use of heparin, I find that there is a common lack of distinction between the minidose regimes as proposed by Dr. Kakkar for the prevention of thrombotic disease and the proper therapy of an already established thrombus or pulmonary embolus. I was tremendously impressed by the results obtained in the latter category by Dr. Basil and his colleagues, as reported in the New England Journal, 1972, from McMaster. I would like to address my question to Dr. Sherman and Dr. Rosenberg. Given a patient with an established venous thrombus, whether that be severe deep vein thrombosis or a pulmonary embolus, what would you consider the optimal way of handling such patients whether you are in a community hospital or in a major medical institution?

SHERMAN

The optimal way will depend upon the given patient. A patient with an established thrombophlebitis or pulmonary embolus within the early post-operative period is handled much more gingerly than a patient with a myocardial infarct. I prefer to start such a patient on intravenous heparin immediately, assuming no contraindications, and at some time during the first week to switch over to coumedin. I think it is interesting, that if you go back to the earliest literature, one of the few control studies of the efficacy of anticoagulant and pulmonary embolus, that of Barett and Jordan, all they gave was four doses of heparin for eight days. There is a paucity of data on the duration of coumarin therapy. My inclination now is to continue anticoagulation for six weeks to three months, but no more than three months because I am not impressed that the data suggests that the average patient has any benefit from therapy after three months.

WRIGHT

That program of about three months is standard on our service, too.

The policy of some physicians to use anticoagulants for only seven to eight days is really dangerous. About three months is much safer for a person who has an identifiable thrombophlebitis, especially with pulmonary emboli. But we have patients who continue to have reoccurrences. In some of these people, as I pointed out in the case of our well-known patient, all types of studies fail to reveal why this occurs. If this happens we are apt to keep them on indefinitely and most of them do very well.

MCGHEE

If you are carrying patients to six weeks at least or to a longer period of time, that implies oral anticoagulant therapy. I will not take the patient off intravenous heparin until he is adequately controlled on coumarin drugs for a period of time. Is this shared by the panel?

WRIGHT

We all agree on that.

BROZOVIC

Dr. Rosenberg, in your publications you reported that you could not recover your heparin from the antithrombin heparin complex by protamine. Do you have any explanation for the mechanism? How can we reverse heparin action by protamine in our patients?

ROSENBERG

What I reported was that the thrombin was not released from antithrombin by the addition of protamine , and that kind of a possible clinical rebound might have been significant. I do not know how protamine works. I do not think it works by complexing with heparin and disassociating it from the inhibitor. It is much more complex than that.

DAVIS

I would like to ask the opinion of the panel in regard to the use of large amounts of glucose in this situation. Everyone of these patients in acute stress is mobilizing free fatty acids. These are powerful procoagulants and as soon as you start running in glucose, free fatty acids go down. I use the expression that glucose is the most valuable anti-procoagulant to have. Some people are being maintained and are preventing, for instance, pulmonary embolism or fat embolism by large amounts of glucose alone. If you do not want to use heparin, is it useful?

WRIGHT

You are now referring to glucose, not dextrose?

DAVIS

Right!

SHERMAN

My only comment would be that this would be an ideal type of study for someone to try to see if it were true by using Dr. Wessler's animal model.

JOIST

One point which has not been addressed in this meeting is the question of the usefulness of heparin or anticoagulant therapy in general in preventing the spread of metastatic cancer. I wonder if anybody would like to comment on that?

WRIGHT

This is a very wide open subject at present. I am familiar with these reports, but I do not have any personal experiences with it. Certain patients with cancer are, of course, very prone to thrombophlebitis and we have reported a fairly large series of patients in whom the thrombophlebitis was the first clinical manifestation of cancer.

YIN

I have two questions. Dr. Rosenberg stated that there might be a danger of heparin preventing clot dissolution. I do not know if this is stated from the facts or this is speculation. I would also like to ask Dr. Kakkar if he has observed any difference between clot dissolution among patients developing thrombi in leg veins as to whether or not they were on minidose heparin.

ROSENBERG

I have been drawing analogies from the protein chemistry studies to the whole animal which is always dangerous. I am just raising a possibility that one should look for it. It is certainly biochemically true; whether it is true in the whole animal or not remains to be proven.

KAKKAR

Yes, there is tremendous difference in the rate of thrombolysis depending on whether heparin is present or absent. If you look at the fibrin which is formed under the influence of low dose heparin and the fibrin which is formed in the absence of heparin and study cross-linkages, there is a difference. If you study the effect of heparin therapy by the crude studies we have done, there are changes there. Further studies in this area are for the second International Symposium.

JAQUES

Dr. Wright, at the beginning of this panel you referred to a visit by Dr. Best to New York in 1938, and Dr. Cifonelli on Monday showed a slide with analyses in great detail of a particular heparin preparation which he got from the late P. A. Levine and said he had never been able to find out where that heparin came from. I am pleased to set the record straight. In 1940 on my way to the Federation Meeting, Dr. Best gave me a bottle of 50 or 100 grams of heparin to deliver to P. A. Levine. So a portion of the world's supply of heparin at that time that you did not get for your studies ended up at the Rockefeller Institute.

PARTICIPANTS IN
THE INTERNATIONAL SYMPOSIUM ON HEPARIN, ST. LOUIS, MO.
MAY 13-15, 1974

ALKJAERSIG, Norma
Department of Enzymology
Washington University School of Medicine
St. Louis, Missouri 63110

ATKINS, Edward D.
H.H. Wills Physics Laboratory
University of Bristol
Bristol, England

AVIOLI, Louis V.
Department of Medicine
The Jewish Hospital of St. Louis
St. Louis, Missouri 63110

BANG, Nils U.
Lilly Laboratory for Clinical Research
Marion County General Hospital
Indianapolis, Indiana 46202

BARLOW, Grant H.
Abbott Laboratories
North Chicago, Illinois 60064

BARTENBACH, David E.
Abbott Laboratories
North Chicago, Illinois 60064

BAUE, Arthur
Department of Surgery
The Jewish Hospital of St. Louis
St. Louis, Missouri 63110

BAUGHMAN, D. Joe
Ortho Diagnostics Inc.
Raritan, New Jersey 08869

BLAKE, Diane
Department of Biochemistry
University of Illinois at Urbana-Champaign
Champaign, Illinois 61801

BORSODI, Anna D.
Department of Biological Chemistry
Washington University School of Medicine
St. Louis, Missouri 63110

BRADSHAW, Ralph A.
Department of Biological Chemistry
Washington University School of Medicine
St. Louis, Missouri 63110

BRAUNSTEIN, Kenneth
Medical University of South Carolina
Charleston, South Carolina 29401

BRINKHOUS, Kenneth M.
Department of Pathology
University of North Carolina
Chapel Hill, North Carolina 27514

BROZOVIC, Milicia
Medical Research Council
National Institute for Biological Standards and Control
Holly Hill, London NW 3 6 RB
England

CHAPA, Liberato
Vascular Surgical Services
State University of New York
Downstate Medical Center
Brooklyn, New York 11203

CHIU, H.M.
Department of Pathology
McMaster University Medical Centre
Hamilton, Ontario, Canada

CIFONELLI, J.A.
Department of Pediatrics
University of Chicago
Chicago, Illinois 60637

CONRAD, H.E.
Department of Biochemistry
School of Chemical Sciences
University of Illinois at
Urbana-Champaign
Champaign, Illinois 61801

COOTS, Macie
Coagulation Laboratory
University of Cincinnati Medical
Center
Cincinnati, Ohio 45229

COYNE, Erwin
Cohelfred Laboratories
Chicago, Illinois 60618

DANISHEFSKY, Isidore
Department of Biochemistry
New York College of Medicine
Valhalla, New York 10595

DAVIS, Andrew
Department of Enzymology
Washington University
School of Medicine
St. Louis, Missouri 63110

DAVIS, Herbert L.
Department of Surgery
The University of Nebraska
Medical Center
Omaha, Nebraska 65105

DOMBROSE, Fred A.
Department of Biological
Chemistry
Washington University
School of Medicine
St. Louis, Missouri 63110

DUGDALE, Marion
Hematology Division
The University of Tennessee
Memphis, Tennessee 38163

ENGELBERG, Hyman
Internal Medicine and
Cardiology
465 North Roxbury Drive
Suite 1003
Beverly Hills, California
90210

ESMON, Charles
Department of Biochemistry
University of Wisconsin
Madison, Wisconsin

ESTES, J. Worth
Boston University School of
Medicine
Boston, Massachusetts 02118

FALB, Richard
Biology and Medical Sciences
Department
Battelle Memorial Institute
Columbus Laboratories
Columbus, Ohio 43201

FAREED, Jawed
Pharmacology Department
Loyola University
Stritch School of Medicine
Maywood, Illinois 60153

FLETCHER, Anthony P.
Department of Enzymology
Washington University
School of Medicine
St. Louis, Missouri 63110

FORMAN, Walter B.
Hematology Service
Veterans Administration Hospital
Cleveland, Ohio 44106

GARSKA, Carol
Organon Company
West Orange, New Jersey 07052

GENTRY, Patricia A.
Department of Biomedical Sciences
Ontario Veterinary College
University of Guelph
Guelph, Ontario, Canada

GEORGE, James N.
Division of Hematology and Oncology
The University of Texas
Health Science Center
at San Antonio
San Antonio, Texas 78284

GERTLER, Menard M.
Cardiovascular Research
New York University
Medical Center
New York, New York 10016

GITEL, Sanford
Department of Medicine
The Jewish Hospital of St. Louis
St. Louis, Missouri 63110

GLASER, Luis
Department of Biological Chemistry
Washington University School of
Medicine
St. Louis, Missouri 63110

GLAZIER, Richard
Department of Medicine
University of Wisconsin
Madison, Wisconsin 53706

GLUECK, Helen I.
Coagulation Laboratory
University of Cincinnati
Medical Center
Cincinnati, Ohio 45229

GOLDSMITH, Edward I.
New York Hospital
Cornell Medical Center
New York, New York 10021

GOLDSTEIN, Jack
The New York Blood Center
New York, New York 10021

GOODMAN, Terry
New York Hospital
Cornell Medical Center
New York, New York 10021

GORMSEN, Johs
Sundby Hospital
Copenhagen S, Denmark

GOTT, Vincent
Department of Surgery
The Johns Hopkins University
Baltimore, Maryland 21205

HANDIN, Robert I.
Department of Medicine
Hematology Division
Peter Bent Brigham Hospital
Boston, Massachusetts 02115

HECKER, Sydney P.
Palo Alto Medical Clinic
Palo Alto, California 94301

HORNER, Alan A.
Department of Physiology
University of Toronto
Toronto 5, Canada

HENRIKSEN, Ruth Ann
Department of Biological Chemistry
Washington University
School of Medicine
St. Louis, Missouri 63110

ITTYERAH, Roy
Department of Biological Chemistry
Washington University
School of Medicine
St. Louis, Missouri 63110

JAQUES, Louis B.
Department of Physiology
Hemostasis-Thrombosis Research Unit
University of Saskatchewan
College of Medicine
Saskatoon, S7N OWC, Canada

JEANLOZ, Roger W.
Laboratory for Carbohydrate Research
Harvard Medical School
Massachusetts General Hospital
Boston, Massachusetts 02115

JOIST, J.H.
Hemostasis and Thrombosis Laboratory
Washington University School of Medicine
St. Louis, Missouri 63110

KAKKAR, V.V.
Department of Surgery
King's College Hospital Medical School
Denmark Hill
London, S.E. 5, England

KIMBALL, Daniel B.
Hematology Oncology Service
Department of the Army
Walter Reed Army Medical Center
Washington, D.C. 20012

KLIEN, Harvey
Naional Lung and Heart Institute
Division of Blood Diseases and Resources
National Institutes of Health
Bethesda, Maryland 20014

KOEHLER, Karl A.
Departments of Pathology and Biochemistry
University of North Carolina
Chapel Hill, North Carolina 27514

KWAAN, Hau C.
Hematology Section
Veterans Administration Research Hospital
Chicago, Illinois 60611

LASKER, Sigmund E.
New York Medical College
Flower and Fifth Avenue Hospitals
New York, New York 10029

LATOUR, Gean Gilles
Montreal Heart Institute
Montreal, Canada 81T 1C8

LEVIN, Jack
Hematology Division
Department of Medicine
The Johns Hopkins Medical Institutions
Baltimore, Maryland 21205

LINDAHL, Ulf
Institute of Medical Chemistry
University of Uppsala
S-751-22 Uppsala 1
Sweden

LITTLEFIELD, Arthur
Organon Company
West Orange, New Jersey 07052

LOWENBERG, Robert I.
Vascular Surgery - Angiology
The Lowenberg Clinic, P.A.
Austell, Georgia 30001

LUTCHER, C.L.
Medical College of Georgia
Veterans Administration Hospital
Augusta, Georgia 30904

LYONS, Roger M.
Department of Hematology
Washington University School of Medicine
St. Louis, Missouri 63110

MAJERUS, Philip W.
Department of Hematology
Washington University
School of Medicine
St. Louis, Missouri 63110

MARCINIAK, Ewa
University of Kentucky
Albert B. Chandler Medical Center
Department of Medicine
Lexington, Kentucky 40506

MARCUS, Aaron J.
Hematology Section
New York Veterans Administration Hospital
New York, New York 10010

MARX, Gerard
The New York Blood Center
New York, New York 10021

MC GEHEE, William G.
Hematology Department
University of Southern California
School of Medicine
Los Angeles, California 90033

MESSMORE, Harry L.
Loyola University Medical Center
Clinical Hematology
Maywood, Illinois 60158

MINK, Irving B.
Roswell Park Memorial Institute
Buffalo, New York 14203

MIOTTI, Angelica B.
Division of Hematology
The Jewish Hospital and
Medical Center of Brooklyn
Brooklyn, New York 11238

NELSON, Thomas E., Jr.
Department of Pharmacology
Southern Illinois University
School of Dental Medicine
Edwardsville, Illinois 62025

NESS, Paul
National Lung and Heart Institute
Division of Blood Diseases and Resources
National Institutes of Health
Bethesda, Maryland 20014

OLIVECRONA, Thomas
Department of Chemistry
Umeå University
Section of Physiological Chemistry
S-901-87 Umeå
Sweden

OLSON, Robert E.
Department of Biochemistry
St. Louis University
School of Medicine
St. Louis, Missouri 63104

OWEN, Whyte G.
Department of Pathology
University of Iowa
Iowa City, Iowa 52240

RATNOFF, Oscar D.
Department of Medicine
Lakeside Hospital
Case Western Reserve University
Cleveland, Ohio 44106

ROSENBERG, Daniel
Department of Hematology
Washington University
School of Medicine
St. Louis, Missouri 63110

ROSENBERG, Robert D.
Departments of Medicine and Surgery
Beth Israel Hospital
Harvard Medical School
Boston, Massachusetts 02115

SAWYER, Philip N.
Vascular Surgical Services
State University of New York
Downstate Medical Center
Brooklyn, New York 11203

SEEGERS, Walter H.
Department of Physiology
Wayne State University
School of Medicine
Detroit, Michigan 48201

SHANBERGE, J.N.
Department of Pathology
Mount Sinai Medical Center
Milwaukee, Wisconsin 53233

SHERER, Peter
Division of Blood Diseases
and Resources
National Lung and Heart
Institute
National Institutes of Health
Bethesda, Maryland 20014

SHERMAN, Lawrence
Department of Pathology
Washington University
School of Medicine
St. Louis, Missouri 63110

SHERRY, Sol
Department of Medicine
Temple University
School of Medicine
Philadelphia, Pennsylvania
19140

SHIVELY, Jack
Department of Biochemistry
School of Chemical Sciences
University of Illinois at
Urbana-Champaign
Champaign, Illinois 61801

SILBERT, Cynthia K.
Department of Medicine
Veterans Administration Hospital
Boston, Massachusetts 02130

SILBERT, Jeremiah E.
Department of Medicine
Veterans Administration Hospital
Boston, Massachusetts 02130

SILVERGLADE, Alex
Riker Laboratories, Inc.
Northridge, California 91324

SIMON, Ernest
Blood Diseases Branch
Naional Institutes of Health
Bethesda, Maryland 20014

SIMON, Toby L.
National Lung and Heart
Institute
National Institutes of Health
Bethesda, Maryland 20014

SMITH, Carroll M.
Armour Pharmaceutical Company
Kankakee, Illinois 60901

STENGLE, James M.
Thrombosis and Hemorrhagic
Diseases Branch
Division of Blood Diseases
and Resources
National Institues of Health
Bethesda, Maryland 20014

STONE, Bill
Lettermen Army Medical Center
San Francisco, California 64129

STUART, R.K.
Hematology Service
Department of Medicine
University Hospital
London, Ontario N6G 2K3 Canada

THERRIAULT, Donald
Division of Blood Diseases
and Resources
National Lung and
Heart Institute
National Institutes of Health
Bethesda, Maryland 20014

THOMPSON, Arthur R.
Division of Hematology
Harborview Medical Center
Seattle, Washington 98104

TODD, Margaret E.
Technicon Instruments Corp.
Tarrytown, New York 10591

TOLLEFSEN, Douglas
Department of Biochemistry
Washington University
School of Medicine
St. Louis, Missouri 63110

VECCHIO, Thomas J.
The Upjohn Company
Kalamazoo, Michigan 49001

WAGH, Premanand V.
Veterans Administration
Hospital
Little Rock Hospital Division
Little Rock, Arkansas 72206

WALDMAN, Alan
The New York Blood Center
New York, New York 10021

WALLACE, Herbert W.
Department of Surgery
The Graduate Hospital
Philadelphia, Pennsylvania
19146

WELCH, Michael
Department of Radiology
Washington University
School of Meidicne
St. Louis, Missouri 63110

WESSLER, Stanford
Department of Medicine
The Jewish Hospital of St. Louis
St. Louis, Missouri 63110

WINE, Alan
Lakeside Hospital
Case Western Reserve University
Cleveland, Ohio 44106

WRIGHT, Irving
450 East 69th Street
New York, New York 10021

YIN, E. Thye
Department of Medicine
The Jewish Hospital of St. Louis
St. Louis, Missouri 63110

YUAN, Leon
Department of Biochemistry
Loyola University
School of Dentistry
Maywood, Illinois 60153

SUBJECT INDEX

www.ingramcontent.com/pod-product-compliance
Ingram Content Group UK Ltd.
Pitfield, Milton Keynes, MK11 3LW, UK
UKHW051131260726
13967UKWH00010B/2977

* 9 7 8 1 4 6 8 4 0 9 4 7 5 *